AF412000

Laboratory Automation in the Chemical Industries

Laboratory Automation in the Chemical Industries

edited by

David G. Cork
Takeda Chemical Industries, Ltd.
Osaka, Japan

Tohru Sugawara
ChemGenesis, Inc.
Tokyo, Japan

MARCEL DEKKER, INC.　　　　　NEW YORK · BASEL

ISBN: 0-8247-0738-9

This book is printed on acid-free paper.

Headquarters
Marcel Dekker, Inc.
270 Madison Avenue, New York, NY 10016
tel: 212-696-9000; fax: 212-685-4540

Eastern Hemisphere Distribution
Marcel Dekker AG
Hutgasse 4, Postfach 812, CH-4001 Basel, Switzerland
tel: 41-61-261-8482; fax: 41-61-261-8896

World Wide Web
http://www.dekker.com

The publisher offers discounts on this book when ordered in bulk quantities. For more information, write to Special Sales/Professional Marketing at the headquarters address above.

Current printing (last digit):
10 9 8 7 6 5 4 3 2 1

PRINTED IN THE UNITED STATES OF AMERICA

Preface

Automation is now used in such a wide variety of ways in the chemical industries that when we were asked to edit a book on the subject it seemed it would be a huge task, which could easily be developed into a series of volumes. The last decade has seen an explosion of interest in automation as a tool for enabling the strategies expounded in combinatorial and parallel synthesis. Faster synthesis has increased the demand for automation at other stages, for example, to enable new parallel chromatography or other rapid purification strategies, to facilitate high-speed and high-throughput analysis steps, and even to help with the reaction planning and optimization stages. The following trends have also been important in increasing the use of automation: (1) availability of tried and tested, affordable equipment, including disposable parts and tools, (2) demand for reduced costs and higher throughput, (3) advances in technology to allow miniaturization and the handling of larger numbers, and (4) greater computing power for instrument control, data management, and so on.

We have gathered chapters that give a comprehensive coverage of automation related to chemistry in industry, with the exception of the screening stage, which has been the subject of several books. The work of most industrial chemists who are now involved with laboratory automation can be broadly divided into two categories: chemistry discovery and process chemistry. We have used this division to structure the chapters of the book.

The chapters in Part I, related to chemistry discovery, cover aspects of automated synthesis, reaction optimization, and parallel purification techniques. This part concludes with a chapter on the prospects for automation in the field. Automated synthesis in chemistry discovery has been developing for at least two decades, and with tremendous impetus in the past five years, as needs and pressures have mounted. The first two chapters reflect this background, covering several of the commercially available apparatus and robotic systems now being used in the industry and then examining in detail the in-house automated apparatus, units, and devices that have been specifically developed for chemistry in a pharmaceutical research laboratory.

Automation of the reaction investigation and optimization stages in chemistry research has been progressing at a pace similar to that for synthesis. We include a chapter that describes in detail how software for planning and scheduling experiments has been developed. Automation has also been advantageously applied to the parallelization of purification processes, notably HPLC but also liquid–liquid extraction and some supported-reagent scavenging strategies. Chapters on both of these aspects are included.

Part II broadly encompasses automation related to process chemistry, incorporating five chapters that cover laboratory information management systems (LIMS), design and optimization of chemical processes, calorimetry, and chemical analysis. Automation of chemistry, in both the discovery and process stages, has, by necessity, been accompanied by advances in automation related to information management. Some kind of LIMS is now essential for any industrial laboratory in which automated chemistry is used. Another area of process chemistry in which automation has been of use is the design and development of new chemical processes that are specifically suited to automated chemistry systems. Notably, one-pot processes offer greater advantages, and some of the recent research is described in one chapter.

Automation in its various guises and at the various stages of process development is comprehensively covered in a chapter on optimization of chemical processes. Calorimetry is often overlooked at the chemistry discovery phase but becomes vitally important when developing processes. The book examines commercially available systems, and several case studies are presented to highlight the key benefits that automated calorimetry offers. Automated systems for serial or sequential chemical analysis are well established, but the needs of high-throughput systems have driven serial analysis to be largely replaced by parallel systems that rely on advances in automation. The features and performance characteristics of these parallel systems are described in the final chapter.

The contributors—industrial and academic experts in the field—have created a comprehensive book that serves as a unique guide to what and how automation is actually being used in the chemical industries. Many authors submitted material related to their practical experience in implementing automated apparatus for high-throughput synthesis, purification, and reaction or process optimization, and most have included sections on future prospects as well as on the current state-of-the-art techniques in their fields.

As work continues at full pace on the human genome and the number of ''orphan'' targets (for which chemical leads must be found) increases, the role of automation in the chemical industries will become more important. This book will be useful for academics and students, as well as researchers and practitioners working within the pharmaceutical and chemical industries.

David G. Cork
Tohru Sugawara

Contents

Part II. Laboratory Automation in Process Chemistry

Contributors

April Ciampoli *ArQule, Woburn, Massachusetts*

David G. Cork *Takeda Chemical Industries, Ltd., Osaka, Japan*

L. Andrew Corkan *North Carolina State University, Raleigh, North Carolina*

Hai Du *North Carolina State University, Raleigh, North Carolina*

Nick Evens *Avecia Pharmaceuticals, Grangemouth, Stirlingshire, Scotland*

Nick Hird *Takeda Chemical Industries, Ltd., Osaka, Japan*

Patricia Y. Kuo *North Carolina State University, Raleigh, North Carolina*

Jonathan S. Lindsey *North Carolina State University, Raleigh, North Carolina*

Bill MacLachlan *GlaxoSmithKline Pharmaceuticals, Harlow, England*

John E. Mills *Johnson & Johnson Pharmaceutical Research & Development, L.L.C., Spring House, Pennsylvania*

Junzo Otera *Okayama University of Science, Okayama, Japan*

Christine Paszko *Accelerated Technology Laboratories, Inc., West End, North Carolina*

Janice A. Ramieri *Biotage, Inc., Charlottesville, Virginia*

David T. Rossi　*Pfizer Global Research and Development, Ann Arbor, Michigan*

Michele R. Stabile-Harris　*ArQule, Woburn, Massachusetts*

Tohru Sugawara　*ChemGenesis, Inc., Tokyo, Japan*

Brian H. Warrington　*GlaxoSmithKline, Harlow, England*

Kexin Yang　*North Carolina State University, Raleigh, North Carolina*

Laboratory Automation in the Chemical Industries

1
Robotic Workstations and Systems

Nick Hird
Takeda Chemical Industries, Ltd.
Osaka, Japan

Bill MacLachlan
GlaxoSmithKline Pharmaceuticals
Harlow, England

1 INTRODUCTION

Although it is a relative latecomer to organic synthesis, automation is now being increasingly incorporated into most areas of the modern synthetic chemistry laboratory. The driving force of increased productivity within the pharmaceutical industry has led to a new generation of instruments specifically designed for high-throughput synthesis. Prior to the advent of combinatorial chemistry [1], the use of automation for organic synthesis was a minor specialty, carried out mainly by academic groups for the optimization of reaction condition [2,3]. Hardware and software on all instrumentation was homemade, which not only required considerable development but also limited its use to experts. However, in the last decade, automation of organic synthesis has been widely recognized as a key technology by which productivity can be improved, and considerable resources have been directed into this area. In particular, the requirement for instrumentation to perform high-throughput, parallel chemical synthesis has resulted in the development and commercialization of a number of novel technologies [4]. Nevertheless, although the number of instruments available may seem large, automated high-throughput organic synthesis is still in its infancy. While the ability to carry out reactions in parallel is the chief advantage that automation offers over manual chemistry, instruments that can routinely and reliably outperform a skilled synthetic chemist in terms of chemical ability are not yet available.

The development of synthesis automation has been a long and difficult struggle. In other areas of drug discovery such as high-throughput screening and analysis, instruments that successfully replace manual operations have been in use for a number of years. In these applications, the processes are usually repetitive with little variation, involving the handling of aqueous solutions with similar properties, and thus the engineering requirements are relatively modest. In marked contrast, automating synthetic chemistry requires precise control of a wide range of reaction environments, as well as handling a vast array of reagents and solvents with different compatibilities and physical characteristics. This constitutes a considerable engineering challenge. In the initial stages of development of synthesis automation [5,6]. It was convenient to ignore a number of these factors, and as a result the early instrumentation was limited in its use for general organic synthesis and therefore not widely accepted. It has now been realized that the needs of the ''traditional'' chemist dictate the performance requirements of the instrument and not the reverse, and consequently the use of automated synthesis workstations is becoming firmly established in laboratories. While in many cases usage of the instruments does require specialist training, it is probably the simpler tools [7] that are making the greatest inroads into enabling chemists to increase their productivity. Large-scale mechanization is also taking place, leading to the concept of ''compound factories'' [8], which will be discussed later in this chapter. For any kind of automation, it is important to realize that the success or failure of a particular technology lies in how readily it can be assimilated by the operator. In many cases the impact of otherwise useful automation has been limited due to the high degree of skill required by the users. For synthesis workstations, it is essential that users gain sufficient confidence in the operation and performance of the instrument to choose to use it in preference to manual synthesis. This not only requires a well-configured and well-engineered instrument that can be easily used, but also the provision of a high level of technical support from the manufacturer to the user to solve problems as they arise.

2 WORKSTATION DESIGN

By far the most common approach to automating synthetic organic chemistry has been the concept of a synthesis workstation [9]. The key requirements for a successful automated synthesizer are fairly obvious: high chemical performance and throughput, ease of operation and maintenance, reliability, and cost. The workstation needs to automate reagent addition, reaction incubation, work-up, and isolation of the products. In some cases additional tasks may be carried out such as reagent preparation, on-line analysis, evaporation, and purification. However, in general the greater the number of tasks that can be performed by a workstation, the lower the throughput will be. Furthermore, the requirements of solid phase chemistry and solution phase chemistry are considerably different, and often need separate components.

Several kinds of automation principles [10] have been employed in the design of workstations for high-throughput organic synthesis. By far the most common design is to combine an overhead gantry robot with a reaction block (which is normally held in a fixed position) containing the reaction vessels. The reaction block can usually be heated, cooled, and maintained under an inert atmosphere and has some means of agitation. The overhead gantry xyz robotic arm is used to carry out liquid handling and other tasks necessary to perform the synthesis. Such systems have the advantage that they are well suited to performing parallel operations and can easily be constructed into compact units suitable for standard laboratories.

Another approach uses a mobile functional robotic arm, which can mimic the manual actions of a chemist carrying out a synthesis, such as picking up a reaction vessel, adding reagents, or shaking. In these systems, the reaction vessels are usually mobile and transferred by the robot arm to several different components laid out within reach of the arm to carry out the individual tasks of the synthetic process. Since such systems tend to process on the individual reaction vessel level, they can closely mimic the manual process, enabling excellent chemical control, although this means that they are less easily adapted to high-throughput processing.

Nonrobotic fluidics-based approaches have also been used for automated synthesis (see Chap. 2). In this case, instead of using moving robotic arms to carry out liquid handling, the reaction vessels are directly connected to input and output fluid lines. Thus by using inert valves to control the fluidic pathways, liquid transfer can be carried out without the need for robotics. The key advantage of this approach is that fluid transfers can easily be carried out in parallel, greatly enhancing the throughput of the instrument. Furthermore, because liquid handling is carried out in a completely inert environment, high levels of chemical performance are possible. However, the number of vessels that can be used (and thus the throughput) is limited by the complexity of the fluid line network, and problems may arise in handling insoluble materials.

Irrespective of the type of automation used, the most influential determinant of the chemical performance of any synthesizer is its reactor design. Most systems use reactors made from inert materials such as glass or Teflon, which have high chemical resistance and can be subjected to wide temperature variation. The majority of designs incorporate a frit to allow filtration for solid-phase chemistry, although in some cases separate solid- and solution-phase vessels can be used interchangeably on the same instrument. In most high-throughput applications, it is more expedient to use disposable rather than reusable consumables, and much effort has been expended to develop high performing but disposable reaction vessels, which has led to simplification of vessel design.

Workstations by definition are self-contained systems in which all the functional components necessary to carry out the synthetic tasks are integrated within a single instrument. These units are convenient for the user, as not only can they be located within a standard laboratory, the integrated nature of these instruments generally makes them easier for chemists to maintain and operate. The major disadvantage of multi-functional workstations is that generally only one of the functions can be employed at any one time, resulting in inefficient instrument usage, particularly if high-throughput synthesis is required. Furthermore, the closed architecture of these systems limits their flexibility, for example to add new functionality or to upgrade the system.

Due to this need for flexibility, modular systems are becoming increasingly popular. Modularity enables separate processes involved in compound synthesis to be segregated to individual processors that can be operated simultaneously. These modules consist of several simple, often monofunctional, devices that can be linked together to perform each task and are often built with an open architecture giving flexibility and customization possibilities. However, the integration of several components can present mechanical and operating difficulties (particularly with control software) for the user, and the size of the instrument can require purpose-built containment facilities. Modular use can be obtained from unit instruments by using off-line incubation, for example with removable reaction blocks, but this usually requires manual intervention. Thus, although modular systems are more complex and difficult to develop, there is a trend towards modular design for efficient high-throughput instruments.

Another key aspect of any workstation is the control and data management software [11], which fundamentally dictates the usage of the instrument. Originally instruments were operated by expert programmers, but now great effort is expended to make synthesizers accessible to most chemists. Point-and-click pictorial software with a comprehensible set of instructions is now standard. In addition, error tracking and diagnostics are becoming increasingly available on new instruments to enable a greater degree of self-reliance for the chemist user.

3   CURRENTLY AVAILABLE SYSTEMS

Several instruments for automated high-throughput organic synthesis are now (Autumn 2000) commercially available (Table 1). These instruments vary considerably in throughput capacity, chemical performance, size, and cost, and some are discussed in detail below. While this is not intended as a comprehensive list, it does serve to illustrate the major types of workstations currently available to perform organic synthesis. However, it should be stressed that only by repeated use can the true synthetic capability of a system be fully assessed. Particular emphasis is given to the Mettler-Toledo Myriad® system, to consider its key technologies and the practical use of the instrument.

Table 1 Commercially Available Automated High-Throughput Synthesisers

Instrument	Reaction vessels	Agitation	Chemistry	Automation	Probes for liquid handling	Type	Manufacturer
Benchmark Omega 496	8–96	orbital	solution/solid	xyz arm	2	unit	Advanced ChemTech
Chemspeed	40–112	orbital	solution/solid	xyz arm	1	unit	Chemspeed
Combinatorial chemistry workstation	96	stirring	solution/solid	xyz arm	8	unit	Zymark
Coherent system	120	stirring	solution/solid	xyz arm	1	unit	Personal Chemistry
Iliad	96–384	orbital	solution/solid	xyz arm	1	unit	Charybdis Technologies
Magellan	96	orbital	solid	xyz arm	1/4	unit	Zenyx
Discoverer	24	stirring/gas	solution/solid	xyz arm	1	unit	Mettler-Toledo
Myriad MCS	192	stirring/gas	solution/solid	xyz arm	1	modular	Mettler-Toledo
Nautilus	24	rocking	solid	valves	—	unit	Argonaut Technologies
Neptune	48–192	stirring	solution	xyz arm	2	unit	Mettler-Toledo
Quest 205	20	vertical	solution/solid	valves	—	unit	Argonaut Technologies
Quest 210	10	vertical	solution/solid	valves	—	unit	Argonaut Technologies
Solaris 530	48	orbital	solution/solid	xyz arm	8	unit	PE Biosystems
Sophas	36–864	orbital	solid	xyz arm	4	modular	Zinsser
Trident	192	orbital	solution/solid	xyz arm	—	modular	Argonaut Technologies
Vantage	96	orbital	solution/solid	xyz arm	6	unit	Advanced ChemTech
Zymark solution synthesizer	40–50	stirring	solution	cylindrical arm	1	unit	Zymark

3.1 Charybdis Synthesis System

The Calypso system, produced by Charybdis Technologies [12], is probably the best developed of a number of microtiter plate-based synthesis systems that are available for high-throughput synthesis. It uses a PTFE Teflon reactor block (Fig. 1) in 24, 48, and 96 well formats with separate modules for filtration and inerting. In addition, a multi-temperature block using glass reaction vessels that can be heated or cooled (-80 to $180°C$) by recirculating fluid and withstand up to 30 psi is available for more challenging chemistry. Although originally designed for manual operation, the Calypso blocks have now been incorporated into a number of robotic workstations for automated use. For these systems a modular approach has been used, resulting in a number of separate units for synthesis, washing, cleavage, and purification, thus allowing for very high throughputs. Synthesis can be carried out on the Illiad PS^2 personal synthesizer and the Illiad PS^2 Gemini Edition synthesizer (Fig. 2), which use two overhead xyz robot arms for reagent delivery.

In addition to the Charybdis system, there are other microtiter plate-based synthesis platforms including the Sophas instrument from Zinsser [13] and the

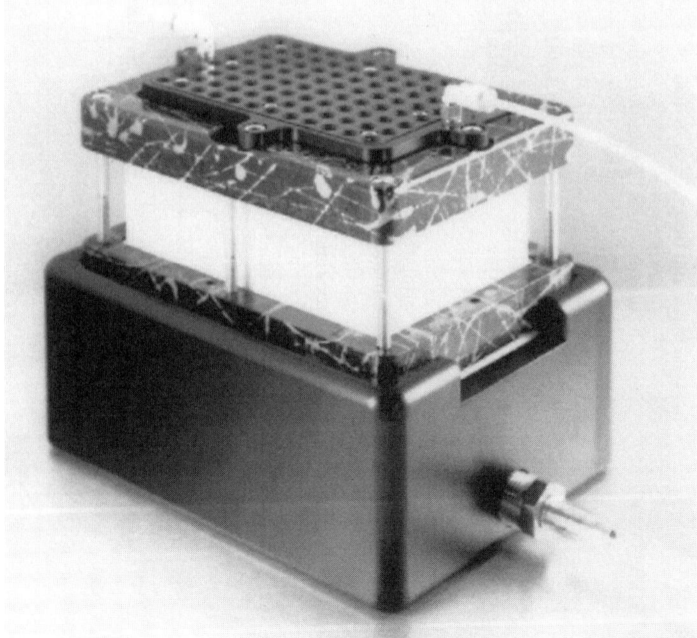

Figure 1 Calypso system reaction block.

Figure 2   Illiad PS^2 Gemini Edition synthesizer.

Flexchem system from Robbins [14]. The Sophas (solid phase synthesizer) uses disposable glass or reusable PTFE tubes that are inserted into the microtiter-plate footprint aluminum reaction block. Because it has parallel liquid handling with a four-probe head and can robotically transport the reaction blocks around the instrument deck, it is capable of very high throughput, with the potential to carry out 864 reactions simultaneously. However, the heating mechanism (heat transfer from temperature-controlled plates upon which the reaction blocks are placed) is inherently less efficient than those based on direct heating by recirculating fluids.

Although the simplicity of microtiter-based systems provides cost-efficient platforms for synthesis that can be easily used by most chemists, they are probably best suited for high-throughput rather than for high-performance chemistry.

3.2 Chemspeed

The Chemspeed [15] systems are multifunctional workstations, originally developed by chemists at Roche (Basel, Switzerland), which unlike most other workstations are capable of automating the complete compound preparation process. Chemspeed (Fig. 3) has been conceived as a complete personal synthesis tool for use by individual chemists, and thus it is possible to perform synthesis, work-up, purification, evaporation, and analysis, all within the same compact instrument. In theory, the chemist needs only to load reagents and starting materials onto the instrument in order to obtain analyzed, dried-down products. Currently two automated Chemspeed synthesizers are available, the AWS 1000 and AWS 2000, which are of similar design but differ in capability. Both these instruments are based around a Gilson 233 liquid handler, into which reagent racks and up to seven reaction blocks can be placed.

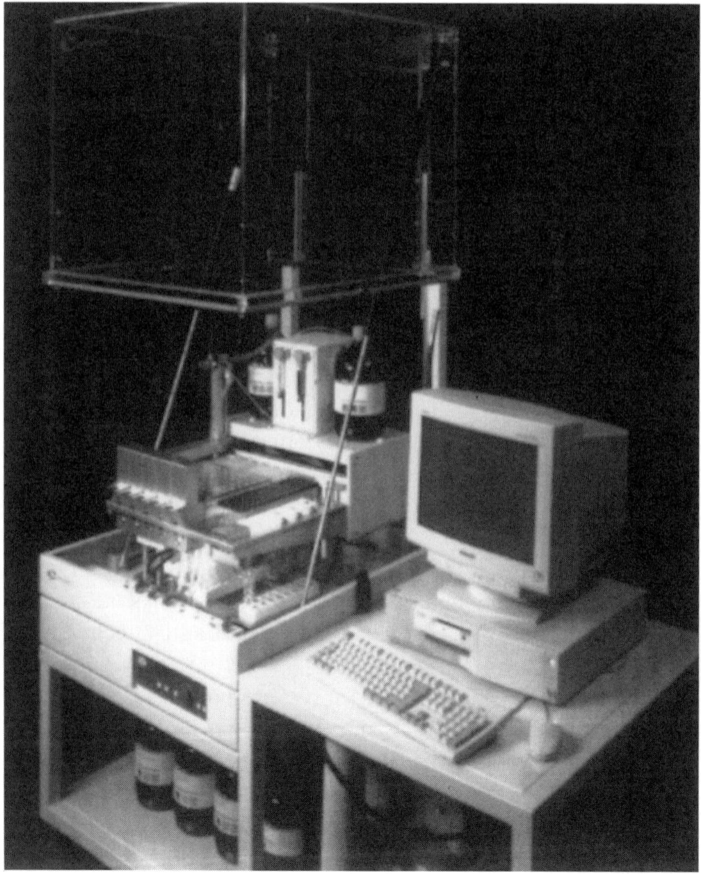

Figure 3 Chemspeed automated workstation (AWS).

The key to the flexibility and the multifunctional nature of this instrument is the reaction block and reactor design. The reaction block (Fig. 4) consists of a manifold that holds the reaction vessels in a flexible linkage from the top only, allowing the top of the vessel to be held in position while the bottom is agitated by rapid low-amplitude vortexing. This alternative agitation mechanism not only permits simultaneous agitation and reagent addition but also has the major advantage of permitting on-line evaporation, a capability that few synthesizers have. Clearly this is useful for multistep synthesis, where solvent changes are often required to allow subsequent reactions to be carried out. The manifold contains a valve system enabling it to switch between an inert gas and a vacuum, and the vessels are closed using ceramic valves under positive pressure, avoiding the need for septa. The reaction blocks can hold five sizes of glass reaction vessel (5, 13, 27, 50, 75, 100 mL), and therefore synthesis can be carried out on a wide range of scales, with the possibility of performing high-throughput synthesis and large-scale synthesis on the same platform. As well as different sizes, several different types of reaction vessel are available for specific performance requirements. An integral external jacket on the vessel is used to control temperature, and the jacket of each vessel is linked sequentially by tubing, which is attached to a recirculating cryostat. The temperature range is dictated by the choice of cryostat, but can range from -70 to $+145°C$ using a Huber Unistat 390. For solid phase synthesis and filtration, another reaction vessel, in which two reaction vessels have been linked via a filter, is used. Reflux is obtained by using a cold

Figure 4 Chemspeed reaction block.

finger inserted into each reaction vessel. Although this design offers flexibility and performance, the reaction vessel is complex and intricate, suggesting that assembly in the reaction block and cleaning after reaction may be time consuming.

The Chemspeed systems can be customized to individual requirements, for example to perform workup techniques such as liquid–liquid extraction, solid-phase extraction, and the use of scavenger resins. Other options include linkage to on-line analytical HPLC and TLC and also preparative HPLC. Another aspect of these systems is that they can be easily mounted on trolleys with their own integral fume hood, giving a portable synthesis workstation that can be easily transported between laboratories. The inherent flexibility of this system makes it ideal for use by individual chemists with rapidly changing requirements. For example, it has been used to carry out phenol alkylation, as shown below for library synthesis, as well as for the gram scale synthesis of the precursors (Scheme 1) [16].

However, the integrated nature of the Chemspeed may not only reduce its speed of processing but also the intricate design may be problematic and limit its use in sustained high-throughput synthesis. As well as the AWS series, the same components have been incorporated into a manual version, which can be used in conjunction with the AWS systems.

3.3 Coherent Synthesis System

Very recently the novel concept of applying microwave technology as the mode of heating in an automated synthesizer has resulted in the development of the Coherent Synthesis System from Personal Chemistry [17] (Fig. 5). The use of microwaves to accelerate chemical reactions has lately received much attention [18], but wider application has been hampered by the practical problems of generating a predictable and reproducible reaction environment. To address this prob-

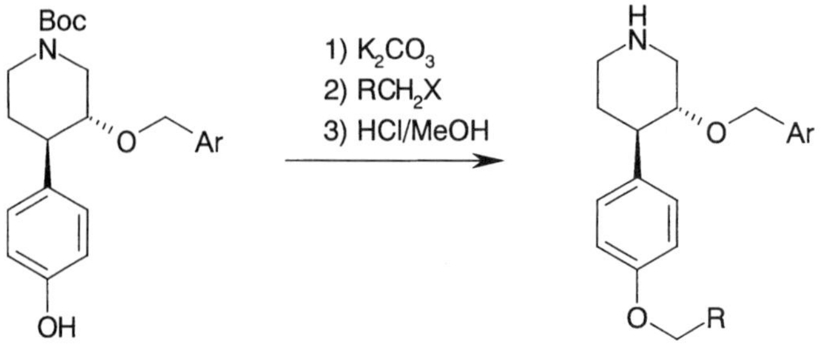

Scheme 1

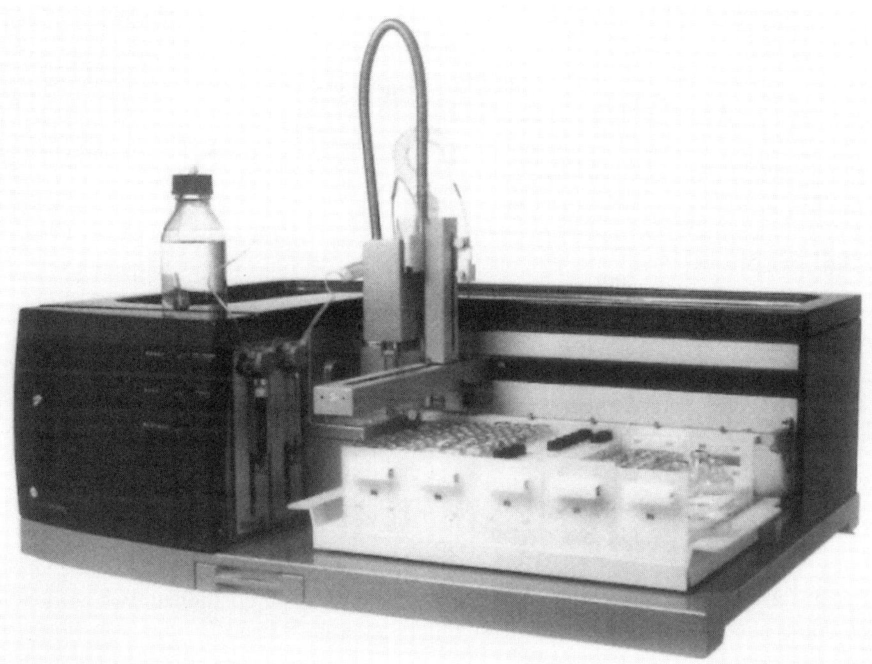

Figure 5 Coherent synthesis system.

lem, the Coherent system uses advanced numerical modeling of the microwave fields and resultant heating pattern to enable uniform heating of a range of volumes and polar environments. Furthermore, this technology has been miniaturized to allow 100 reactions from 0.5–5 mL scale to be carried out. The instrument consists of an xyz robot to perform liquid handling and transfer the reaction vials from the deck of the robot into the microwave source. Both reaction vessels, which are simple septum sealed vials, and reagents are located on the instrument deck. The chief benefit of this system is the dramatic acceleration of reaction rate that can be achieved, so that high-throughput synthesis is obtained primarily by reduction in reaction time. Typically, reaction times of hours can be reduced to minutes by microwave acceleration, in principle allowing the generation of approximately 50,000 reaction products per year with this instrument. However, an additional significant automation resource would be required to access these products via reaction work up.

Microwave acceleration can also be applied to solid-phase synthesis. Although the Coherent system offers great potential as a high-throughput synthesizer, by rate acceleration, yield improvement, and access to chemistries not feasible using nonmicrowave methods, it is too soon to assess the general applicability of this approach, particularly to new chemical methods. Personal Chemistry are compiling microwave-assisted chemistry experience into a database, which will lead to a wider understanding of this new type of synthesis. However, it is likely that this system will play a niche role to complement existing technologies rather than replace them.

3.4 Neptune

The Neptune workstation from Mettler-Toledo Bohdan® [19], which was based on the RAM synthesis system, is a multifunctional unit capable of performing synthesis on a number of different platforms, including the MiniBlock™ reactors, the RAM solid- and solution-phase reaction blocks, and also a high-pressure reaction block. This gives the instrument great flexibility and allows access to a wide variety of users. For example, high-throughput synthesis can be carried out with the Miniblock system, and lower throughput, higher performance synthesis can be carried out using the RAM synthesizer blocks.

The Neptune itself (Fig. 6) is the workstation, consisting of a robot, an open-format deck, upon which a variety of racks can be positioned, two vortex mixers, and a balance. Unlike most other systems, instead of a simple liquid

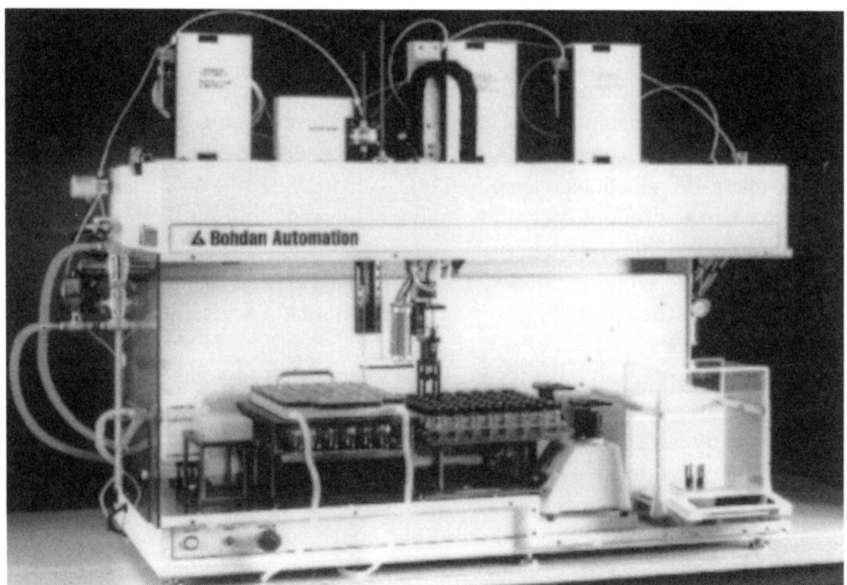

Figure 6 Neptune workstation.

handling probe, the overhead xyz gantry robot drives a powerful multifunctional arm. This arm contains two different cannula probes for separate aqueous and organic liquid handling (particularly suited for efficient liquid–liquid extraction), a vessel gripper unit, and a septum piercing unit that holds vessels down as the needle is withdrawn from the septum. Thus the arm has the ability not only to distribute liquid materials from septum-sealed containers but also to move vessels around the system, for example to the vortexers for agitation. The vortex unit gives excellent mixing and is superior to the more commonly used magnetic stirring or orbital shaking when used for liquid–liquid extraction or reagent dissolution procedures. The other performance requirements are determined by the individual characteristics of the reaction block system used.

For the RAM solution system (Fig. 7a), temperature control can be carried out using electrical heaters or recirculating chillers. Two independent lower and upper reaction vessel heating zones are available, both for reflux heating and for

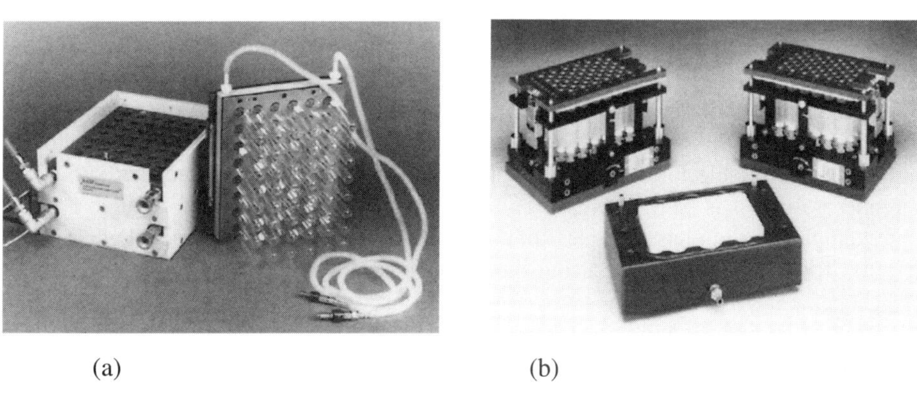

(a)   (b)

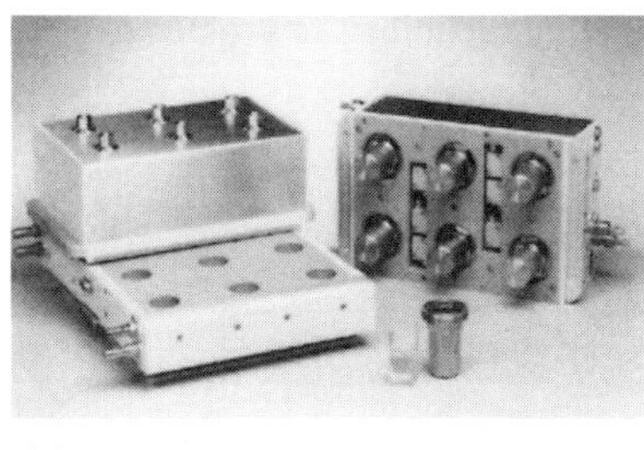

(c)

Figure 7 Neptune workstation reaction blocks. (a) RAM reaction block. (b) Miniblock system. (c) High-pressure reaction block.

low-temperature incubation. In addition, an on-deck stirrer unit is available to enable simultaneous mixing and reagent addition for sensitive chemistry. Although on-deck agitation is not available for the RAM solid phase synthesizer or the Miniblock system, both can be mixed off-line by placing the blocks on an orbital shaker, as can the RAM solution synthesis block. The RAM synthesizer uses two different glass vessels and reaction blocks for solution- and solid-phase synthesis, both in a 48 vessel format. The solution vessel is a simple disposable tube, whereas a reusable frit-containing tube is used for solid phase synthesis. Both RAM systems offer a high level of chemical performance, and the Astra group has described [20] how the capability of the RAM synthesizer makes it a particularly effective tool for lead optimization within medicinal chemistry programs, aided by its user-friendly pictorial software.

The Miniblock system (Fig. 7b), which was developed by Bristol Myers Squibb and commercialized by Mettler-Toledo Bohdan [21], is a simple but versatile reaction block system that can be used both manually and robotically. The key advantage to this system is that individual disposable filter-tube reactors are incorporated into a block design that will permit a high degree of chemical performance. Furthermore, a range of sizes and formats are available (48 $\times$ 4.5 mL, 24 $\times$ 10 mL, 12 $\times$ 20 mL, 6 $\times$ 40 mL), and a series of off-line modules for heating and cooling, shaking, inert atmosphere maintenance, and filtration are available. The Miniblock system was specifically designed for its ease of use, to make it available to all chemists without any requirement for specialist skills, and as such it is being increasingly used by a number of research groups. For example, a solid-phase synthesis of cinnolines by palladium-catalyzed alkynation and Richter-type cleavage using the Miniblock system has been reported [22] (Scheme 2).

Although the Miniblock system is similar to the microtiter plate-based synthesis systems described above, it is likely that because it uses individual reactors, a higher level of chemical performance can be obtained.

In addition, the Neptune can be used with a 12-position high-pressure reaction block, in which 15 mL scale reactions can be carried out up to 200 psi, with a temperature range of 0 to 100°C. This is a capability few other automated

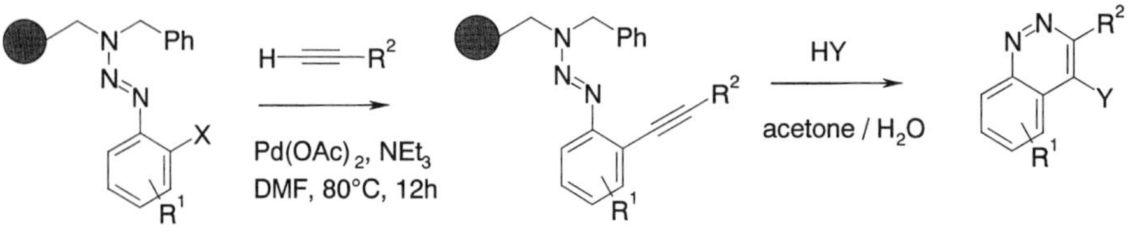

Scheme 2

instruments can offer, and although of limited use in high-throughput synthesis, it is particularly suited to polymer and new material research activities.

For all the reaction blocks that can be used on the Neptune, it is possible to obtain modular operation, since all the reaction blocks can be easily removed while still maintaining environmental control. Thus not only may throughput be increased by transferring the reaction blocks to external incubators after reagent processing, but also it allows the system to be concurrently used for different applications.

3.5 Quest

In addition to the Trident automated high-throughput synthesis system, Argonaut [23] have launched the Quest series of instruments. These are the first of a new generation of simplified semiautomated synthesis workstations that have been specifically designed as easy-to-use personal synthesis tools. The Quest series are characterized as being compact and lightweight units mounted on a rotating stand that is portable and can fit into standard laboratory fumehoods. One of the chief advantages of the Quest systems is that they do not require a PC, allowing their easy operation by chemists without need for specialist training.

The Quest 210 series (Fig. 8) contains two back-to-back banks of reaction vessels, each bank containing 10×5 mL or 10 mL disposable Teflon reaction vessels. The reaction vessels are held in an aluminum jacket, to provide temperature control from -40 to $130°C$ using a recirculating chiller for subambient cooling and electrical heating for superambient heating. Although it is not possible to add reagents automatically, solvent addition, liquid extraction, and filtration (including directly into SPE cartridges) can be performed. An interesting feature of the Quest is that a novel vertical mixing mechanism is used, in which Teflon coated magnets inside the reaction vessel are moved up and down by a compressed gas–powered, vertically oscillating external magnetic bar. The mixing rate can be adjusted by altering both the amplitude and the velocity of oscillation, and the merit of this technique is that it gives agitation of the complete contents of the vessel, essential for good mixing of both phases during liquid–liquid extraction. Furthermore, vertical mixing avoids the problem of resin crushing, which can be problematic when stirrer bar mixing is used. It is also possible to carry out vessel-to-vessel transfers, which is useful for multistep synthesis where one bank of vessels can be used to generate starting materials for synthesis performed in the second bank of vessels. Evaporation is possible by simultaneous heating and inert gas bubbling, although this mechanism (as compared to vacuum-based evaporation) is likely to be slow and limited to volatile solvents. Two versions of this instrument are available, the Quest 210 SLN, specifically for solution-phase synthesis, and the Quest 210 ASW, which is adapted for automated solvent washing and is suitable for solution- and solid-phase chemistry.

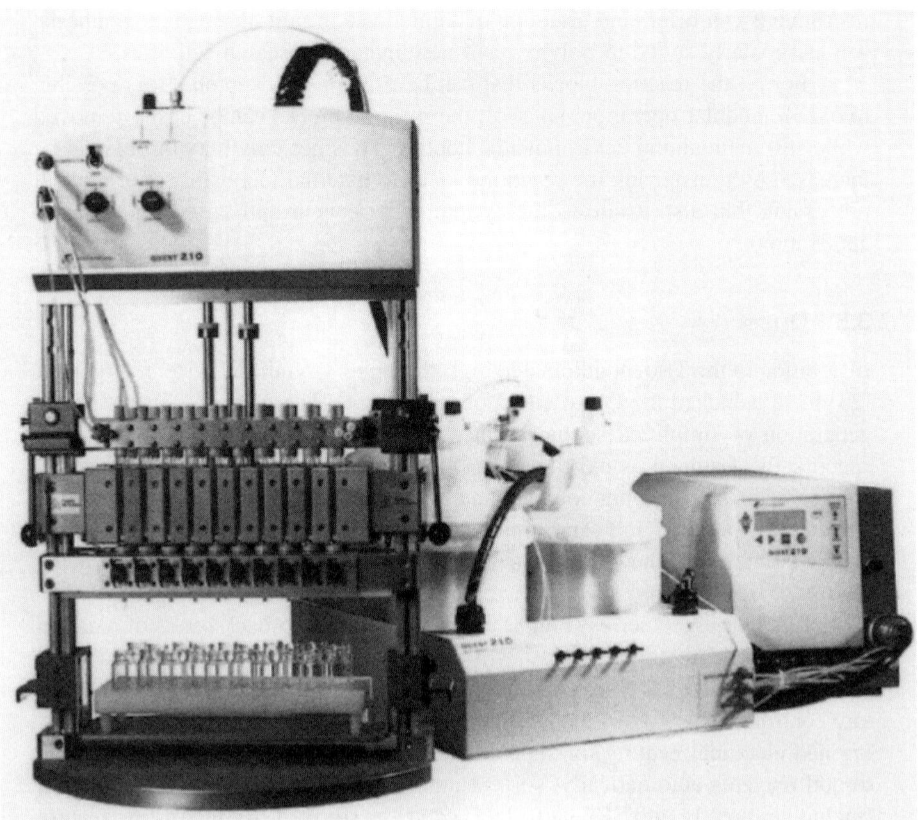

Figure 8 Quest 210 personal synthesizer.

As well as the Quest 210, there is the Quest 205, which is virtually identical except that the reaction block contains 10×100 mL vessels. The larger reaction volume available makes this instrument suitable for preparation of starting materials, resynthesis of active compounds, and use with microreactors, such as IRORI microkans [24]. The Quest 205 and 210 have been used in a coordinated manner for synthesis of 2-aminothiazoles (Scheme 3) [25]. In this case the 2-aminothiazoles were prepared on a 0.5 g scale using the Quest 205, and on-line evaporation was used to precipitate the hydrobromide salts, which were collected by filtration.

The products were then acylated on a 30 mg scale using the Quest 210, where the free base was first generated by reaction with resin-bound base and then automatically transferred to another reaction vessel containing the acid chlo-

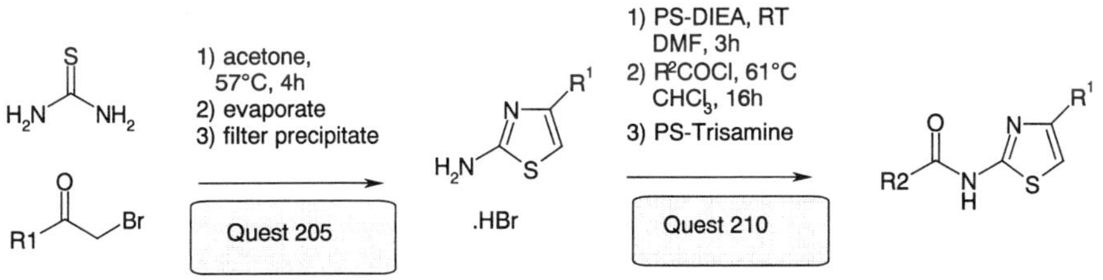

Scheme 3

ride. This illustrates how the large reaction vessels Quest 205 can provide feedstocks for the Quest 210, which has a greater number of smaller reaction vessels.

In addition to the Quest series, there are a growing number of even simpler, completely nonrobotic parallel reactors for use as "personal productivity tools." Argonaut have incorporated many of the components of the Quest instrument into a manual 10-position reactor called the Firstmate. Other such systems include the simple and convenient 12-position Carousel Reaction Station (Fig. 9) from Radleys [26] and the 24-position reactor from J-Kem [27], which can both perform reflux heating.

The Carousel (originally developed by GlaxoWellcome, Stevenage, UK) uses magnetic stirring agitation, while the J-Kem requires an orbital shaker platform, although it has the additional benefit of being able to carry out low-temperature reactions. As part of an evaluation by the Polymer Supported Reagents group at Cambridge University, the Carousel was used successfully to carry out the Buchwald reaction shown below (Scheme 4) [28]. Recently, a variant of the Carousel Reaction Station for subambient chemistry has become available.

These types of reactor may be viewed as space-efficient parallel synthesis apparatus to replace the round-bottom flask and heating and cooling immersion bath apparatus that chemists have been traditionally using for carrying out single reactions. Because these instruments are inexpensive and their operation obvious, they are likely completely to supersede individual reactor apparatus. Furthermore, manual parallel tools for other parts of the synthesis process are also appearing, such as the SynCore Evaporator system from Buchi [29] for simultaneous solvent evaporation from multiple reaction vessels.

3.6 Solaris

The Solaris 530 synthesizer (Fig. 10) is a reengineered version of the CombiTec instrument originally developed by Tecan and now marketed by PE Biosystems [30]. It is a self-contained unit built within an enclosed cabinet, which can be

Figure 9   Carousel reaction station.

Scheme 4

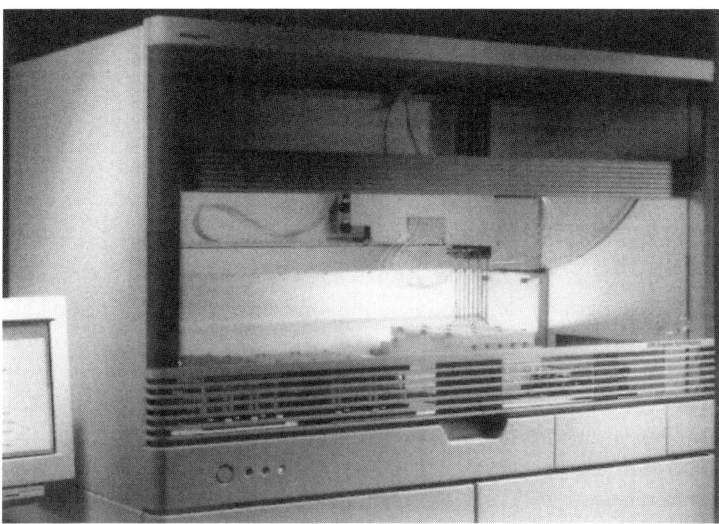

Figure 10 Solaris 530 organic synthesis system.

easily connected to a laboratory venting system. The upper part of the 2 m tall unit contains the worktable and overhead gantry robot, below which are located solvent reservoirs, waste containers, valves, syringes, and a vacuum pump. A key merit of this instrument is that it uses an eight-probe Genesis robot, giving a high degree of parallelization to liquid handling operations. Thus four probes are used for reagent addition and four probes are used for aspiration, and these can be used simultaneously to increase the speed of liquid handling. In addition, the four dispensing probes are coaxial needles, allowing inert atmosphere liquid transfers. For single-probe handlers, the slowness of liquid handling, especially for diversity agents and in resin washing, can often determine the throughput of the instrument. Reactions are carried out in 10 mL round-bottomed glass flasks that are screwed into an inert polymer reaction block and sealed by a double septum. The lightweight reaction block contains 48 flasks and can be removed from the system for off-line operations such as incubation. The aspiration probe accesses each flask via a fritted straw contained within the flask, enabling top filtration to be carried out, and thus both solid- and solution-phase chemistry can be performed in the same reaction vessel. Agitation is affected by controllable orbital shaking, which prevents simultaneous agitation and addition of reagents. The reaction block has a working temperature range of -30 to $+150°C$, and an off-line temperature control system allows reflux heating to be carried out. Diversity reagents are stored in a septum-sealed rack, which can contain 144 12 mL bottles, and common reagents are stored in a 48-position 60 mL bottle rack. Because these, and the 48-position output rack, are located on the instrument deck, they can all be accessed by the multiprobe liquid handler. A number of reports have demonstrated the high chemical performance of this synthesizer, for example in solid-phase chemistry such as the Wittig reaction [31] and also the Mitsunobu [32] reaction shown below (Scheme 5).

Recently Zymark [33] have launched the Combinatorial Chemistry Workstation, which is also a Genesis liquid handler–based synthesizer and was previously developed by Scitec, Switzerland. Unlike the Solaris, the CCW system can use all eight probes in parallel for liquid handling operations, enabling fast parallel processing. The key design feature of this synthesizer is the 96 vessel reaction block with a novel sliding shutter sealing mechanism. The reaction block itself is constructed from a nonthermally expanding material, which contains 96 holes for the disposable glass tube reaction vessels. The properties of the block material ensure that good thermal contact is made between the block during heating and cooling, which is carried out by using a recirculating fluid.

The reaction block is enclosed with a top plate, in which 12 sliding shutters close access to the vessels. Each shutter can be independently activated, to allow simultaneous access to eight vessels at a time (Fig. 11). A constant stream of inert gas is piped into the top plate, thus ensuring that an inert atmosphere is maintained throughout liquid handling when the shutter is open and during reaction. In addition, the upper part of the vessel is cooled by a separate recirculating fluid, enabling reflux heating, and agitation is provided by magnetic stirring. The Scitec block thus offers a particularly elegant solution to inert atmosphere containment, not only avoiding the use of septa and enabling unrestricted access, but also allowing the use of a cheap, disposable reaction vessel. The shutter mechanism is also employed to maintain an inert atmosphere in the reagent racks, which also can be maintained at nonambient temperatures. Another novel feature is the filter probe, which can be employed for solid-phase chemistry, giving efficient resin washing by top filtration. Although not currently widely used, this workstation offers considerable advantages for high-throughput synthesis with very rapid liquid handling and a disposable reaction vessel. However one draw-

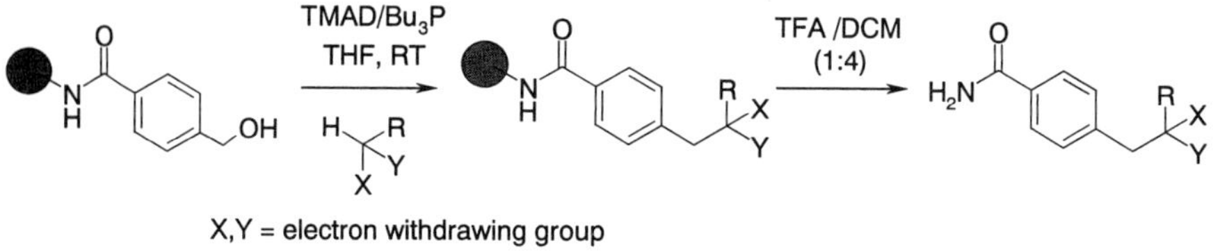

Scheme 5

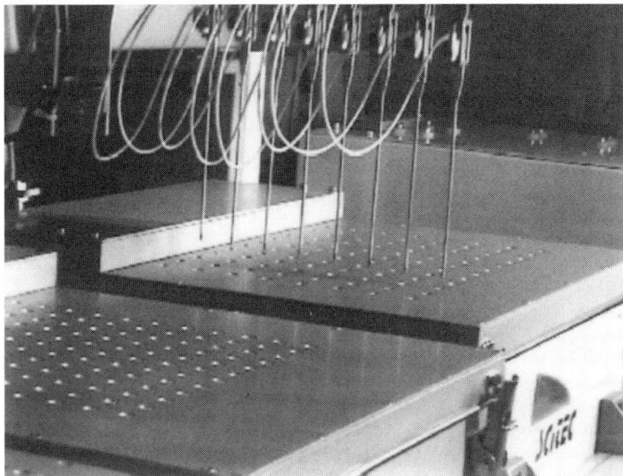

Figure 11 Zymark combinatorial chemistry workstation (detail).

back is that it is not easily adapted to modularity, as the reaction block cannot be conveniently removed for off-line incubation.

3.7 Trident

The Trident [34] system (Fig. 12) was developed as a high-throughput synthesizer by Argonaut [23] in consortium with a number of pharmaceutical companies. This instrument differs from other designs by its use of fluidic technology to carry out liquid handling, with minimal use of robotics. The Trident system uses individual glass round-bottomed reaction vessels that are sealed by a Teflon cap, and these vessels are arranged in 48 vessel cassettes. The capacity of the instrument is four cassettes, i.e., 192 vessels. Reagent delivery is mediated via a multiple-dispensing robotic head, in which a slider valve sealing mechanism between the reaction vessel cap and the cassette maintains an inert atmosphere during reagent delivery and within the reaction vessel. The inert valve distribution system has been specifically designed to operate with organic reagents and is used to meter and deliver reagents and solvents and to collect product from the reactors. Fluid entry and exit is carried out via a fritted straw inside the closed glass reaction vessel. The input of diversity reagents and the output of reaction products are managed by Gilson liquid handlers, but all other reagents can be processed in a closed system. The advantage of this technology is that atmospheric exposure of the reagent, which can occur with robotic liquid handlers, is minimized during handling, thus enabling high chemical performance. The accessible temperature range is -40 to $+150°C$, obtained by use of heated or cooled gas, with liquid

Figure 12 Trident automated library synthesis system.

nitrogen as the refrigerant source. Agitation is provided by variable single-axis oscillation, and each cassette can be independently heated and shaken. Although the Trident system can be operated as a self-contained unit, a number of additional units have also been developed for use in conjunction with Trident. These off-line modules, which can be used independently of Trident, include the Trident Workstation for processing of a single 48 vessel cassette, the Trident Processing Station for workup such as liquid–liquid extraction, and purification using scavenger resin. Since Trident uses sealed reactors, cassettes can be easily transferred between these units without exposure of the reaction mixtures, and hence the throughput of Trident may be significantly enhanced by coordinated use of these off-line modules. In addition to Trident, a 24 reaction vessel instrument called the Nautilus [35], which incorporates the same technology, is available for reaction condition optimization and small library synthesis.

Recently the Trident synthesizer has been used in the high-throughput synthesis of chiral libraries by derivatization of manually synthesized chiral templates [36] (Scheme 6).

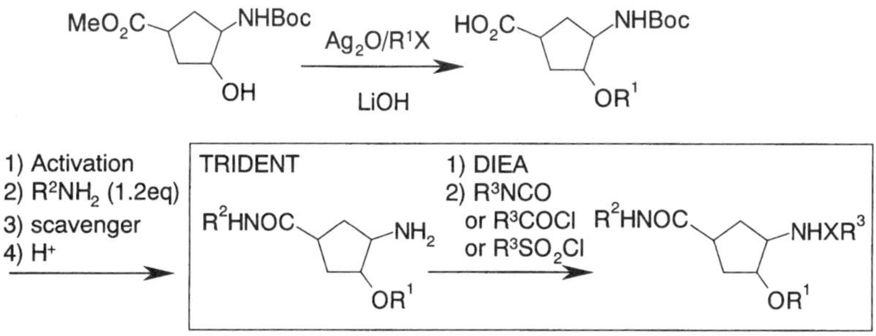

Scheme 6

3.8 Zymark Solution Synthesis System

The Zymate synthesis system (Fig. 13) was co-developed by Zeneca [37] (Alderley Edge, UK) and Zymark [33], and it has been widely used by a number of groups for several years. Unlike all the other workstations described, instead of a Cartesian robot, a cylindrical robot arm provides the automation.

This arm lies at the center of the system (Fig. 14), and various simple devices for carrying out individual processes in the synthesis, such as reagent

Figure 13 Zymark solution synthesizer.

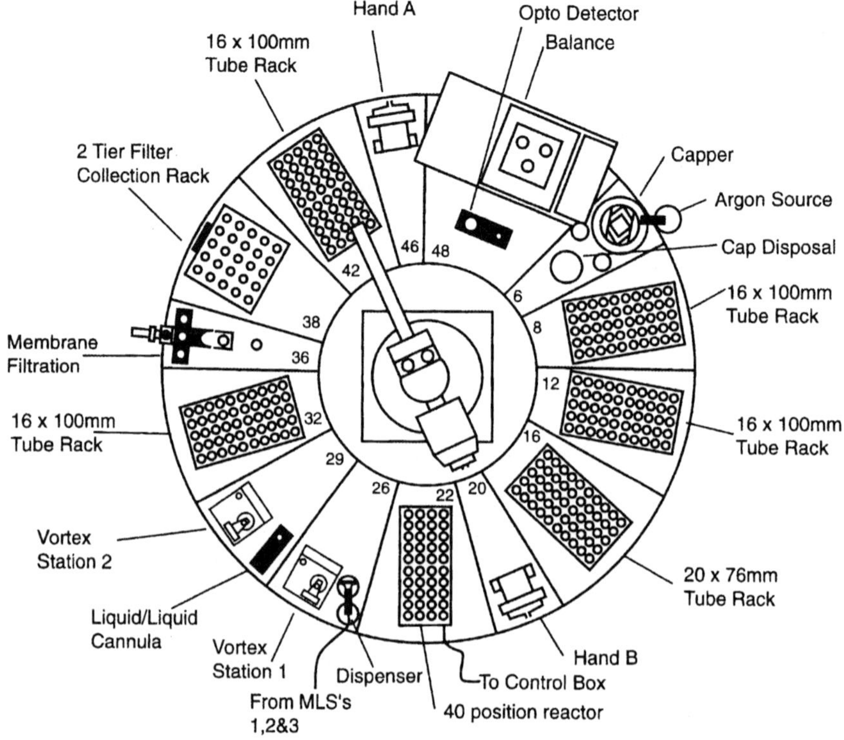

Figure 14 Zymark solution synthesizer layout.

addition, temperature incubation, liquid–liquid extraction, and filtration, are located on a circular table around the arm. The system works at the individual reactor level, and the robot arm is used to move the reactor to the appropriate modules as determined by the software program. This individual processing permits close mimicking of the conditions used in manual synthesis and can give high levels of chemical performance. The reactor is a simple 25×150 mm glass tube, which can be sealed either by a screw cap or a precut septum and enables large working volumes (ca 50 mL) to be used. This is a major advantage of this instrument, since it can perform gram-scale synthesis on 40 to 50 reactions at a time, making it useful for reagent as well as final product synthesis. This scale of reaction is also suited for synthesis of compounds for secondary evaluation or for agrochemical research where large amounts of sample are required for testing. However, because each operation is carried out on the individual reactor scale, parallel processing is not possible, and careful scheduling of the module use is required to obtain efficient high-throughput synthesis.

Because modules in the system can be arranged in any order around the circular table, it gives the system a high degree of flexibility, as the individual components can be replaced or modified as required. In this way, unlike with most workstation systems, it is easily possible to customize the synthesizer to the user's own requirements, for example to include on-line purification [38]. It is also possible to expand the capacity, and thus the throughput, of the system by mounting the arm on a linear track to enable a greater number of racks and modules to be addressed by the robot hands [39]. However, versatility comes at a price, as there are often issues of integrating the components, and the EasyLab control software is open to expert users only. Furthermore, the size of the instrument requires custom-built ventilation cabinets, adding considerably to the costs involved.

4 MYRIAD SYNTHESIS SYSTEM

The Myriad Core System (MCS) is a modular high-throughput, high-performance synthesizer that was originally developed by a consortium of pharmaceutical companies (BASF, Chiroscience, Merck, Novartis, Pfizer, SmithKline Beecham, and Takeda) and The Technology Partnership and is now marketed by Mettler-Toledo Myriad [40]. The fully automated MCS (Fig. 15), which was developed with the aim of providing large libraries of compounds, operates with 192 individual reaction vessels that are independently processed as four sets of 48 vessels and is the first generation of a number of synthesis systems using [41] the same technology. The Myriad Discoverer, which is a synthesis workstation that incorporates key components of the MCS, will be described later in this chapter.

4.1 MCS Description and Key Technologies

The MCS (Fig. 16) is divided into three areas: the processing module, the incubator module, and an input/output module. Individual reaction vessels are placed in minitrays and transported between the modules by a conveyor belt, using tray lifters positioned on each module to place and remove trays from the conveyor. The processor module (PM) is used for reagent additions and workup procedures and consists of an ambient robot processing area (APA), a sub-ambient incubator (SAI), and a solvent processing area (SPA) where solvent addition and solid phase resin washing are carried out. A Huber Unistat 360 recirculating chiller is used in the SAI and is capable of working at $-60°C$. Reagents are stored in removable drawers located between the APA and the SAI. The incubator module

Figure 15 Myriad core system.

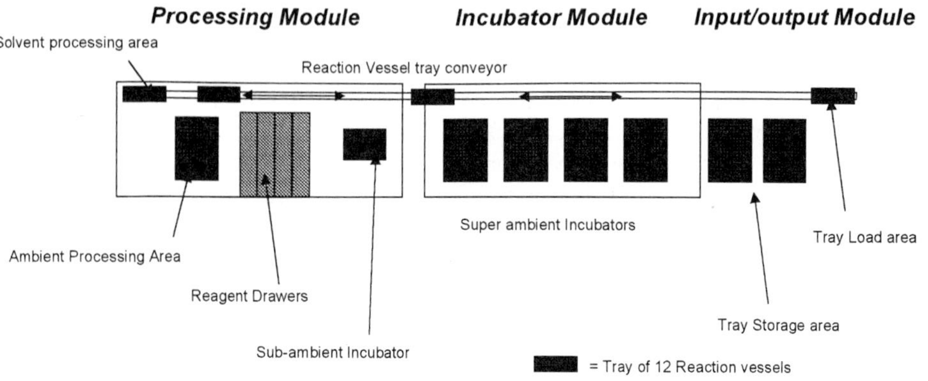

Figure 16 Myriad core system layout.

consists of four banks of electrically heated aluminum blocks into which up to 48 reaction vessels in minitrays can be inserted. Each bank can be independently maintained at different temperatures, and magnetic stirring or gas agitation can be provided. The minitrays are bar-coded, and their progress through the system is continuously monitored and stored on a database.

4.2 Reaction Vessels

Myriad synthesizers employ two different vessels, for solution-phase and solid-phase synthesis, which are both constructed from borosilicate glass and sealed with the same novel twist cap device. The solid-phase vessel (Fig. 17a) is a U-tube design containing a porosity-2 glass frit to retain the resin in the main chamber. The capacity of the main chamber is 10 mL with a working volume of 3 mL, the size of the side arm. The solution-phase vessel (Fig. 17b) has the same physical diameter as the solid-phase vessel but has the U-tube fused to provide a single 17 mL chamber with a maximum working volume of 13 mL. In addition, a 3 mL working volume solution vessel in which only the main chamber is used (i.e., the U-tube has not been fused) is also available.

The twist cap is a novel vessel sealing mechanism constructed from a chemically inert polymer that can be opened by a 90° rotation to expose the main and side chambers. Cap opening on a Myriad synthesizer occurs under an inert gas shroud, thus maintaining an inert atmosphere during reagent and solvent addition or removal. After the vessel access operation is complete, the twist cap is rotated 90° back to the closed position, and the inert gas shroud is removed. The twist cap mechanism thus allows for septum-free access of the reaction vessel and complete inert handling of chemistries without requiring an inert environment

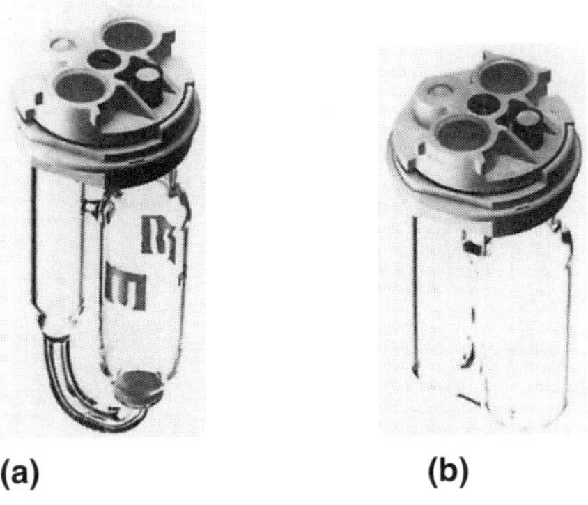

(a)   (b)

Figure 17 Myriad reaction vessels. (a) Solid phase vessel; (b) solution phase vessel.

for the entire robot area. In addition, an inert atmosphere can be maintained inside the reaction vessel by periodic addition of inert gas to the side chamber of the reaction vessel via the gassing pin without the need to open the twist cap.

The reaction vessels are located in minitrays that hold 12 vessels (Fig. 17c) and are the basic processing unit of Myriad. The minitrays enable the reaction vessels to be conveniently transported to all the modules in the system while still allowing the reaction vessels to be processed individually

4.3 Reagent Delivery

Liquid handling is carried out by a multifunctional robot head, which uses chemically inert positive displacement pipettes. These function like disposable syringes and enable accurate dropwise delivery of volatile, viscous, and slurry reagents. The robot head is used to perform a number of operations on the reaction vessel (Fig. 18). The pipette tips are taken up by the robot head and held within an inert gas shroud, allowing all the liquid handling to be carried out in an inert atmosphere. These are disposed of after use, thus avoiding the need for time-consuming washing and preventing cross-contamination. Furthermore, since the pipettes are made from a fluoropolymer, handling of corrosive materials and air-sensitive materials is possible. The rate of withdrawal and dispensing of chemicals can be accurately controlled to ± 10 μL/s to allow both accurate handling of volatile

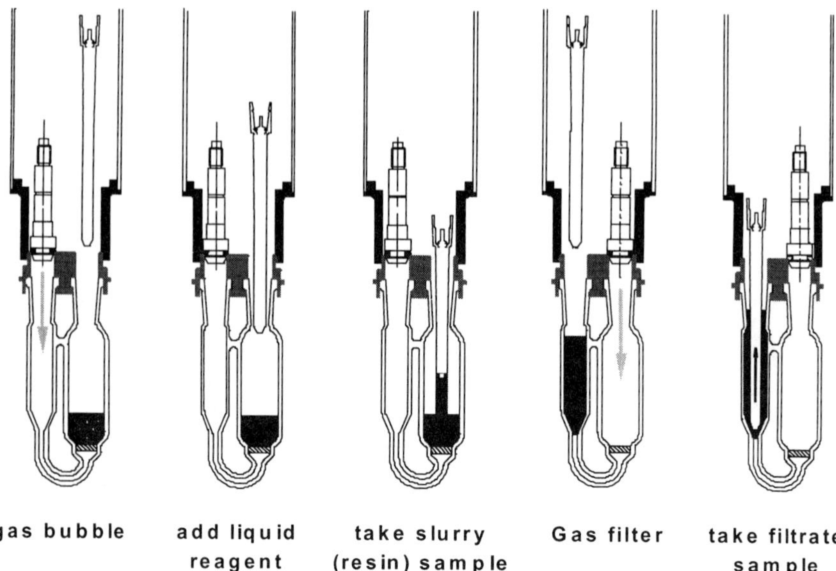

Figure 18 MCS robot head operation.

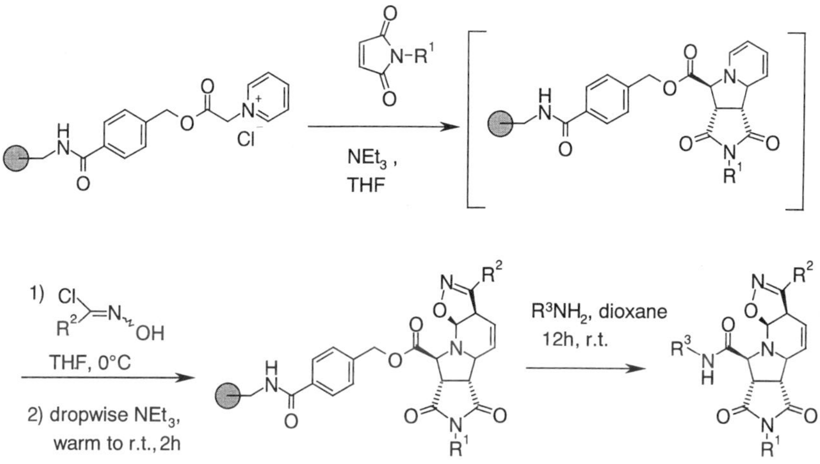

Scheme 7

and viscous materials and dropwise addition of sensitive reagents. This precision and control of reagent delivery, coupled with the ability to add reagents to a cooled and stirring reaction vessel, contributes to the capability of the Myriad systems to carry out high-performance chemistry in a manner analogous to a skilled practical chemist. Such techniques were used successfully to perform a Tsuge reaction on solid phase (Scheme 7), which required the generation of a reactive intermediate in a controlled concentration, to prepare isoxazole-fused tetrahydroindolizines [42].

Another beneficial consequence of using the positive displacement pipettes, which have an aperture of 0.8 mm, is the ability to transfer suspensions or resin slurries.

Other reagents are stored under inert conditions in 7, 14, or 28 mL vials that are sealed by chemically inert split septa, which allow access for the pipette tips and reseal after tip removal. As well as reagent handling, the disposable pipettes can be used to deliver and collect chemicals from either chamber of the reaction vessel. The robot can be used for reagent addition at room temperature in the APA and at sub-ambient temperature in the SAI.

4.4 Solvent Processing

Solvent addition and removal are carried out under an inert atmosphere in a separate unit called the SPA (solvent processing area, Fig. 19). Typically, the major use of the SPA area is work-up of solid-phase reactions by washing the resin to remove excess reagents. The SPA employs a 12-headed unit enabling simultane-

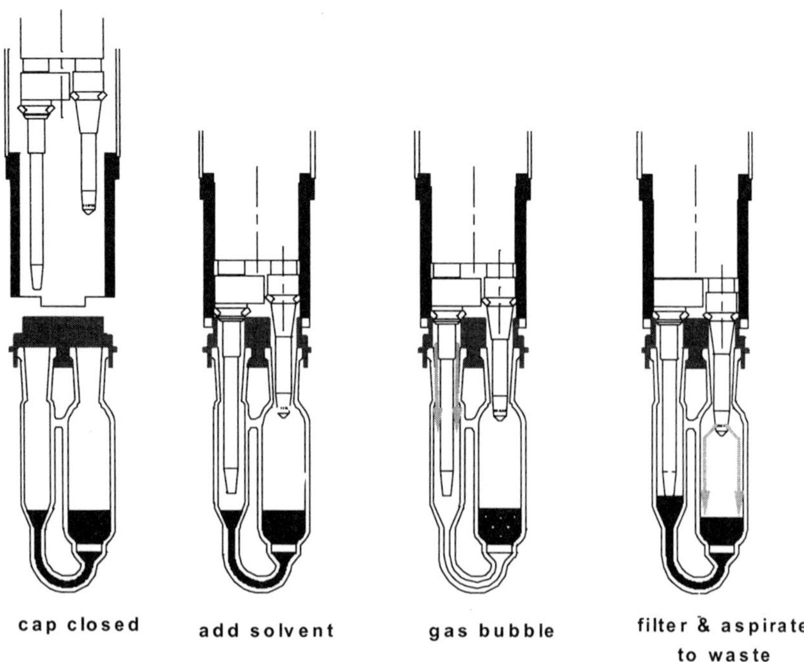

Figure 19 MCS solvent processing.

ous processing of the entire minitray. First the manifold moves over the tray and makes a seal on the twist cap, and then simultaneously it opens the vessels under inert gas using a similar shroud as that used by the robot head. Solvent is introduced into the main chamber of the reaction vessel using an orthogonal spray to wash resin that may have become adhered to the sides of the vessel during the reaction. Introduction of an inert gas through the side arm or magnetic stirring may be used to ensure good mixing of the solvent and resin to obtain an efficient wash process. Aspiration of the solvent occurs from the side chamber via a push-pull mechanism in which inert gas is used to push all of the solvent into the side arm and a vacuum is applied through a needle to the side chamber. The presence of a blocked frit is detected using a pressure sensor. The inert processing possible with the SPA is shown below (Scheme 8), in which a reactive dianion intermediate can be washed prior to reaction with various alkyl halides [43].

4.5 Software

Myriad software is separated into two main components according to function: MyrEd (Myriad Editor) for programming, and MyrMan (Myriad Manager) for

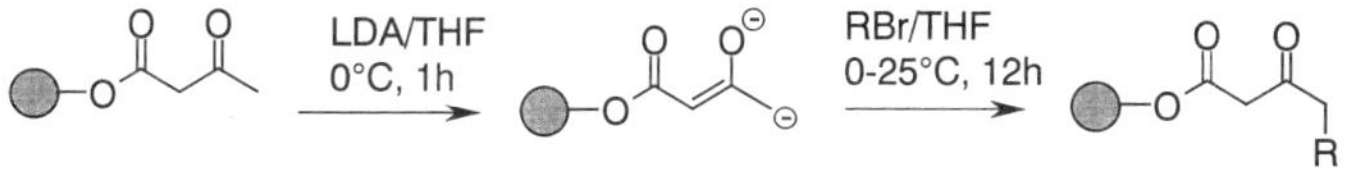

Scheme 8

instrument operation. The chemistry process is written in MyrEd using a simple intuitive graphical interface, which chemists can use after minimal training. The experimental procedure is defined using a standard set of processes, in which each process corresponds to a specific MCS module, such as the APA or Incubator. Within each process, the available operations of that module, in drag-and-drop icon form, can be selected to carry out the relevant task of the experiment. The processes are then joined pictorially in the order in which they are to be carried out to give the complete experiment and are linked to Tray In and Tray Out Processes showing the number of reaction vessels to be processed. The whole program is called a Job, which not only defines the synthesis but also generates a file (called the shopping list) with all the resources (reagents, solvents, pipettes) required. Each MyrEd Job can process a maximum of four trays (48 vessels) and is automatically validated to ensure it contains no errors and is suitable for use on the instrument using MyrMan, which has a similar graphical interface to MyrEd. The Myriad Core System is capable of operating four different jobs (192 vessels) simultaneously. The progress of each Job and the state of each module are represented pictorially, as well as the currently available quantity and location of system resources such as solvents and reagents. MyrMan has an error-checking capability, ensuring that only experiments for which appropriate resources are available, including waste containment, are performed. All data relating to Myriad is stored on a database, enabling easy tracking of current and previous experiments. In addition, MyrMan has an error recovery capability to enable malfunctions to be detected and resolved if they arise during an experiment. For a lower level of instrument control, for example in the case of serious faults or engineering, a more advanced level of software is available but it is limited to expert users. Thus by separating the software into components for programming, routine operation, and engineering, the use of Myriad becomes accessible to non-expert chemists without compromising the sophistication essential for an instrument of this kind.

4.6 Other Myriad Modules

4.6.1 Myriad Discoverer (PS)

The Discoverer (Fig. 21) (originally called the Myriad Personal Synthesizer) evolved from the MCS and, although it was initially designed to enable chemis-

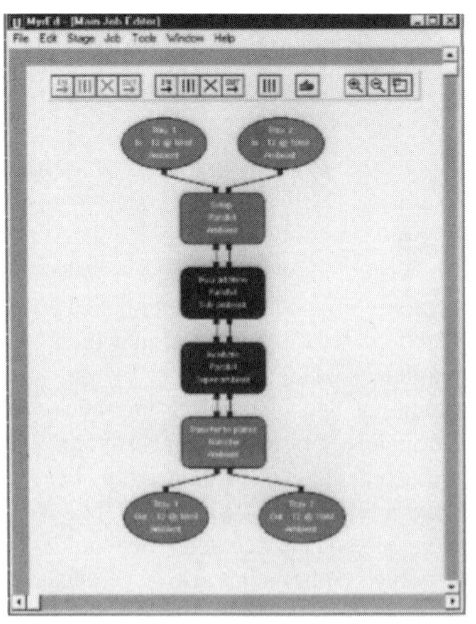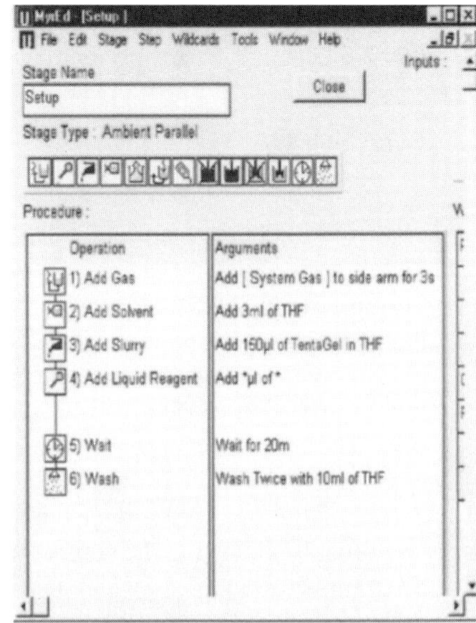

Figure 20 Myriad software.

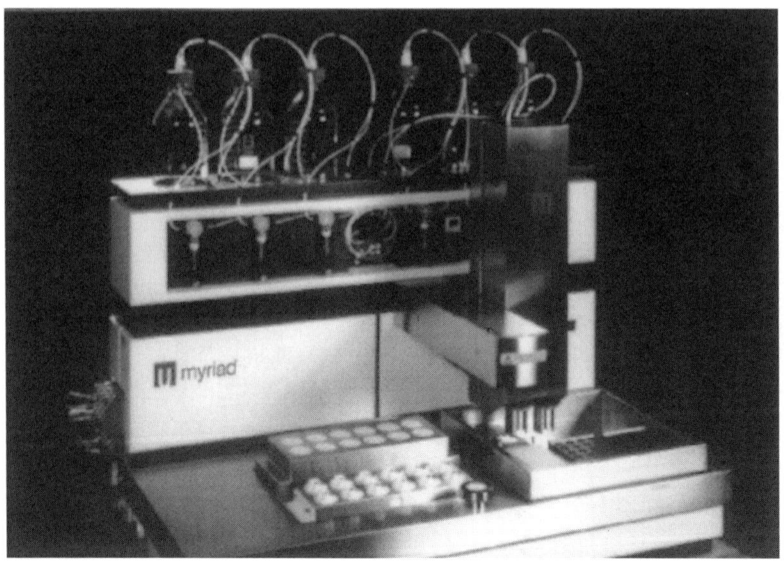

Figure 21 Myriad Personal Synthesizer.

tries to be developed on the Discoverer and then transferred for larger scale library production to the MCS, it can also be successfully used as a stand-alone synthesizer. The Discoverer has 24 reaction vessels and runs within a temperature range of -60 to 150°C. MyrEd is common to both the Discoverer and the MCS, and therefore procedures are fully transferable. Both systems also use identical reaction vessels, reagent vials, and pipetting delivery systems, therefore enabling reaction conditions developed on one system to run without further optimization on the other.

4.6.2   Myriad Discoverer Plus

The Discoverer Plus (Fig. 22) is a compact unit designed to increase the throughput of the PS. It has 48 reaction vessel positions, thus increasing the throughput of the Discoverer to 72 reactions. Four independent temperature zones can be used with heating capability up to 150°C, and magnetic stirring, gas agitation, and inert gas can be applied in all vessel positions. Operation of the remote incubator is simplified by use of a built-in controller and display.

4.6.3   Myriad ALLEX (Automated Liquid–Liquid Extractor)

ALLEX is a dedicated automated liquid–liquid extractor workstation developed to address the most common workup procedure in solution-phase chemistry. The

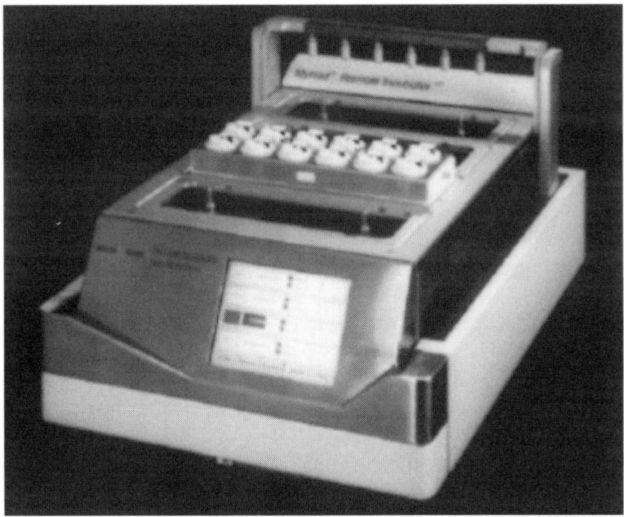

Figure 22   Myriad remote incubator.

Figure 23   Myriad ALLEX.

unit is composed of a Bohdan chassis that is fitted with Myriad sensor technology that can detect the boundary between two immiscible liquids (Fig. 23). The liquid interface sensor works by sensing electrical differences between the two solvents and then determining the cut-off point to give two layers. If no boundary is detected, this indicates that the sample may have formed an emulsion, which is then registered in the software for later reprocessing at the determination of the chemist. The detector and settling chamber can be washed with multiple solvents between separation processes to prevent cross-contamination between samples. This washing procedure using deionized water also allows the use of brine and other saturated salt solutions, as residues can be washed away between separations, preventing any blockages from occurring.

5   COMPOUND FACTORIES

The development of automation for high-throughput compound synthesis has predominantly progressed to provide tools to help improve individual productivity. These tools range from automated workstations to simpler devices such as nonautomated parallel reactors that have been specifically designed for use by synthetic chemists in a laboratory environment to accelerate research. While the skilled laboratory organic chemist will never be replaced by a machine, clearly the only practicable way to achieve routine high-throughput synthesis is by mechanization of most, if not all, the tasks in the synthesis process. This progressive mechanization not only increases the space required to house the necessary instruments but also leaves less practical work for the skilled organic chemist to perform. Hence efficient high-throughput synthesis progresses from a research to a manufacturing activity, and it is with this in mind that the concept of a "compound factory" has arisen [8]. By analogy to the large high-throughput screening systems that are currently in use, it is also possible to construct systems for ultra-high-throughput synthesis. Such systems can be built by the combination of existing workstations to create production lines for synthesis. This approach has been pioneered by companies such as Ontagen [44], Panlabs [45], and ArQule [46], which have used customized as well as commercially available instrumentation. As the number of reliable workstations for all aspects of the synthetic process increases, construction of compound factories will become accessible to groups lacking specialist engineering skills.

The Myriad system has been designed with the factory concept in mind, enabling a number of specialist modules to be linked to perform all the tasks required for compound preparation. Linkage of these modules by common control software can enable a factory environment to be established (Fig. 24).

It is envisaged that the Myriad Discoverer will be used as a rehearsal instrument to develop chemistry for the MCS, which, because of the high compatibility between the two instruments, is likely to present few transfer problems. At SmithKline Beecham (UK), for example, Myriad Discoverer systems are widely distributed in medicinal chemistry research laboratories for the development of synthetic protocols towards potential drug candidates, and the MCS instruments are located in a production environment for high-throughput preparation [47]. In this way it is possible to establish a compound factory facility that is responsive to the requirements of medicinal chemistry groups.

A key requirement to this approach is of course the facility in which it is to be housed, as the requirements of space and facilities are not met by most laboratory designs. Moreover, the operation of this system no longer requires skilled chemists, but only well-trained operators, to keep the process operating on a routine if not 24-hour basis. Clearly there are parallels to the scale-up pilot plant facilities already present in pharmaceutical companies. With the availability of ultra-high-throughput screening systems and the compound factories, there is no reason (except history) to segregate these activities, and the "factory" concept is logically extended to develop a facility to include synthesis and screening as a continuous process. This is obviously the goal of the microtechnologies [48],

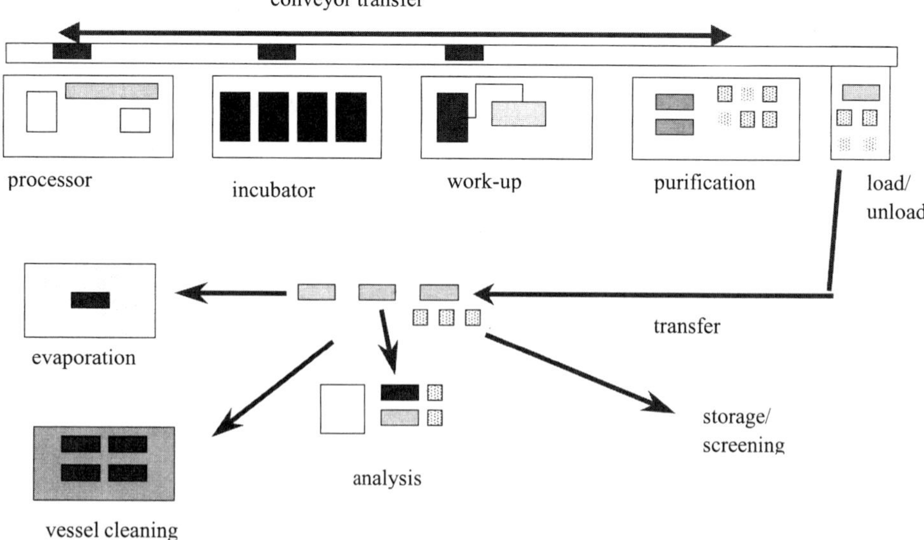

Figure 24   Myriad compound factory.

although major new technology developments will be required to produce operational systems. However, in the near future, the use of compound factories may be integrated into the operation of pharmaceutical companies and become major accelerators of the drug discovery process. Indeed, at SmithKline Beecham (UK) plans are nearing completion to construct a specially designed building to accommodate not only high-throughput chemistry but also high-throughput screening and genetic technologies. These platform technologies are key components of the drug discovery process, which rely on extensive automation to achieve operational efficiencies. The successful integration and optimization of this automation can be obtained in a high-tech factory environment, which is not limited by traditional laboratory design. Such an environment would have to be flexible to manage the rapid changes in the size and the format of automation. Laboratory furniture, air-handling, hazard containment, power supplies, gas services, and safety features would also need to be adaptable. To achieve these requirements, a large robotics hall with a minimum of structural impediments, such as pillars, low ceilings, and nonfunctional walls, is envisaged in which easily accessible services are arranged overhead in a grid covering the hall and furniture and fumehoods are movable. Thus a rapid change in operational equipment can be easily achieved with only minimal need for engineering support.

6 CONCLUSION AND FUTURE PROSPECTS FOR AUTOMATED WORKSTATIONS

The role of automation in compound synthesis will undoubtedly become increasingly important as more reliable and robust instrumentation is developed. At the moment, the use of robotic workstations in organic synthesis laboratories, while not uncommon, is far from routine. Often the decision of whether to use a particular instrument is determined by whether its capabilities are perceived to be appropriate to the required synthesis. In other words, the synthetic task is usually preselected as being suitable for robotics. This in itself is not surprising, and indeed the early history of combinatorial chemistry is replete with examples specifically selected "easy" syntheses, designed to yield large numbers of compounds. In other words, for a technology to be genuinely applied, its use should not restrict the nature of tasks it seeks to automate. The challenge for the future is to develop instruments until the use of automation is independent of the nature of the synthetic problem. No practical chemist thinks twice about using a round-bottomed flask to perform a synthesis; automation workstations need to achieve the same level of confidence among chemists.

The impact of automation on high-throughput synthesis is unlikely to be measured by numbers alone, but by whether it can deliver the compounds to the same standard as hand-crafted ones and in massively greater numbers.

The implementation of any technology can only be justified if it answers a simple question, "Does it make the process better?" If this criterion is not satisfied, the technology, however advanced, becomes irrelevant. It seems inconceivable that the first efforts to mechanize organic synthesis will not be built upon to enhance and accelerate discovery chemistry. However, the real successes will come to the organizations that balance the technology with their own needs and select the most appropriate technology, not necessarily the most advanced technology, for their research. Another key factor for the future will be the cost/compound analysis. This not only includes the capital cost of the instrumentation and its servicing and maintenance but also the cost of consumables. The latter is particularly important for high-throughput applications, as is seen with high-throughput screening, where the cost of reagents can become the limiting factor rather than the throughput capability of the instrument.

In conclusion, this chapter has summarized some of the current platforms for high-throughput synthesis, with particular focus on the Myriad systems. It is unlikely that any one synthesis system will dominate, but probable that a few kinds of automation platform will be accepted as standard over the next few years. For a start, efficient mechanized high-throughput synthesis will require several automated platforms to provide not only the final products but also feedstock materials for the high-throughput instruments. It seems highly improbable that a single instrument can have the widely differing performance requirements to carry out both high-throughput and large-scale synthesis. As the process of organic synthesis becomes progressively more mechanized, the workforce by necessity will acquire new skills. The sophistication of the synthetic chemist has rapidly increased in virtually all areas, with most having practical experience of, for example, computational modeling and automated analytical techniques.

Paradoxically, until recently, the least altered task was the actual practice of synthesis, although the use of automation and personal productivity tools is becoming commonplace. A decade ago a PC was a fairly rare sight on a chemist's desk, but now it is an essential item (indeed many chemists have to be prized away from them to enter the laboratory!). Perhaps we can speculate that a decade from now, personal robotic synthesis workstations will also be part of the standard armory of every chemist.

ACKNOWLEDGMENTS

The authors would like to thank the many colleagues throughout the industry who have kindly helped in the preparation of this article, in particular John Bergot (PE Biosystems), Jim Harness (Bohdan), and Jeff Labadie (Argonaut Technologies).

REFERENCES

1. Gallop, M. A. et al. J. Med. Chem. 37:1233–1251, 1994.
2. Lindsey, J. S. Chemometrics and Intelligent Laboratory Systems. Laboratory Information Management 17:15–45, 1992.
3. Sugawara, T., Cork, D. G. Laboratory Robotics and Automation 8:221–230, 1996.
4. Hird, N. W. Drug Discovery Today 4:265–274, 1999.
5. Christensen, M. L., Peterson M. L., Saneii, H. H., Healy, E. T. In: Peptides: Chemistry and Biology (Kaumaya, P.T.P, Hodges, R. S., eds.), 1996, pp. 141–143.
6. Selway, C. N., Terrett, N. K. Bioorg. & Med. Chem. 4:645–654, 1996.
7. Merritt, A. T. Drug Discovery Today 3:505–510, 1998.
8. Archer, R. J. Assoc. Lab. Autom. 3:52–55, 1998.
9. Hardin, J. H., Smietana, F. R. High Throughput Screening (Devlin, J. P., ed.). pp. 251–261, 1997.
10. Dewitt, S. H., Czarnik, A. W. Current Opinion in Biotechnology 6:640–645, 1995.
11. Warr, W. J. Chem. Inf. Comput. Sci. 37:134–140, 1997.
12. Charybdis Technologies, 5925 Priestly Drive, Suite 101, Carlsbad, CA 92008, USA. www.charybtech.com.
13. Zinsser Analytic, D-60489 Frankfurt, Eschborner Landstrasse 135, Germany. www.zinsser-analytic.com.

14. Robbins Scientific Corporation, 814 San Aleso Avenue, Sunnyvale, CA 94086-1411, USA. www.robsci.com.
15. Chemspeed Ltd., Rheinstrasse 32, CH-4302 Augst, Switzerland. www.chemspeed.com.
16. Guller, R. 2nd International Conference on Microplate Technology, Laboratory Automation and Robotics, Montreux, Switzerland, May 17–21, 1999.
17. Personal Chemistry, Hamnesplanaden 5, SE-753 19 Uppsala, Sweden. www.personalchemistry.com.
18. Loupy, A. et. al. Synthesis:1213–1234, 1998.
19. Bohdan Automation, 562 Bunker Court, Vernon Hills, IL 60061, USA. www.bohdan.com.
20. Cheshire, D. EuroLabAutomation'98, 26–30 September, Oxford, UK, 1998.
21. Felder, R. J. Assoc. Lab. Autom. 4, 46–47. Goodman, B. A. (1999) ibid. 48–52, 1999.
22. Brase, S., Dahmen, S., Heuts, J. Tetrahedron Letters 40:6201–6203, 1999.
23. Argonaut Technologies, 887 Industrial Road, Suite G, San Carlos, CA 94070, USA. www.argotech.com.
24. Xiao, X. et al. Biotechnol. Bioeng. (Comb. Chem.) 71:44–50, 2000.
25. Yun, Y. K., Leung, S. S. W., Porco, J. A. Biotechnol. Bioeng. (Comb. Chem.) 71: 9–18, 2000.
26. Radleys Discovery Technologies Ltd., Shire Hill, Saffron Walden, Essex, CB11 3AZ, UK. www.radleys.co.uk.
27. J-Kem Scientific Inc., 6970 Olive Blvd., St. Louis, MO 63130, USA. www.jkem.com.
28. Baxendale, I.R. Testing results for the Radleys Carousel reaction station. Available from Radleys Discovery Technologies Ltd.
29. Buchi Labortechnik AG, Postfach, CH-9230 Flawil, Switzerland. www.buchi.com.
30. PE Biosystems, www2.perkin-elmer.com/530.
31. Salvino, J. M., Kiesow, T. J. ISLAR'97 Proceedings, 1998, pp. 99–106.
32. Chaturvedi, S., Otteson, O., Bergot, J. Tet. Letts. 40:8205–8209, 1999.
33. Zymark Corporation, Zymark Center, Hopkinton, MA 01748-1668, USA. www.zymark.com.
34. Deegan, T. ACS meeting, Boston, 25 September 1998, 1998.
35. Porco J. A. et al. Molecular Diversity 2:197–206, 1997.
36. Beroza, P., Suto, M. J. Drug Discovery Today 5:364–372, 2000.
37. Main, B. G., Rudge, D. A. ISLAR'93 Proceedings, 1994, pp. 425–434.
38. Rudge, D. A., Crowther, M. L. ISLAR'97 Proceedings 1998, pp. 264–274.
39. Walter, G. LabAutomation 2000, Palm Springs, CA, 24–26th January, 2000.
40. Mettler-Toledo Myriad, 2 Saxon Way, Melbourn, Royston, Hertfordshire, SG8 6DN, UK. www.mtmyriad.com.
41. Hird, N., MacLachlan, B. ISLAR'98 Proceedings CD, 1999.
42. Brooking, P. et al Synthesis 11:1986–1992, 1999.
43. MTM internal data.
44. Cargill, J. F., Maiefski, R. R. Laboratory Robotics and Automation 8:139–148, 1996.
45. Garr C. D. et al. Biomolec. Screening 1:179–186, 1996.
46. Gustafson, G. R. et al. Tetrahedron 54:4051–4065, 1998.
47. MacLachlan, B. Lab Automation 2001, Palm Springs, CA, 27–31st January, 2001.
48. Vetter, D. Drug Discovery Today 3:253–254, 1998.

2

Nonrobotic Automated Workstations for Solution Phase Synthesis

Tohru Sugawara
ChemGenesis, Inc., Tokyo, Japan

David G. Cork
Takeda Chemical Industries, Ltd., Osaka, Japan

1 INTRODUCTION

It is possible to categorize the various automated apparatus that have been developed for synthesis based on their design, with two main categories, those using robotic transfers and those using flow lines for transfer between fixed reactors [1]. In this chapter we will describe some of the unique features of the flow line–based automated apparatus that have been developed in the research laboratories at Takeda Chemical Industries Ltd., Osaka [2] and present some examples of their use for syntheses. These will show applications of automated synthesis, analysis, extraction, and purification of compounds.

The latest apparatus have been designed to function as automated workstations with dedicated purposes. They are used for preparing libraries of compounds by combining a series of reactants, in all or selected permutations. The common series of operations used in solution phase synthesis are automated to allow continuous and unattended preparation [3]. The apparatus are mainly used for medium- to large-scale synthesis, for preparing focused libraries of compounds, or producing series of reactants for subsequent use in smaller scale combinatorial library synthesis [4].

2 DEVELOPMENT OF FLOW LINE–BASED AUTOMATED APPARATUS: DEVICES, UNITS, WORKSTATIONS, AND SYNTHESIS SYSTEMS

The development of any automated synthesis apparatus can be broken down into the parts that make up the whole. Complete systems may be made up of workstations with specific objectives, which in turn are composed of units for performing the different steps involved in synthesis. Ultimately, the development of the units relies on the availability of devices and parts [1,5]. Below we look at some of the main devices and units that have been developed for our flow line–based automated apparatus.

2.1 Synthesis Unit

Takeda's automated apparatus consists of units for performing various tasks, carrying out the reactions, supplying the reagents, reactants and solvents, performing two-phase extraction, separating compounds by chromatography, monitoring reactions by HPLC, and washing the whole apparatus [2–3,6–8]. A photograph and schematic diagram of the unit layout of a system is shown in Fig. 1 and a photograph of three flow line–based automated workstations, RAMOS, ASRA, and SOAP, is shown in Fig. 2. Table 1 lists the automated systems and workstations that have been developed at Takeda.

Key devices for any flow line synthesis system are the solenoid valves. For high chemical inertness Teflon™ is generally the material of choice, but perfluorinated elastomers have been used beneficially for making flexible parts such as the moving diaphragm. For our apparatus we arranged with the manufacturer (Takasago, Japan) to redesign the shape and direction of the holes in the Teflon™ body, as shown in Fig. 3. The holes for the normally closed and normally open connections were made larger in cross section, and they were drilled at an angle, to decrease the number of 90° turns and thus resistance to flow through the valves.

The reaction unit is perhaps the most fundamental unit of any synthesis apparatus, since it determines what chemistry can be carried out. Some kind of reaction vessel with means of agitation, temperature control, and connections for the transfer of materials in and out is required. In developing flow line–based apparatus we have used four reaction flask designs:

    Jacketed flasks
    Round-bottomed flasks
    Column-shaped flasks
    Flanged separable flasks

Reagents and solvents are held in reservoirs and delivered to the reaction flasks through Teflon™ flow lines, the flow being controlled by the Teflon™ solenoid valves.

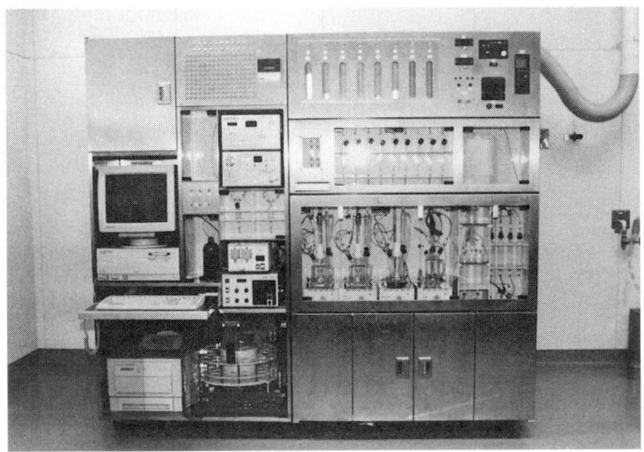

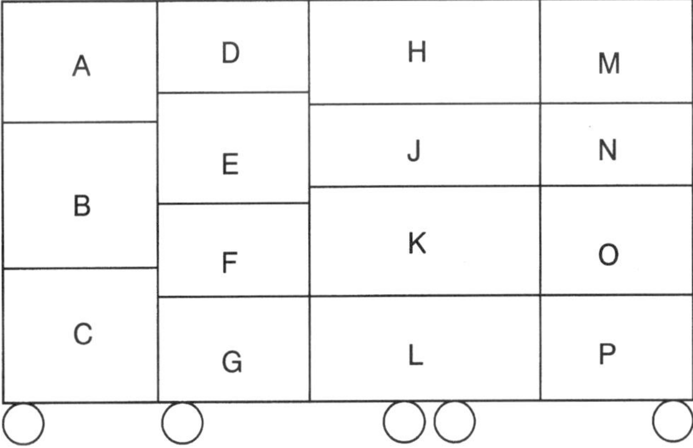

Figure 1 Photograph and schematic diagram of the unit layout of EASY. A = control interface, B = computer, C = keyboard and printer, D = monitor panel and manual switches, E = analytical HPLC, F = preparative HPLC, G = fraction collector, H = solvents and common reagents, J = measuring tubes, diversity reagents, K = four reaction flasks, L = hot and cold baths, M = temperature controllers, N = washing solvents, O = liquid–liquid extraction unit, P = vacuum pump and waste.

2.1.1 Reaction Flask Design

Jacketed Flasks. A schematic diagram of one jacketed reaction flask that is used in EASY is shown in Fig. 4. Stirring is achieved using a mechanical stirrer with a long flexible drive shaft. Temperature control, between about −15 and 80°C, is carried out by circulating hot or cold ethylene glycol through the outer jacket [6–8]. The flask contents can be transferred out using a Teflon™

Figure 2   Photograph of the workstations RAMOS, ASRA, and SOAP.

tube that is held at one side of the flask, outside the sphere of the stirrer, extending to the bottom of the flask, which is designed so that the last drops of solution will collect where the tube is placed.

Round-Bottomed Flasks.   A photograph of the reaction unit in our workstation RAMOS is shown in Fig. 5, and a schematic diagram of the reaction flask is shown in Fig. 6. The reaction flask has a round bottom and conventional ground glass joints, which allow it to be easily removed for manual operation and occasional maintenance. The flask, which can hold a reaction mixture of about 200 mL, is clamped firmly, and the reaction bath is raised and lowered by a motorized jack. The bath was designed to enable a wide range of temperatures to be used, with heating being performed by a sheath heater immersed in the bath and cooling by circulation of a coolant (e.g., aqueous ethylene glycol) through the inside of the bath walls. With silicon oil in the bath a temperature range of ca. 15 to 230°C can be used, while aqueous ethylene glycol gives a range of ca. −25 to 80°C. Operation of the controllers for heating and cooling is managed by the main control computer, to achieve the desired temperature.

In order to transfer solutions from these round-bottomed reaction flasks we needed to devise a means to draw solutions from the bottom of the flask without interfering with stirring. We have found that a mechanical stirrer with two-speed control (high and low) is a versatile and robust means of stirring, and in order to remove solutions from the flask we devised a means to draw solutions up through the central glass stirring shaft, as shown in Fig. 7. The lip seals on the central shaft, at the top of the flask, and at the uppermost joint are crucial to enable vacuum to be applied, to transfer solutions to and from the flask. The seals (R408-2, Fluid Metering, Inc., USA) are made of fluorocarbon specifically formulated for resistance to wear, abrasion, heat, and chemical attack. They possess an exceptional mechanical memory, which allows them to maintain a relatively constant wiping pressure on the stirring rod.

The two-speed capability of the mechanical stirrer allows gentle mixing during reaction on the one hand but rapid emulsification for two-phase extractions or vigorous agitation for flask washing, etc.

Column Shaped Reactors. Solid catalysts and reagents are widely used in organic synthesis, and their importance is growing as new ones are developed and the benefits of easy workup are realized [9,10]. We developed one automated apparatus capable of efficiently handling solid catalysts in a fixed-bed column-type reaction flask [6,11]. The apparatus can be used independently or connected to our other workstations for automated purification and/or further reaction.

The column-shaped reactor is shown in Fig. 8. The reactor is clamped firmly in a jacketed holder through which a coolant can be circulated. Heating can be facilitated in a similar manner or by a thin film (capton) heater that is attached to the surface of the holder in contact with the glass reactor. This combination allows a temperature range of ca. 5 to 100°C to be obtained. Additionally, some cooling is facilitated by blowing air over the column reactor using a fan.

The flask design also includes two screw connectors for tubing and a screw-capped side port for the manual addition of solid material. Filtration of the product from the solid catalyst is achieved by means of a porous glass (e.g., G3 or G4) or Teflon™ filter (e.g., 5 μm) that is fitted at the bottom of the reaction flask.

Agitation of the flask contents is achieved by shaking. A stepping motor (Oriental Motor, Japan), with control unit (Technolab, Japan), is used to operate the shaking of the reaction flask in two modes; a swing of ca. 40° from the vertical and a rocking seesaw motion of ca. 50° from the horizontal, as shown in Fig. 9.

Flanged Separable Flasks. The design of a reaction flask for using powdered supported reagents (SRs) is shown in Fig. 10 and the appearance in Fig. 11. The top of the flask is clamped firmly, but the bottom part is separable, allowing the size to be changed according to the scale and also enabling easy maintenance and cleaning. Motors for a stirrer and a powder addition funnel (PAF) are mounted at the top and side, respectively. The powdered SR is stored in the hopper reservoir, of variable volume, and addition of the reagent into the flask is controlled by turning the Teflon™ screw with the stepping motor. The motor and screw shaft are connected via a zero-backlash universal joint to give smooth turning. The top of the flask has a cooling jacket to enable the upper part

Table 1 Automated Synthesis Systems and Workstations Developed at Takeda Chemical Industries, Ltd.

Generation	Years	Name (Type)	Features
1. 1985–1987	(system)	The prototype	Two reactors, 0–80°C, column chromatography, predictive logic control algorithm, fixed control sequence.
2. 1988–1989	TACOS (system)	Takeda's Automated Computer Operated System	Three reactors, 0–80°C ($\times$2), 25–180°C ($\times$1), large reservoirs and measuring for multi/repetitive use, column chromatography, HPLC reaction monitoring, unit operation control program.
	MAVIS (system)	Multipurpose Automated Versatile Intelligent System	
3. 1990–1993	VACOS (system)	Versatile Automated Computer Operated System	Three reactors, 0–80°C ($\times$2), 25–180°C ($\times$1), reservoirs for single use, thermal sensor for solvent evaporation, conductivity sensor for phase separation, column chromatography, HPLC monitoring, versatile unit operation control.
	MATES (system)	Multipurpose Automated Technical Equipment for Synthesis	
	TARO (system)	Takeda's Automated Reliable Operator	
	EASY (system)	Expert Automated Synthesizer for You	

4. 1994–1997	RAMOS (workstation)	Reliable Automated Multireactor Organic Synthesizer	Three reactors, -15–$180°C$ ($\times 3$), synthesis and phase separation units only, quick-fit reaction flasks and reagent reservoirs, precipitate sensor.
	TAFT (workstation)	Takeda's Automated Flow Technique	
	ASTRO (workstation)	Automated Synthesis with Totally Reliable Operation	
	FUTOSHI (workstation)	A Japanese nickname	One reaction flask of 2 liters, phase separation
	ASRA (workstation)	Automated Supported Reagent Apparatus	One reactor, -15–$180°C$, handling of solid powdered reagents, workup by extraction and filtration from solid.
	ASCA (workstation)	Automated Solid Catalyst Apparatus	One fixed-bed type reactor, 15–$80°C$, solid powdered catalysts, multibatch/cycles possible.
	SOAP (workstation)	Standalone or Option-type Apparatus for Purification	Preparative scale column chromatography, with fraction collector.
5. 1997–	WOOS (workstation)	Windows Operated Organic Synthesizer	One reactor with 1 liter flask, -15–$180°C$, Windows operating system.
	TACOS-II (workstation)	Takeda's Automated Computer Operated System II	Two reactors, -15–$180°C$, centrifugal two-phase separation, panel PC control.

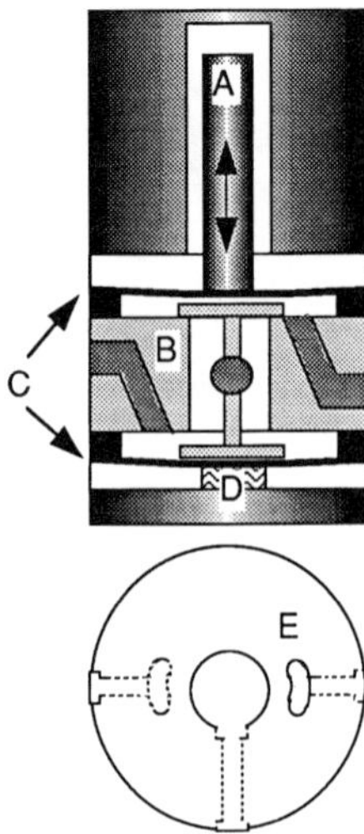

Figure 3 Specially designed Teflon™ solenoid valves for flow line synthesis apparatus. A = solenoid plunger, B = Teflon block with diagonally drilled flow tubes, C = flexible Teflon diaphragms, D = spring, E = enlarged mouth to flow tubes for greater flow rate.

of the flask to be kept dry, even under reflux conditions, and the solid reagent flows smoothly into the reaction mixture under a gentle flow of nitrogen to both the top of the hopper reservoir and the flask. The flask design also includes two screw connectors for tubing adapters and a screwed side arm for a condenser.

Filtration of the product from the solid SR is achieved by means of a porous Teflon™ ball filter that is attached to the tip of a Teflon™ tube connecting to a collection flask (e.g., a 300 mL round-bottomed flask). Solid waste material is removed from the reaction flask through a wide bore Teflon™ tube (5 × 4 mm) after a suitable wash solvent is added and mixed to give a suspension. The suspension is rapidly transported into a waste collection bottle by suction through a pinch valve (ASCO, Japan).

The flask is clamped firmly and the reaction bath is raised and lowered by a motorized jack, as for the round-bottomed flasks described above.

2.1.2 Handling Addition of Solid Reagents

In ASRA [11], the automated addition of powdered reagents from a hopper reservoir into the separable reaction flask is achieved by controlled turning of a Teflon™ screw shaft. A stepping motor, with a unit step of 7.2°, is controlled by two output signals (start and reverse) and the angle of each turn is set by a

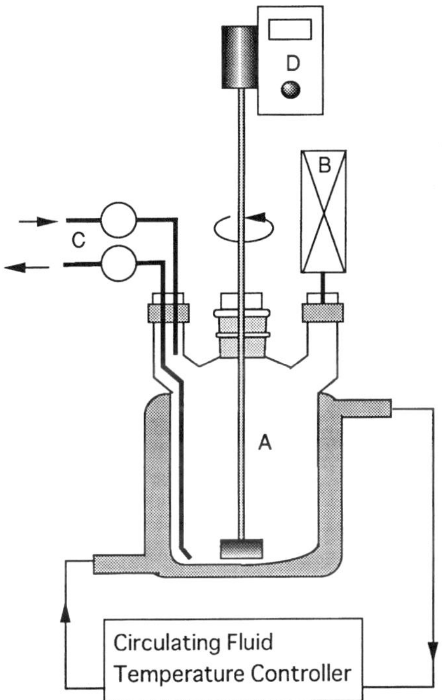

Figure 4 Schematic diagram of a reaction flask used in EASY. A = ca. 300 mL jacketed glass flask, B = water cooled condenser, C = inlet/outlet Teflon tubes, D = mechanical stirrer.

three-digit (3 × 4 bit) number. It was possible to achieve smooth addition of SRs by applying a slow flow of nitrogen gas to both the hopper reservoir and the top of the reaction flask, and using a special screw-turning protocol for each cycle. A control sequence of two steps back followed by three steps forward was effective to prevent jamming of the screw due to powder becoming tightly packed between the screw and the tube walls. Figure 12 shows the pattern of the stepping motor control signal.

The frequency of the stepping motor signals and the angle that the screw must be turned forward each cycle to achieve addition of the reagent over the required time is calculated using a control algorithm. The volume of powder delivered per degree of cycle-turn was determined experimentally using four types of typical powders for a range of cycle times. Bulk densities were measured by determining the volume of a fixed amount of SR, e.g., 1.0 g, in a graduated tube under a similar nitrogen flow as used in the PAF. With the above correlations

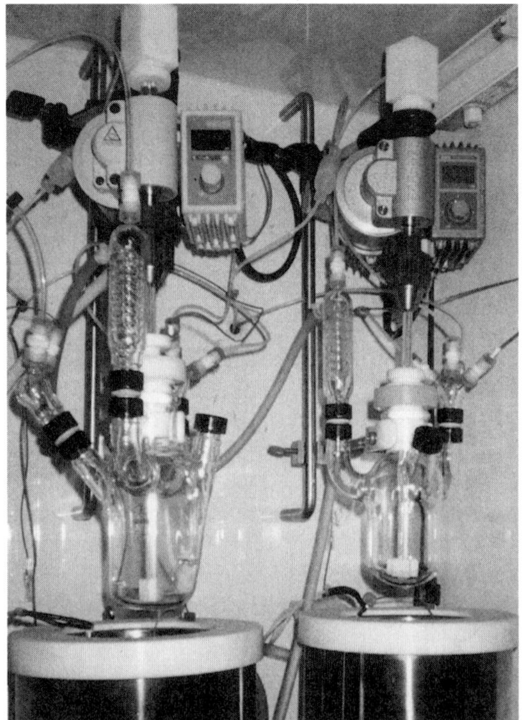

Figure 5 Photograph of the reaction unit in the workstation RAMOS.

it was possible to obtain a control algorithm based on the input of the bulk density, reagent weight, and required addition time [11].

2.1.3 Handling Precipitates

A chemical reaction may start homogenous but unexpectedly a precipitate forms due to a change in reaction conditions or formation of a product with poorer solubility characteristics than the starting material. In automated flow line apparatus the appearance of such a precipitate can lead to disaster, if the solid material causes a blockage in a valve or tube. One way around this problem is simply to avoid reactions that are likely to cause trouble, but this can require extensive experimentation and time-consuming trial runs to guarantee that a new synthesis is not going to present any serious handling problems.

A second option, which we employed for several years, was to place glass filters in the flow lines leading from the reaction flasks, to prevent solid material

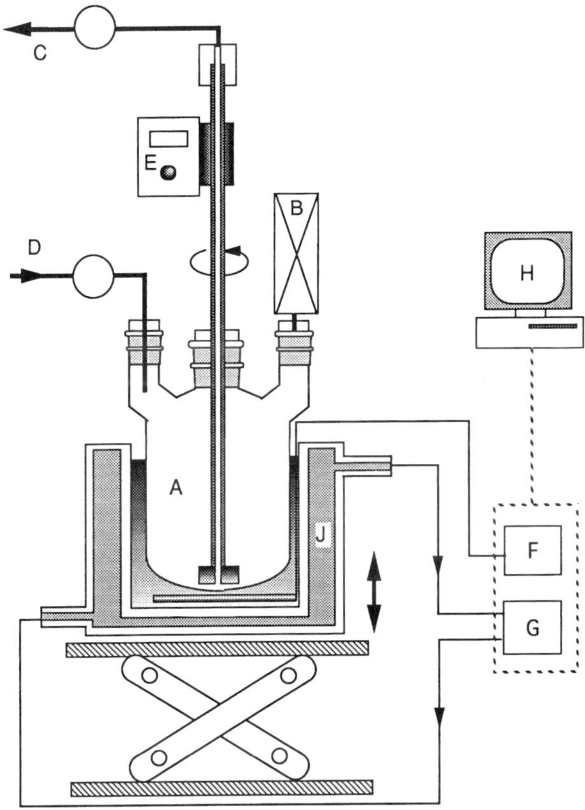

Figure 6 Schematic diagram of the reaction flask used in RAMOS. A = ca. 300 mL glass flask, B = water-cooled condenser, C and D = inlet/outlet Teflon™ tubes, E = mechanical stirrer with hollow rod, F = heater controller, G = cooler controller, H = control computer, J = stainless steel reaction bath.

entering valves. Furthermore, we chose to employ solenoid valves, which used fairly flexible perfluorinated elastomer for the diaphragm. Unlike Teflon™, the perfluorinated elastomer is sufficiently flexible to function correctly even if a small amount of solid is present.

The third way we have handled the problem of precipitates relies on using a sensor to indicate when one is formed. Figure 13 shows a precipitate sensor that was developed to enable the automated handling of reaction mixtures that form unexpected solids. The sensor is made of a glass fiber that transmits to and collects light from the reaction mixture [12]. In order to detect when a solid precipitate is present, the probe irradiates the reaction solution with laser light

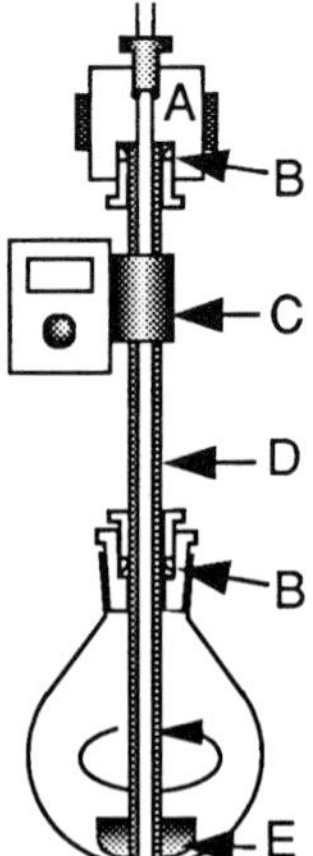

Figure 7 Combined central glass stirring shaft and transfer line. A = fixed Teflon™ connector, B = lip seal, C = motor, D = hollow glass stirring shaft, E = Teflon stirring vane.

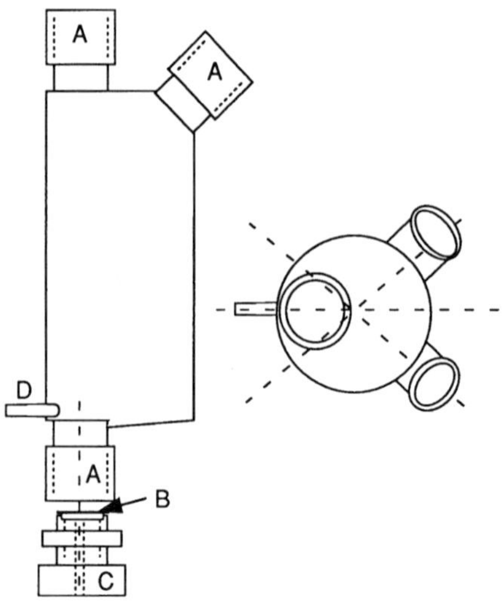

Figure 8 Column shaped reactor. A = screw joint #15, B = filter fitting, C = Teflon™ screw fitting #15, for 3 mm ϕ tube, D = temperature sensor fitting.

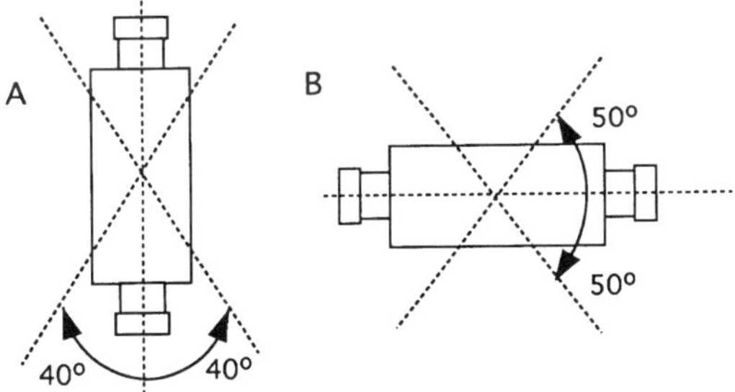

Figure 9 Swing and shake modes of agitation. A = swing mode, useful to agitate the contents over the whole flask, B = shake mode, generally used for agitation during reaction.

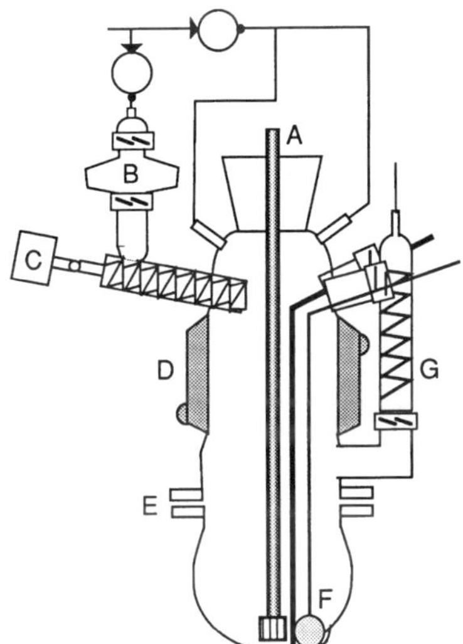

Figure 10 Design of a separable reaction flask for handling supported reagents. A = two-way stirrer shaft, B = adjustable volume solid reagent reservoir, C = stepping motor, D = cooling jacket, E = flanged separable reaction flask, F = Teflon™ ball filter, G = condenser.

Figure 11 Appearance of the separable reaction flask for handling supported reagents.

(780 nm) and measures the reflected light intensity. If a solid precipitate is present it blocks the path of the light, and the reflected intensity drops significantly. The glass fiber is 2.6 mm in diameter and is sharpened so that it will easily pass through a septum.

For this sensor to work reliably it is important that the precipitate be suspended in the reaction mixture rather than forming a layer on the walls of the reaction flask. Thus it is necessary to have efficient stirring, often with a forward-reverse two-way operation, and it can be useful to add several Teflon™ balls to the reaction flask, which tend to detach and break up any precipitate that begins

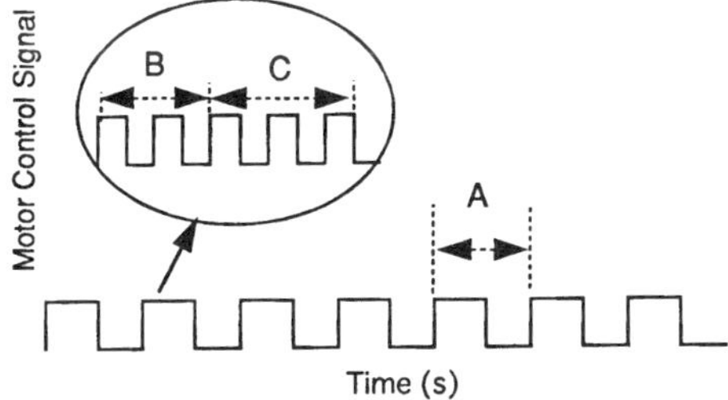

Figure 12 Pattern of the stepping motor control signal. A = cycle time(s), B = two reverse pulses, C = three forward pulses.

to form on the flask walls. Figure 14 shows the flask attachment that enables the precipitate to be filtered.

The attachment consists of a glass filter and a receiver, and the filtrate can then be easily transferred to another vessel, as required. To prevent any unintentional transfer of the reaction mixture to the filtration attachment, it is important to connect the top of both vessels in order to maintain the same pressure in both at all times.

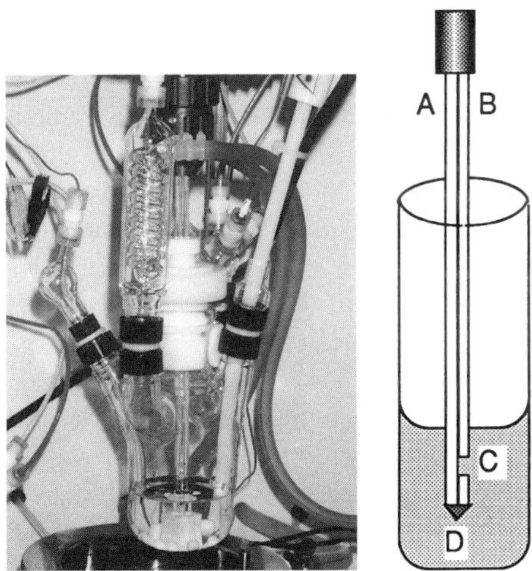

Figure 13 Precipitate sensor. A = source optical fiber (780 nm), B = detection optical fiber, C = gap through which suspended solid particles can flow (ca. 5 mm), D = prism at tip of fibers to reflect laser light.

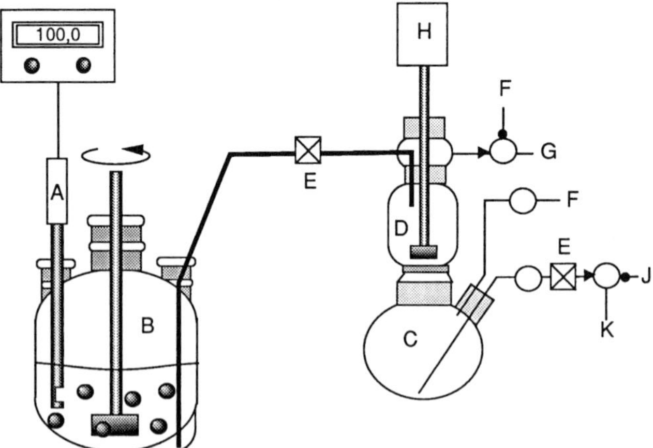

Figure 14   Flask attachment for handling precipitates. A = precipitate sensor, B = reaction flask, C = filtrate receiving flask, D = upper filtration flask fitted with a glass filter at the bottom, E = liquid sensor to verify completion of transfer, F = vacuum, G = open, H = mechanical stirrer, J = next reaction flask, K = separatory funnel.

The collected precipitate can be easily recovered as the glass filter of the attachment is separable. A general procedure for checking and handling precipitate formation is as follows:

> Measure the light transmittance value of the starting mixture in the reaction flask (time = 0, transmittance = 100%).
> Measure the light transmittance value of the reaction mixture before a transfer is attempted (usually < 20% transmittance if a significant amount of precipitate has formed)
> If precipitate is present, then vigorously stir in the presence of Teflon™ balls, which aid suspension of the precipitate.
> Transfer the reaction mixture through a wide-bore (3–4 mm) flow line to the filtration attachment by applying suction, until the end of the transfer is indicated by a sensor on the flow line.
> Transfer the filtrate from the attachment to a reaction flask for subsequent steps of the synthesis procedure or wait for the operator's intervention.

2.2   Automated Extraction

The liquid–liquid extraction process is of great importance for purification in solution-phase organic synthesis, and with the development of phase tags [13] in recent years, it looks as though it will retain its importance in the age of high-throughput and combinatorial technology. The use of fluorous groups for aiding the liquid–liquid separation of the desired compound has been utilized for automated combinatorial synthesis [14].

In order to apply liquid–liquid extraction in an automated manner, the two phases should separate readily without forming an emulsion. Traditionally, it is common practice to aid the separation by (1) adding more of one of the two solvents, (2) adding a small amount of a cosolvent, (3) salting-out by addition of a saturated aqueous salt solution, (4) leaving to stand for a long time, or (5) using a centrifuge. Below we describe two automated approaches that we have developed for liquid–liquid extraction.

2.2.1 Automated Separatory Funnel

A diagram of the automated separatory funnel (SF), which is used for reaction workup, is shown in Fig. 15. An organic and aqueous phase, usually the reaction and wash solutions, respectively, are transferred into the funnel-shaped flask, usually from the reaction flask and one or more of the reagent/solvent reservoirs [15]. Below we outline the steps involved for a two-cycle aqueous bicarbonate wash of an acidic reaction mixture.

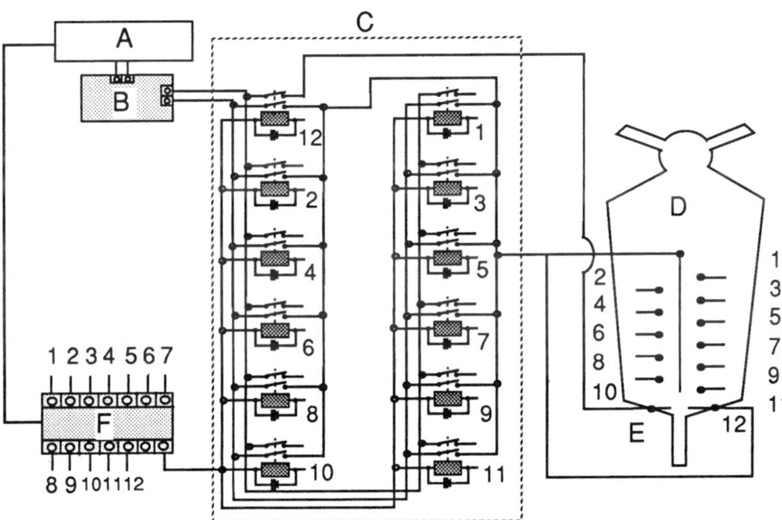

Figure 15 Automated separatory funnel. A = control computer, B = electric conductivity meter, C = relay switches, D = glass funnel fitted with 11 pairs of platinum electrodes at varying heights, E = bottom pair of platinum electrodes, F = computer interface board.

1. Transfer the reaction mixture in ethyl acetate to the SF and add 5% aqueous $NaHCO_3$ solution (30 mL) from a solvent reservoir.
2. Mix the combined phases in the SF by bubbling nitrogen from the bottom.
3. Allow the phases to separate on standing while monitoring the extent of separation by measuring the conductivity at each of the eleven electrodes positioned at 5 mm intervals in the wall of the funnel.
4. When a clear distinction between the top and bottom phases is recognized, solenoid valves on the line from the bottom of the SF are opened to direct the lower organic phase into a collection flask. The flow rate must be kept slow in order to prevent turbulence that will cause remixing of the two layers. When the two-phase boundary is detected at the final pair of electrodes, the solenoid valve above the collection flask is switched to direct the upper aqueous wash to a second collection flask or waste.
5. The organic layer can be returned to the SF for further cycles of washing.
6. The organic layer is dried after the final wash by passing it through a short cartridge of anhydrous sodium sulfate, either into a clean reaction flask for further reaction or into a standard round-bottomed collection flask.

2.2.2 Automated Centrifugal Separator

A difficulty we found when using the automated separation flask described above was that it could take an excessively long time for the mixed phases to separate. Indeed, it was found that some mixtures (e.g., aqueous alkaline/dichloromethane) formed particularly stable emulsions that did not separate even after several hours.

In order to overcome this problem we devised an automated means to separate emulsified extracts using centrifugal force to separate and divide rapidly the individual phases from an emulsified extract. Figure 16 shows the extraction unit of our workstation TACOS II. The unit is also shown in the photograph of Fig. 17. Our centrifugal separating unit offers the ability (1) to shorten the time, (2) to avoid the need to add another reagent or solvent, (3) to handle volumes of about 50 to 300 mL, and (4) to avoid the need for multiple tubes and to balance samples.

The emulsified sample is transferred into a round glass flask that is spun at high speed. As the flask is turned, the emulsified sample spins and the two phases separate under the centrifugal force that is created. The bottom of the flask is connected to the central motor and the top is held by a collar containing ball bearings. The flask can be easily removed for maintenance by unscrewing

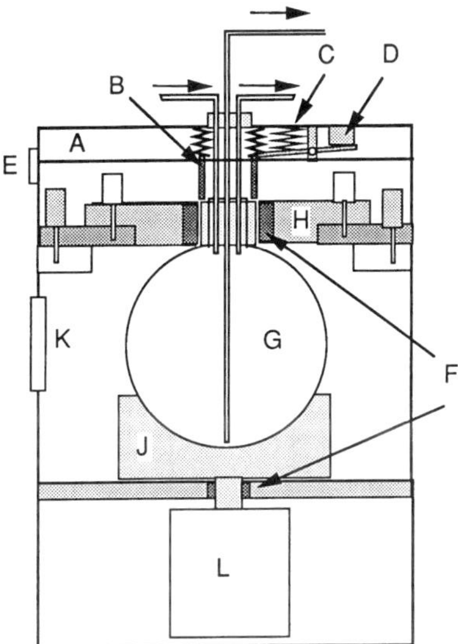

Figure 16 Extraction unit of TACOS II. A = lid, B = viton packing, C = spring, D = solenoid, E = metal fastener, F = ball bearings, G = separation flask, H = holder, J = flask receptacle, K = window, L = motor.

the upper holder. In order to transfer solutions into the flask when it is stationary, the top cap is pushed down onto the mouth of the flask to form a seal. Three tubes are fixed to the cap, one extending to the bottom and the other two into the top of the flask. Using these tubes it is possible to apply vacuum or pressure to accomplish the transfer of solutions. A window is fitted in the front panel of the unit to allow visual monitoring of the separation, which is useful during development work before unattended full automation is attempted. The power to the unit is switched on with one relay and the speed is set at 600, 1800, or 3000 rpm with another two relays. The speed of the motor can be monitored on a digital display. One further relay switch controls the solenoid that pushes the cap down onto the mouth of the flask to form a seal.

After the emulsified sample is transferred to the separating flask and the phases are separated, the sample is transferred out of the flask passing a flow-type conductivity sensor, which can detect the boundary between the phases and be used to control a solenoid valve to direct the phases into separate collection flasks in a fraction collector [16].

For example, an emulsion of salad oil (150 mL) and water (150 mL), which did not separate on standing for 5 h, was separated clearly into two layers after 4 min centrifuging at 1800 rpm. Similarly, an emulsified mixture of ethyl acetate

Figure 17 Photograph of the centrifugal extraction unit of TACOS II.

(150 mL) and water (150 mL) was separated within 1 min under the same conditions.

2.3 Automated Purification and Analysis Units

Automation of the purification process is crucial for making a complete system from synthesis to isolation of the pure compounds. We developed an automated purification workstation [17] for column chromatography that can be operated independently or connected to the synthesis workstations to facilitate more elaborate or extensive automation. The fundamental components of our Standalone or Option Apparatus for Purification (SOAP), are outlined below. A photograph of SOAP is shown in Fig. 2, and the basic layout of the workstation is shown in Fig. 18.

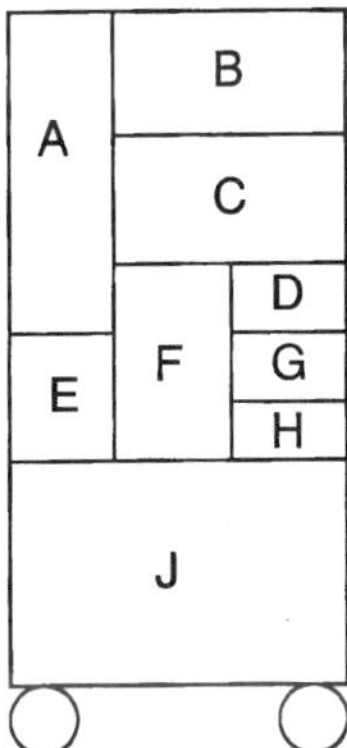

Figure 18 Layout of the workstation SOAP. A = column, B = status display panel, C = solvent reservoirs, D = sample reservoirs, E = HPLC pump and detector, F = sample injector, G = wash solvent reservoir, H = control computer, J = fraction collector.

The hardware consists of reservoirs for eluant solvents, a pump, a sample reservoir, an injection device, columns, an autosampler or connection to a synthesis apparatus, detector, and fraction collector, as shown in the flowline chart of Fig. 19.

2.3.1 Syringe-Type Column Injector

The most difficult step to automate was the charging of the sample onto the chromatography column. To allow the whole of a sample to be injected, without air being included, we developed an automated syringe-type injector [18]. The piston of the syringe is drawn down to transfer the whole of the sample from the sample reservoir into the syringe. Then, to avoid loss of sample in the line and reservoir, some solvent is added to the reservoir, and this is also drawn into the syringe cylinder by means of a vacuum pump. When no further sample is detected by the photosensor on the inlet side of the syringe, the syringe piston is pushed up to charge the sample onto the column. To make sure air is excluded from the column, the photosensors are used to check when liquid enters the outlet tubes attached to the cylinder of the syringe, to control the operation of the two rotary valves so that air is not allowed to flow to the column. A pressure sensor on the syringe piston is used to stop it moving if dangerous pressure levels (e.g., over 2.5 kg/cm^2) occur. When the piston is in the uppermost position, the two tubes leading in and out of the syringe remain connected to each other so that the solvent from the HPLC pump to the column is not interrupted.

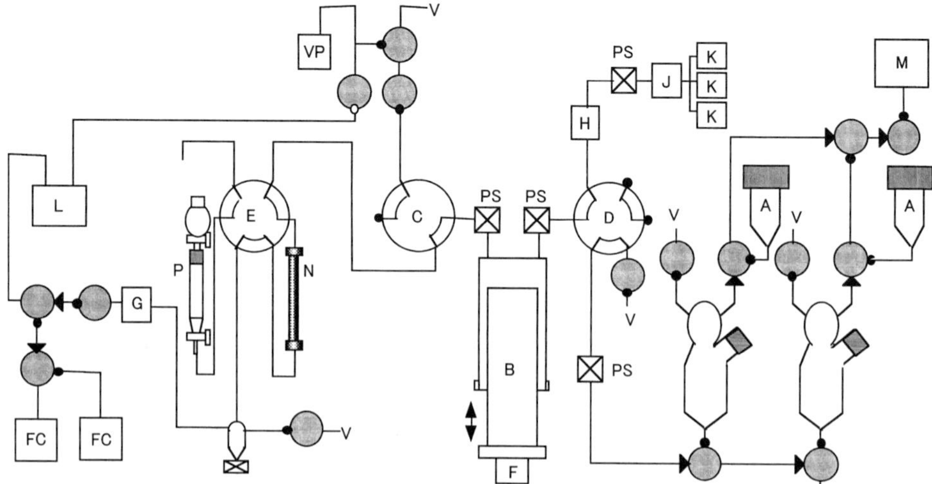

Figure 19   Flow lines of the workstation SOAP. A = sample reservoirs, B = syringe injector, C, D, E = rotary valves, F = pressure sensor, FC = fraction collector, G = UV detector, H = HPLC pump, J = HPLC gradient mixer, K = HPLC solvents, L = waste tank, M = wash solvent, N = HPLC column, P = liquid chromatography column, PS = photosensor, V = vacuum line, VP = vacuum pump.

2.3.2   Degassing the Column Eluant

The eluant from the column passes through a UV-vis detector so that the desired peaks may be collected and the remainder of the eluant passed to waste. It was found that the signal from the detector could be extremely noisy due to gas bubbles passing through the line, and the noise caused serious problems in determining when peaks should be cut. Thus a device was incorporated on the line between the column and the detector to remove the gas bubbles [19]. The flow from the column passes into a small glass tube that has a photosensor just above the bottom, as shown in Fig. 20. When liquid flows from the column it is detected by the sensor and flows straight out to the detector. But when a gas bubble passes from the column into the device, the sensor registers that no solution is present and the solenoid valve to vacuum is opened briefly until the sensor registers liquid again.

A similar device is used to allow change-over of the eluting solvents, on the flow line to the HPLC pump (Fig. 19). During the change-over of solvents, if no solvent is detected in the line to the pump, a slight vacuum is applied to the degassing device to prevent air from getting in the line to the HPLC pump.

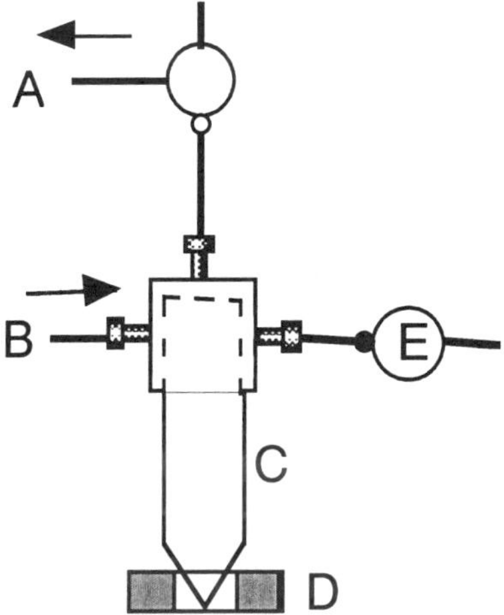

Figure 20 Device to remove gas bubbles. A = to detector, B = from column, C = glass tube, D = photosensor, E = normally closed solenoid valve, F = vacuum pump.

3 APPLICATIONS OF FLOW LINE-BASED AUTOMATED APPARATUS

Several applications of our apparatus have already been published [2–4,20–22]. Here we will briefly describe how our automated workstations were used in the preparation of a series of tetrapeptide derivatives, which are starting materials for (pharmaceutically attractive) pentapeptide derivatives.

3.1 Tetra- and Pentapeptide Library Synthesis

There are two basic strategies to construct libraries of tetra- and pentapeptides. One uses a linear strategy in which the amino acids are added to each other one by one, and the other is a convergent strategy that relies on joining fragment di- and tripeptides. In general the linear strategy has often been used for automated repetitive syntheses that rely on using the same kind of reaction procedure multiple times. But as peptides get longer their solubility characteristics get worse, and handling in an automated apparatus becomes more problematic. The convergent strategy is favorable because series of di- and tripeptides can be made on a large scale and then combined combinatorially to generate a large library, with fewer purification steps and higher overall yield than can be achieved using the linear method.

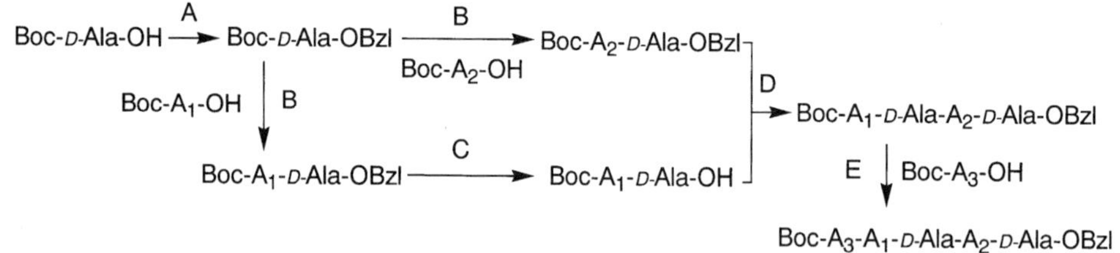

Figure 21 Automated library synthesis of pentapeptide derivatives. A = benzylation with benzyl bromide, 0.5 mol scale repeated nine times using FUTOSHI, B = formation of dipeptide derivatives by the methanesulfonic acid method, 0.4 mol scale using FUTOSHI, C = hydrolysis of the benzyl esters using either ASCA for selective hydrogen transfer with Pd-charcoal or FUTOSHI for saponification with sodium hydroxide, 0.2 mol scale, D = formation of tetrapeptide derivatives by the methanesulfonic acid method, 12 mmol scale using RAMOS, ASTRO, EASY, TAFT, and WOOS, E = formation of pentapeptide derivatives by the methanesulfonic acid method, 1 mmol scale using RAMOS, ASTRO, EASY, and TAFT.

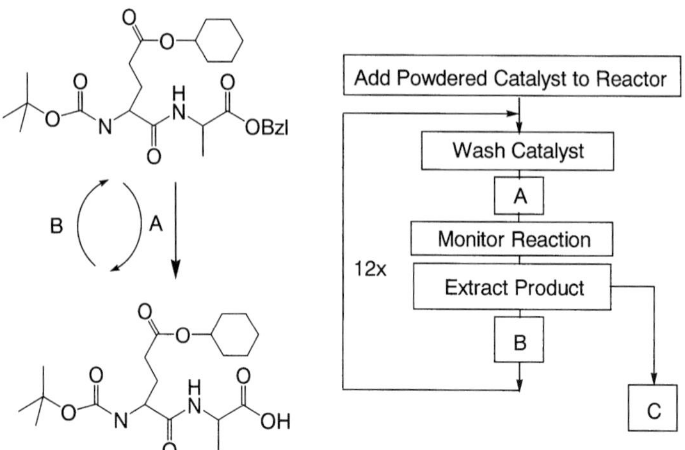

Figure 22 Multibatch selective debenzylation using ASCA. A = catalytic reduction of Boc-Glu(OcHex)-D-Ala-OBzl in MeOH at 25°C, B = reactivation of the catalyst with methanoic acid in MeOH, C = collection of the product, Boc-Glu(OcHex)-D-Ala-OH, solution followed by evaporation to give 50–100 g over 12 cycles.

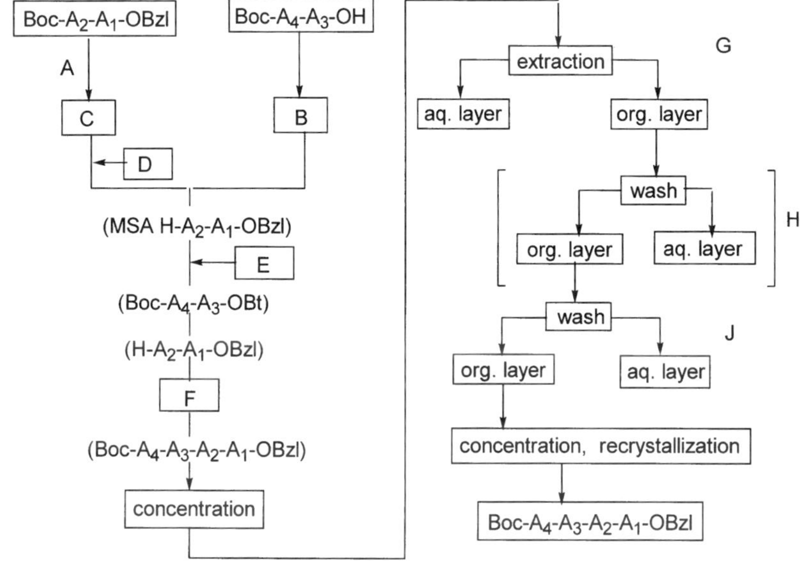

Figure 23 Flowchart for the tetrapeptide synthesis by the MSA method. A = generation of the methane sulfonic acid derivative using 2 M MSA in MeCN (15 mL; 30 mmol) at 25°C, B = activation of the unprotected peptide with HOBt (2.02 g, 13.2 mmol) in DMF (10 mL), C = deprotection of the Boc group (40°C, 1 h), D = neutralization using DIEA in MeCN (20 mL, 18 mmol) at 0°C, E = addition at 0°C of B to D and condensation using WSCD (2.05 g) in MeCN (5 mL), F = reaction at 25°C for 10 h, G = dissolve concentrated product in EtOAc (30 mL) and THF (20 mL), then add 5% $NaHCO_3$ (40 mL) and EtOAc (40 mL), extract and separate the layers, H = add aqueous HCl (0.2 M) to the organic layer (40 mL), extract and separate the layers. Omit this step when His (Bom) amino acid is used, J = add water (40 mL) to the organic layer, extract and separate the layers.

A general method for the automated synthesis of oligopeptide derivatives was developed, in which we used methanesulfonic acid (MSA) to give a milder yet more efficient automated synthesis than is possible with the trifluoroacetic acid (TFA) method [23]. The MSA method not only avoided the need for the highly corrosive acid TFA but also allowed us to avoid using the restricted solvent dichloromethane as well as prevent some intractable emulsions from being formed.

Table 2 Typical Setup and Subroutine Sequence for a Tetrapeptide Synthesis Using RAMOS

RAMOS
(Reaction setup)

RS1: H_2O (40 mL)	RS2: 5% NaHCO3 (40 mL)	RS3: 0.2 M HCl (40 mL)	RS4: Ethyl acetate (70 mL)
RS5: THF (20 mL)	RS6: MeCN (10 mL)	RR1: 2 M MSA/MeCN (15 mL)	RR2: 0.9 M DIEA/MeCN (20 mL)
RR3: 2 M WSCD/MeCN (5 mL)			

Program TETRAPEP
(Reaction process)

1. START	2. RF1-ST-ON	3. RF2-ST-ON	4. RR1-RF1
5. RF1-LF-UP	6. REA1 (40°C, 60 m)	7. RF1-T-ON (0°C)	8. RR2-RF1
9. WAIT	10. RF2-RF1	11. RF6-RF2	12. RF2-MIX
13. RF2-RF1	14. RF2-ST-OF	15. RR3-RF1	16. RF1-T-OF
17. REA1 (25°C, 600 m)	18. CON1 (40°C, 60 m)	19. RF1-LF-DN	20. RS5-RF1
21. RS4-RF1	22. RS2-RF1	23. RF1-MIX	24. BKTETRANK
25. RF1-ST-OF	26. ALARM	27. SR1-DR	28. BKWASHTET
29. BKWASHTET	30. RF1-LF-UP	31. RF2-LF-UP	32. RF1-ST-ON
33. RF2-ST-ON	34. RF1-DRY	35. RF2-DRY	36. RF1-LF-DN
37. RF2-LF-DN	38. RF1-ST-OF	39. RF2-ST-OF	40. END

Program BKTETRANK
(Extraction process)

1. BKSTART	2. RF1-SF	3. SF-BUBB	4. SEP-SR1
5. SF-SR0	6. SR1-SF	7. RS4-RF1	8. RS5-RF1
9. RF1-MIX	10. RF1-SF	11. SF-BUBB	12. SEP-SR1
13. SF-SR0	14. SR1-DR	15. SR0-SF	16. RS3-RF1
17. RF1-SF	18. SF-BUBB	19. SEP-SR1	20. RS1-RF1
21. RF1-SF	22. SF-BUBB	23. SEP-SR1	24. SF-RF3
25. RS4-RF1	26. RS5-RF1	27. RF1-MIX	28. RF1-SF
29. SF-BUBB	30. SF-SR0	31. SR0-BUBB	32. SR0-SF
33. SF-RF3	34. BKEND		

Program BKWASHTET
(Wash process)

1. BKSTART	2. WS-RR1	3. WS-RR2	4. WS-RR3
5. WS-RF1	6. WS-RF2	7. RF1-BUBB	8. RF2-BUBB
9. RF1-SF	10. SF-BUBB	11. SF-SR1	12. SR1-BUBB
13. SR1-SF	14. SF-SR0	15. SR0-BUBB	16. SR0-DR
17. RF2-RF1	18. RF1-SF	19. SF-RF3	20. BKEND

Reservoir for solvents (n = 1–6), RSn; reaction flasks (n = 1–3), RFn; reservoir for reagents (n = 1–9), RRn; reaction, REA; separation reservoir (n = 0–2), SRn; separation flask, SF; wash solvent, WS; stirrer, ST.

3.1.1   Large-Scale Synthesis of Dipeptide-Benzyl Esters and Dipeptide-Carboxylic Acids

Figure 21 shows the route to preparing a library of 504 pentapeptide derivatives. Using our workstation FUTOSHI, we synthesized benzyl protected D-alanine on a kilogram scale by repeating a 0.5 mol scale synthesis nine times (av. 86.4% yield). We then made 50–150 g of nine dipeptide-benzyl esters (Boc-A_2-D-Ala-OBzl, 80–97% yield) and nine dipeptide-carboxylic acids (Boc-A_1-D-Ala-OH, 70–90% yield), which are the starting materials for preparing the tetrapeptide derivatives, with our large-scale synthesis workstations, FUTOSHI and ASCA [24].

ASCA, our workstation that can handle solid catalytic reagents, was used to accomplish a selective multibatch hydrogen transfer debenzylation reaction using ammonium formate and Pd-charcoal, which was regenerated by acid washing between each cycle, as shown in Fig. 22 [4].

3.1.2   Penta- and Tetrapeptide Synthesis by Fragment Condensation

A flowchart for the tetrapeptide synthesis using the so-called methanesulfonic acid procedure is shown in Fig. 23. For this series it was necessary to modify slightly the standard procedure by using a mixed reaction solvent of tetrahydrofuran and ethyl acetate to avoid precipitation of the tetrapeptides from the reaction mixture. It was possible to synthesize a single 12 mmol batch in a 100 mL reaction flask. Table 2 shows the subroutine programs that were sequenced to carry out the experiment in RAMOS, in accordance with the flowchart. Tetrapeptides were obtained in excellent yield (average 77%). Those with tryptophanyl or tyrosyl units were found to have the poorest solubility, and sometimes they precipitated in the apparatus, leading to poor yields. In a single operating cycle, 5–10 g of each tetrapeptide was synthesized.

RAMOS and our similar workstations were then used to prepare pentapeptide derivatives from the tetrapeptide ones, on a 1 mmol scale to give 0.6–1.2 g of each compound. In a typical example, Boc-D-His(Bom)-OH was added to Boc-Leu-D-Ala-Glu(OcHex)-D-Ala-OBzl (using methanesulfonic acid/DIEA followed by HOBt and WSCD) to give 0.71 g of pentapeptide derivative, Boc-D-His(Bom)-Leu-D-Ala-Glu(OcHex)-D-Ala-OBzl.

4   FUTURE PROSPECTS

With the interest in what has broadly become known as combinatorial chemistry, various kinds of automated apparatus have been developed all over the world to aid in the synthesis of large numbers of compounds in an efficient and effective manner. Until around the end of the 1980s, the main stream for the development of automated synthesis apparatus was solution-phase, and the solid-phase method, carried out on beads, was restricted to the synthesis of peptides and nucleotides. More recently, however, many series of chemical compounds (chemical libraries) have been synthesized on beads, and many automated solid-phase synthesis apparatus have been developed for this purpose [25]. For many chemists, the term combinatorial chemistry, with its image of generating all and every kind of mixture, has been replaced by automated synthesis or high-throughput organic synthesis, where individual compounds are synthesized and purified, if necessary, using automated techniques in parallel.

During the last 100 years it has been calculated that more than 3000 reaction types have been developed by about 2000 organic synthesis chemists. In order to develop automated synthesis apparatus for carrying out these various kinds of organic reactions, a system that can utilize the best of both solution- and solid-phase synthetic methods is needed. For example, in the case of solid-phase synthesis, it is generally necessary for high yields, of near 100%, to be obtainable for each reaction step, especially for multistep reactions on beads. But, as it is very difficult to achieve such efficient synthesis for many reactions, we can only carry out a few reaction steps using solid-phase synthesis. Thus it is much more efficient and effective to synthesize an (important) intermediate in the solution phase and only use the solid-phase synthetic method for the final step.

A recent hybrid synthetic method, having the merits of both solution- and solid-phase synthesis, has been developed to eliminate the time-consuming purification step, which is the biggest disadvantage of solution-phase synthesis. This approach uses a bead-scavenger method for rapid purification [26], which offers a solution to the purification problem that has hindered the use of solution-phase reactions for high-throughput library synthesis.

The main approach to the development of many automated synthesis apparatus has been to gradually develop more and more sophisticated units for the established apparatus, to allow as many reactions as possible to be carried out. But besides these hardware improvements, it will also be important to develop apparatus for efficiently and effectively getting the most suitable reaction conditions for each reaction step, for both solution- and solid-phase syntheses. If the optimum reaction conditions in each reaction step can be used, so that purification can be done simply and quickly, and only small amounts of reagents and solvents are wasted, it will be beneficial from both economical and environmental points of view [27].

For future automated synthesis apparatus, it will become important, from an effective capital investment point of view, to develop specific apparatus, such as exclusive modular-type synthesis and purification workstations, and then combine these to establish a so-called integrated laboratory system. The first steps towards standardizing the communication protocols for different workstations are being made, but these should be carefully decided as they will be of increasing importance as more and more automated laboratories are established.

Besides the solid- and solution-phase methodologies, there are two different approaches for developing automated synthesis apparatus. One is the scale-up approach, commonly used in process chemistry, and the other is scale-down, which involves developing miniaturized apparatus. Recent developments in miniaturization have been actively carried out in the fields of not only biochemistry but also synthesis and analytical chemistry. Just as the progress in microelectronics produced the prosperity in the computer industry, the progress in microengineering will have a strong impact on the future development of automated synthesis apparatus. Some of the improvements and modifications that remain to be done for the present automated synthesis apparatus may be more easily achieved by miniaturizing the units.

The technical terms microreactor and laboratory on a chip can now be found more and more in scientific meetings and reports [28]. They represent an interesting technology for the automated synthesis of dangerous poisonous compounds, and they allow potentially explosive and fast reactions to be handled safely. The chips have minute vessels for reactions and small microfluidic channels for transferring small amounts of solutions.

The development of sensors to detect or measure changes that occur during the various steps of a synthesis are crucial for successful, reliable, reproducible, and safe automation. Flow line–based apparatus require sensors that can detect, in real time, most if not all of the following:

Precipitation and dissolution	Decomposition
Coloration	Evaporation
Temperature variation	Viscosity
Boiling and freezing	Conductivity and capacitance
Pressure and volume variation	Acidity and basicity
Concentration changes	Liquid surface and level
Water content	Position
Gas flow	

Several of these sensors can be made simply or are commercially available, such as liquid-level sensors for checking when a vessel becomes full or empty, thermometers to record temperature of the reaction mixture, photosensors or microswitches for registering positions of movable objects, mass flow controllers for regulating gas flows, pH sensors for measuring acidity, and pressure sensors for either checking vacuum levels or preventing dangerous buildups of pressure.

In the future we can expect the further development of these sensors, including the development of more compact and multicomponent sensors such as are envisioned in Fig. 24. By the use of more sophisticated, robust, and versatile sensors, automated apparatus will gradually overcome the remaining obstacles

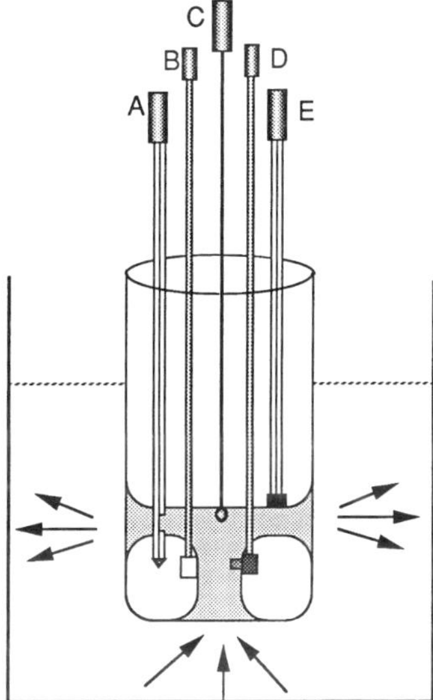

Figure 24 Composite sensor. A = fiber-optic precipitate sensor, B = electric conductivity sensor, C = temperature sensor, D = viscosity sensor, E = microwave sensor.

that currently limit their reliability and performance. When reliability is achieved, any remaining doubts about the usefulness of a particular piece of automated apparatus will finally be dispelled, and it will be accepted as standard laboratory equipment.

REFERENCES

1. Sugawara, T., Cork, D. G. Yuki Gosei Kagaku Kyokaishi 55:466 (1997).
2. Sugawara, T., Cork, D. G. Lab. Robotics and Automation 8:221 (1996).
3. Sugawara, T., Kato, S., Okamoto, S. J. Auto. Chem. 16:33 (1994).
4. Sugawara, T., Cork, D. G. In: Combinatorial Chemistry: A Practical Approach (Fenniri, H., ed.). Oxford Univ. Press, Ch. 13, p. 373 (2000).
5. Hurst, W. J., ed. Automation in the Laboratory. VCH Publishers, New York (1995).
6. Cork, D. G., Kato, S., Sugawara, T. Lab. Robotics and Automation 7:301 (1995).
7. Sugawara, T., Kato, S. Eur. Pat. Appl. EP-510487 (1992).
8. Sugawara, T., Kato, S. U.S. Patent US-0870426 (1992); Jap. Utility Model Appl. 040736 (1992); PCT Int. Appl. WO-9724181 (1997).
9. Drewry, D. H., Coe, D. M., Poon, S. Medicinal Research Reviews 19:97 (1999); Kirschning, A., Monenschein, H., Wittenberg, R. Angewandte Chemie—Interna-

tional Edition 40:650 (2001); Ley, S. V., Baxendale, I. R., Bream, R. N., Jackson, P. S., Leach, A. G., Longbottom, D. A., Nesi, M., Scott, J. S., Storer, R. I., Taylor, S. J. Journal of the Chemical Society. Perkin Transactions 1 23:3815 (2000).

10. Clark, J. H. Catalysis of Organic Reactions by Supported Inorganic Reagents. VCH Publishers, New York (1994); Clark, J. H., Cullen, S. R., Barlow, S. J., Bastock, T. W. J. Chem. Soc. Perkin Trans. 2:1117 (1994).

11. Sugawara, T., Cork, D. G., Kato, S. Jap. Pat. Appl. 06-339492 (1994); Sugawara, T., Cork, D. G., Kato, S. Jap. Pat. Appl. 07-058872 (1995); Cork, D. G., Kato, S., Sugawara, T. Jap. Pat. Appl. 07-131654 (1995).

12. Kato, S., Ikunishi, S. Jap. Pat. Appl. 11-85674 (1999).

13. Yoshida, J., Itami, K., Mitsudo, K., Suga, S. Tetrahedron Lett. 40:3403 (1999).

14. Curran, D. P. Angew. Chem. Int. Ed. Engl. 37:1174 (1998).

15. Sugawara, T., Kato, S., Shintani, M. Jap. Utility Model Appl. No. 040735 (1992); Sugawara, T., Kato, S. Jap. Utility Model Appl. No. 043600 (1992).

16. Sugawara, T., Kato, S. Jap. Pat. Appl. 09-134771 (1997).

17. Sugawara, T., Kato, S. Jap. Utility Model Appl. No. 043603 (1992); Jap. Utility Model Appl. No. 267176 (1995).

18. Sugawara, T., Kato, S. Jap. Utility Model Appl. No. 265833 (1995); Jap. Utility Model Appl. No. 303095 (1995).

19. Sugawara, T., Kato, S. Jap. Utility Model Appl. No. 261546 (1995).

20. Hayashi, N., Sugawara, T. Chem. Lett. 1613 (1988); Hayashi, N., Sugawara, T. Tetrahedron Computer Methodology 1:237 (1989); Hayashi, N., Sugawara, T., Shintani, M., Kato, S. J. Auto. Chem. 11:212 (1989); Hayashi, N., Sugawara, T., Kato, S. J. Auto. Chem. 13:187 (1991).

21. Kuwahara, M., Kato, S., Sugawara, T., Miyake, A. Chem. Pharm. Bull. 43:1511 (1995).

22. Sugawara, T., Kobayashi, K., Okamoto, S., Kitada, C., Fujino, M. Chem. Pharm. Bull. 43:1272 (1995).

23. Gross, E., Meienhofer, J., eds. The Peptide—Analysis, Synthesis, Biology. Vol. 3. Academic Press, New York (1981).

24. Kuroda, N., Hattori, T., Fujioka, Y., Kitada, C., Cork, D. G., Sugawara, T. Submitted to Chem. Pharm. Bull. 49:1147 (2001).

25. Wilson, S. R., Czarnik, A. W., eds. Combinatorial Chemistry—Synthesis and Application. John Wiley, New York (1997); Hird, N. W. Drug Discovery Today 4:265 (1999).

26. Drewry, D. H., Coe, D. M., Poon, S. Med. Res. Rev. 19:97 (1999); Hebermann, J., Ley, S. V., Scott, J. S. J. Chem. Soc. Perkin Trans. 1 10:1253 (1999); Hu, Y., Baudart, S., Porco, J. A. J. Org. Chem. 64:1049 (1999).

27. Cork, D. G., Sugawara, T., Lindsey, J. S., Corkan, L. A., Du, H. Lab. Robotics and Automation 11:217 (1999); Wagner, R. W., Li, F., Du, H., Lindsey, J. S. Org. Process Res. Dev. 3:28 (1999); Du, H., Corkan, L. A., Yang, K. X., Kuo, P. Y., Lindsey, J. S. Chemometrics and Intelligent Laboratory Systems 48:181 (1999).

28. Ehrfeld, W., Hessel, V., Löwe, H. Microreactors. Wiley-VCH, New York (2000).

3

An Automated Microscale Chemistry Workstation Capable of Parallel Adaptive Experimentation

Hai Du, L. Andrew Corkan, Kexin Yang, Patricia Y. Kuo, and Jonathan S. Lindsey
North Carolina State University, Raleigh, North Carolina

For several decades work has proceeded on the development of automated chemistry workstations aimed at augmenting the efforts of the experimentalist engaged in laboratory synthetic chemistry research. In the early years, progress in this area was stymied by primitive mechanical devices, computers, and software. The increasing sophistication and commercial availability of these components, fascinating scientific opportunities, and the economic imperatives of industry have fueled the construction of advanced workstations that now are suitable for addressing a range of chemistries [i]. With more elaborate hardware has come the need for software that enables chemists to propose experiments in a meaningful and straightforward way. We have worked extensively on the development of such experiment planners for use with workstations aimed at microscale chemistry. Our work has been done in conjunction with Dr. David Cork and Dr. Tohru Sugawara at Takeda Chemical Industries (Osaka, Japan), where an identical copy of our workstation resides. In the U.S., we have focused on experiment planners for parallel and adaptive experimentation, developing new algorithms and testing these approaches on appropriate problems in synthetic chemistry [ii]. At Takeda, extensive hardware modifications to the workstation have been made to broaden the scope of accessible chemistry [iii]. This chapter, which provides a comprehensive description of an experiment planner [iv], is reprinted in its entirety with permission from Elsevier Science.

1 INTRODUCTION

The advent of modern automated chemistry workstations holds the promise of improved quality and increased quantity of scientific experimentation in selected domains of chemistry [1–6]. For this promise to be realized, the automation system must offer substantial advantages compared with what an individual scientist can achieve working manually. To exceed human capabilities a workstation must be capable of working relentlessly, precisely, strategically, and autonomously in pursuit of the goals stated by the scientist.

Over the past decade we have been working to develop the hardware and software of such workstations, with a particular focus on systems for fundamental investigations of chemical reactions as well as optimization of reaction conditions for synthetic transformations [7–14]. In particular we have emphasized four broad themes (Fig. 1). First, we have developed a hardware and software architecture that enables a large number of experiments to be performed simultaneously. Such parallel experimentation enables an aggressive and rapid assault on scientific problems. Second, we have developed features for automated decision-making concerning the course of experimentation. This strategic or adaptive experimentation enables the workstation to respond to data, focusing experimentation in pursuit of the scientific goals and eliminating futile lines of inquiry. The focused nature of adaptive experimentation is complementary to the exhaustive approach of parallel experimentation. Third, we have developed a host of book-keeping features that enable the workstation to keep track of available resources and make decisions about implementing experiments accordingly. These resource management features are a key element of autonomous experimentation. Fourth, we have developed a menu-driven user interface that enables the scientist easily to compose descriptions of the experiments for implementation on the workstation. This experiment-planning software provides the link between a scientist's intellectual objectives and the cold mechanics of the system hardware. In this chapter, we describe further developments in these approaches for automated experimentation.

An effective automation system must be able to take input in the form of high-level scientific goals and implement experiments accordingly with automatic management of issues related to sample handling, instrument control, resources, timing, scheduling, and data analysis. In many respects the individual hardware components of an automated chemistry workstation are similar to those that a scientist uses manually (reaction vessels, analytical instruments, syringes), with the exception that the robot performs the jobs of the operator in moving samples and materials among the various hardware subsystems. The robot may manipulate materials with greater precision than is done manually, may perform individual operations faster than is done manually, and may work for prolonged periods. However, any added advantage of robotics must arise from the features

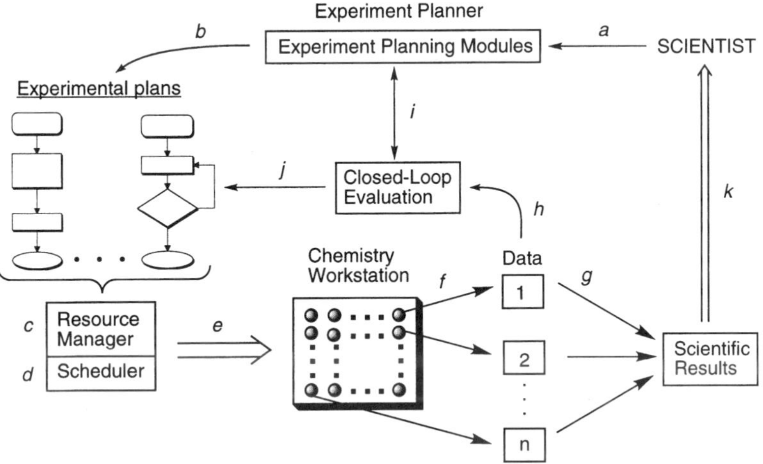

Figure 1 Flow diagram for automated experimentation. (a) The scientist works with one of five experiment planning modules to compose experimental plans describing the operations to be performed. In closed-loop experiments, scientific objectives are stated. Information about available resources (chemicals, containers) must also be provided. (b) From the experiment planning modules a plan or set of plans emerges. Each plan consists of a list of robotic commands and can include conditionals whose output depends on experimental data. (c) The experimental plans are passed to the Resource Management module for look-ahead tabulation of the total resource demands. The schedule is separated into executable experiments and experiments awaiting resources. Other safeguard features are invoked automatically (see text). (d) The experimental plans that meet resource requirements are passed to a scheduler where the plans are rendered in parallel to the extent possible. (e) The schedule of executable experiments is passed to the automated chemistry workstation capable of performing experiments in parallel. (f) Data are generated from the analytical instruments. (g) The data from common experiments are combined in an output file constituting the results (open-loop experimentation). (h) Closed-loop experimentation involves decision making without user intervention about ongoing or planned experiments. (i) The data are evaluated in the context of the scientific objective as stated as part of the experimental plans. (j) Depending on the results of the evaluation, ongoing experiments can be terminated or altered, pending experiments can be expunged from the queue, or new experiments can be spawned. (k) The results from open-loop and closed-loop experiments are available at all times for review by the scientist.

in software that enable high-throughput (as occurs with parallelism), adaptive experimentation (which requires automated decision making), and intelligent use of resources to reach scientific goals. Thus the impact of an automated chemistry workstation depends very much on the sophistication and versatility of the software for experiment planning. The experiment planner and the lower level experiment implementation software taken together are referred to as the experiment manager software.

We previously described a first-generation hardware system and accompanying experiment manager software [7–9] and a second-generation hardware system and experiment manager software [10,11]. The second-generation software was designed for parallel or strategic experimentation. One experiment planner was used for parallel experimentation (without decision making), and a second experiment planner was used for experiments involving automated decision making (without parallelism). The third-generation system described herein retains the hardware design of the previous system, but all components have been selected for high throughput during parallel experimentation. Many new features have been incorporated into the experiment manager software in order to achieve both high throughput and adaptive capabilities. This system has been applied to the optimization of the reaction conditions for the synthesis of *meso*-tetramesitylporphyrin [15].

In this paper we provide a comprehensive description of the third-generation workstation, emphasizing the experiment manager software and issues relevant for composing experimental plans for automated experimentation. The experiment planner is composed of distinct modules for planning general purpose (GP), decision-tree (DT), factorial design (FD), composite-modified simplex (CMS), and multidirectional search (MDS) experiments. This paper describes scheduling, resource management, and decision making issues relevant to the different types of experiments accessible via the five experiment planning modules. The GP experiment planning module is described in detail. Companion papers describe the DT [16], FD [17], and MDS [18] experiment planning modules with illustrative scenarios. This paper concludes with an outline of experiment planning and implementation problems that have not yet been addressed.

2 CHEMISTRY WORKSTATION HARDWARE AND SOFTWARE

The workstation is designed for microscale chemistry. The design generally follows that of the second-generation system. A computer-aided design analysis indicated that a substantial improvement in throughput could be obtained with a new robot, a metering pump for solvent delivery, a syringe pump for sample manipulation, and flow injection analysis [13]. Accordingly, we have upgraded to a new robot with new syringe pumps for sample manipulation, though faster syringe pumps have been used in place of the anticipated solvent metering pumps, and we have not yet incorporated flow injection UV-Vis analysis. The reaction station has been expanded from 15 to 60 reaction vessels, and the number of sample vials increased from 72 to 264. The previous carousel-based product workup station has been replaced with racks of vials. The use of absorption spectroscopy as the only analytical tool presents obvious limitations but is suitable for selected domains of chemistry and is sufficient in such domains for testing new strategies and approaches for automated experimentation. Additional analytical instruments and hardware modules can be added later within the same conceptual framework established using absorption spectroscopy for analysis of reactions.

An artist's rendition of the third-generation workstation is shown in Fig. 2. A Cartesian robot (Intelligent Actuator, Inc., model FS with G-type controller) is equipped with two independent Z-axes that are fitted with side-port needles (Precision Sampling). This robot was selected based on numerous features, including the open Cartesian design, high speed, fast acceleration, precision (DC-servo controlled with internal position sensors), extendibility (longer tracks in the X-direction can be used), versatility (several independent Z-axes can be mounted), and ruggedness (a significant load on the Y-arm can be tolerated). Though some other Cartesian robots have faster maximum speeds, the distances traveled in the small work envelope prevent the maximum speed from being attained; thus acceleration is more important. The robotic work envelope (XYZ =

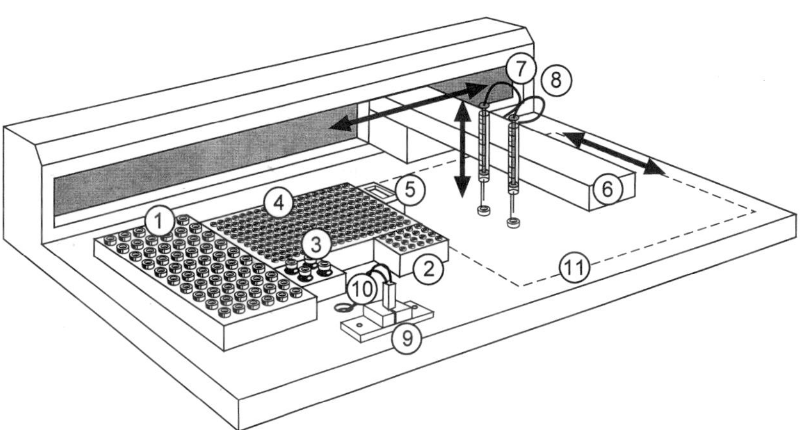

Figure 2 Automated chemistry workstation. (1) 60-vessel reaction station. Each vessel is individually stirred, and all vessels are themostated in a common reservoir. (2) Reagent rack 1. (3) Reagent rack 2. (4) Rack of 264 sample vials (only 190 shown here for clarity). (5) Wash station. (6) Cartesian robot arm. (7) Solvent delivery syringe on one Z-axis. (8) Sample transfer syringe on a second Z-axis. (9) Cuvette assembly for absorption spectroscopy. The fiber-optic lines are not shown. (10) Solvent inlet and drain line for cuvette. (11) Space for future expansion. The syringe pumps, reaction station heater–cooler bath, and solvent bottles are placed beneath the table (not shown).

70 × 40 × 12 cm) encompasses the 60-vessel reaction station, absorption spectrometer cuvette, reagent rack, and sample vial rack. Each of the 60 reaction vessels (vials fitted with septum caps) in the reaction station is stirred magnetically (Variomag stir plate, Florida Scientific, Inc.). A common reservoir in the reaction station provides temperature control via a flowing liquid and a heater–cooler bath (Neslab RTE-140). Solvent delivery is provided by a bank of four syringes (Cavro, 3000 step increments) feeding a four-way valve leading to the needle on the Z-axis. Reagent and sample manipulation are provided by a 1 mL syringe pump (Cavro, 3000 step increments). The absorption spectrometer (Ocean Optics, Inc.) consists of a 3 mL cuvette that is interrogated via optical fibers leading from the lamp and to the photodiode array detector (mounted on a card in the PC). The cuvette is loaded with solvent via a syringe pump and is drained via a separate syringe pump. The PC (Dell DIMENSION XPS p120c, PentiumPro) is equipped with 64 Mb RAM and a 1 Gb hard drive. The workstation hardware was assembled by Scitec, Inc.

This workstation hardware is two to three times faster than the second-generation workstation in performing most operations. In particular, delivery of 10 mL of solvent takes only 53 s and transferal of 50 μL reagent takes only 28 s, compared with 119 s and 77 s, respectively, in the previous system. These delivery times include all robot motion and syringe dispense times. (The liquid dispensing rate can be modified by changing the values in the syringe-driver configuration file.) The high speed of the workstation enables an increase in parallel experimentation (vide infra).

The software has been written within Windows 95, replacing the platform PC-MOS/386 (MOS) in the second-generation system. The experiment manager software is written in C++ (Symantec), and the hardware drivers for the automation equipment are written in Visual Basic. Each window in the user interface is numbered to correspond to a user's manual for workstation operation [19]. The latter also is available as an on-line help feature. The software is described in more detail below.

3 METHODS OF EXPERIMENTATION

Automated chemistry workstations can function in the open-loop mode or closed-loop mode (Fig. 3). In the open-loop mode, experiments are implemented faithfully in accord with a preexisting experimental plan. No decisions are made on the basis of the data collected. Consequently, modifications cannot be made by the workstation to an individual ongoing experiment, and selections cannot be made among a set of possible experiments in order to focus experimentation. In the closed-loop mode, data analysis and evaluation features enable decisions to be made by the workstation based on data obtained from prior or ongoing experi-

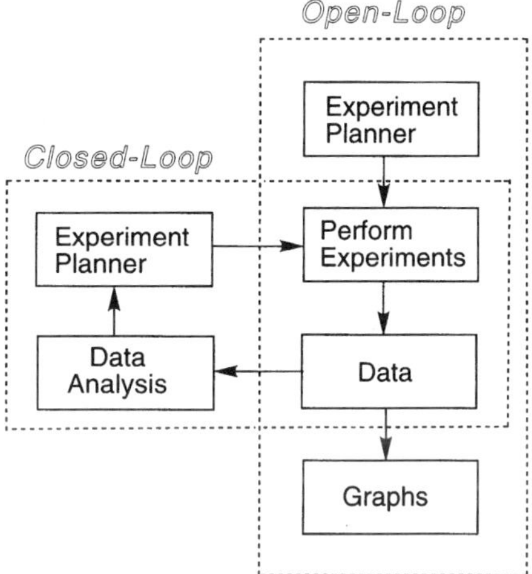

Figure 3 Flow diagrams for closed-loop and open-loop experimentation. Open-loop mode: experiments are implemented in accord with preset plans and data analysis is done off-line by the scientist. Closed-loop mode: decisions about ongoing, pending, or future experiments are made automatically based on the data obtained. Many cycles of experimentation can be performed in pursuit of the scientific objectives.

ments. Decisions can be made leading to modification of ongoing experiments, termination of futile lines of experimental inquiry, or spawning of new sets of experiments. These definitions of open-loop and closed-loop exclusively concern whether decisions are made based on workstation-acquired chemistry data, not on feedback related to proper functioning of hardware subsystems.

The ability of an automation system to modify its actions based on chemistry data greatly expands its flexibility and scope of application. While closed-loop experimentation imbues an attractive level of sophistication to an automation system, a host of experimental planning and implementation issues arise that must be addressed. The implementation issues concern whether decisions made about one experiment will cause adverse effects on other ongoing experiments. Three types of experiments are illustrated in Fig. 4. (A) In open-loop experiments, all commands are implemented as planned without any decision making during experimentation. (B) In one type of closed-loop experiment, several commands (e.g., analysis procedures) are subject to termination based on the nature of the data acquired. (C) In another type of closed-loop experiment, a conditional statement causes branching and a new set of commands is spawned. These three types of experiments require different scheduling approaches and resource management approaches (Table 1).

A) Open Loop:

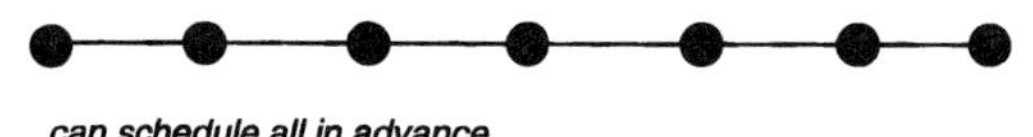

B) Closed-Loop with early termination:

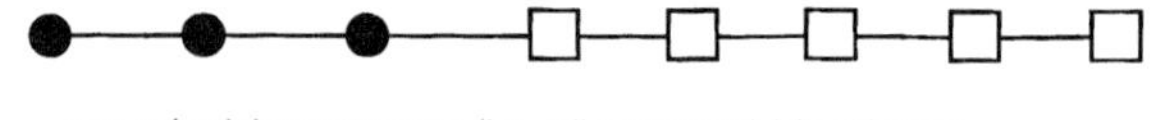

C) Closed-Loop:

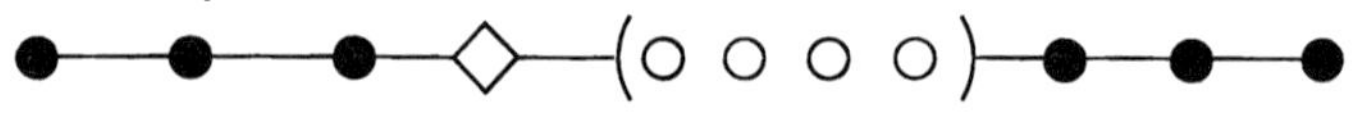

Figure 4 Experiments composed of commands having different types of decision making elements. Legend: ●, command that will be performed; □, command that is subject to an early termination decision; ◇, branch point; ○, command that is contingent on a branch point. (A) An open-loop experiment. None of the commands elicits decision making. All commands can be scheduled in advance. (B) A closed-loop experiment with the possibility of early termination of any of the last five commands (□) based on the experimental data collected in previous step(s). A worst-case schedule can be forecast by assuming that all commands are to be performed. (C) A closed-loop experiment with a conditional statement (◇), which elicits branching into another set of procedures (○). Scheduling can only be performed reliably up to the branch point. At present, experiments of this type must be implemented alone in order to avoid potential conflicts with other experiments.

Table 1 Implementation of Different Experiment Types

Decision-making mode during an experiment	Scheduling	Opportunistic rescheduling	Resource management
No decisions (open-loop)	parallel	not applicable	global look-ahead[a]
Early termination (closed-loop)	parallel	yes[b]	global look-ahead[a]
Conditional branching (closed-loop)	serial	yes[b]	command-by-command[c]

[a] Look-ahead at all commands in the experimental plan(s).
[b] Opportunistic rescheduling is not applicable to ongoing experiments.
[c] Examine each command upon implementation.

Table 2 Experiment-Planning Modules

Module	Individual experiment type[a]	Global decisions[b]	Ordering	Resource analysis look-ahead	Scheduling[c]	Use of templates[d]	Evolutive[e]
General purpose (GP)	OL	no	yes	global[f]	parallel	no	no
Decision-tree (DT)	OL or CL	yes	no	one command	serial	no	possible
Factorial design (FD)	OL or CL(ET)	no	no[g]	global[f]	parallel	yes	no
Simplex (CMS)	OL or CL(ET)	yes	no	one cycle (1 rxn)[h]	serial[i]	yes	yes
Multidirectional search (MDS)	OL or CL(ET)	yes	no[g]	one cycle (1 batch)[h]	parallel	yes	yes

[a] OL = open loop experiment; CL = closed-loop experiment; CL(ET) = early termination of an ongoing experiment.

[b] Global decisions refers to whether decision making concerning new experiments takes into account data from other experiments.

[c] Parallel refers to more than one experiment proceeding at the same time. Serial refers to only one experiment proceeding at a given time, so multiple experiments must be implemented in succession.

[d] Templates describe a general experimental plan in which specific values are inscribed for the variable parameters.

[e] An evolutive process is one where the course of the individual experiment, or a search process involving multiple experiments, is altered based on data collected during experimentation.

[f] Global resource look-ahead refers to examination of all commands in all queued experiments.

[g] In FD or MDS, all experiments have the same basic composition of commands, so experiments are scheduled in the order created by the experiment planning module (default method of ordering).

[h] Local look-ahead analysis.

[i] The $n + 1$ experiments in the first cycle (corresponding to the first simplex) can be run in parallel; all subsequent experiments are performed serially.

We have treated these three types of experiments in the following manner. A batch of open-loop experiments can be scheduled for parallel implementation, and the resource requirements can be assessed in a global look-ahead process in advance of experimentation. Those experiments for which resources are insufficient are designated ''waiting'' and are not placed in queue until resources become available [20]. For closed-loop experiments where the only decisions that can be made involve early termination of individual experiments, a batch of

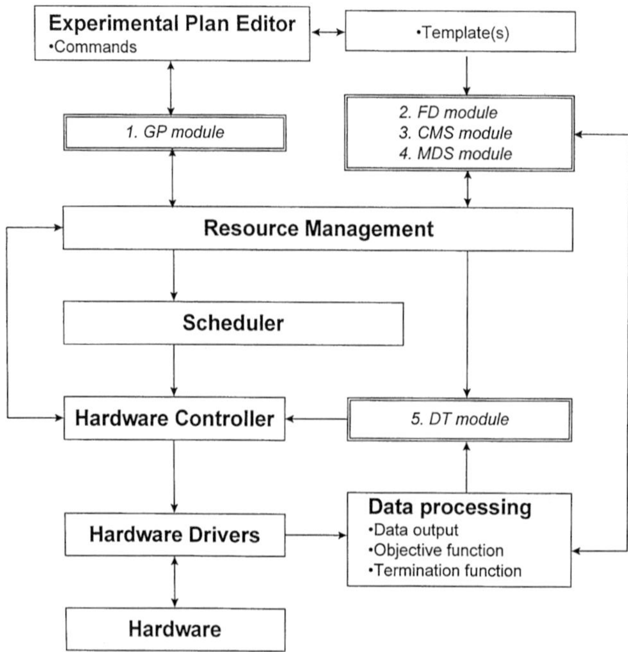

Figure 5 Key components of the experiment manager software. The five experiment planning modules are shown in italics in double-lined boxes. The experimental plan editor is used to compose plans for GP (general purpose), FD (factorial design), CMS (composite-modified Simplex), and MDS (multidirectional search) experiments, while a related set of functions is used in the DT (decision-tree) module. The Resource Management module is accessed by the hardware controller as each impending command is to be passed to the hardware for implementation. The Resource Management module also is accessed by each of the experiment planning modules prior to and during experimentation (see text). The hardware controller only has access to the next impending command. The five experiment planning modules have access to all remaining commands in queue. Information flows back and forth between the hardware controller and the Resource Management module.

resource-sufficient experiments can also be scheduled for parallel experimentation. When terminations occur, opportunistic rescheduling of any pending experiments can be performed without altering the course of ongoing experiments. Any closed-loop experiment where conditional branching can occur must be implemented as a stand-alone experiment. A set of such closed-loop experiments must be implemented with the individual experiments in series. Resource consumption cannot be predicted in advance for such closed-loop experiments and instead is performed during implementation on a command-by-command basis. Alternative approaches can be adopted for parallel implementation of experimental plans that involve branching, but the approach taken herein ensures that decision making in one experiment will not adversely affect other experiments. These three types of experiments appear in a variety of different guises.

Scientists routinely perform experiments in order to reach different types of objectives. To accommodate these diverse objectives we have developed five distinct experiment planning modules (Table 2). The planning modules differ depending on whether the experiments are (1) implemented in the open-loop mode or closed-loop mode, and (2) scheduled serially or in parallel. Three modules (CMS, FD, and MDS) enable statistically designed experiments. Each of these modules is available at the top level of the user interface. The software organization is outlined in Fig. 5. This diagram illustrates the hierarchy of the modules (constituting the user interface) and the hardware. The experiment planning modules serve two roles: (1) they serve as the interface where the scientist plans experiments; and (2) they participate in managing workstation operation during experimentation.

4 EXPERIMENT PLANNER

4.1 Experimental Plan Editor

Commands

Each experiment planning module makes use of commands that constitute the fundamental hardware operations of the robotic workstation. An experimental plan editor is used to compose experimental plans. The experimental plans can be used directly in the GP experiment planning module, or converted into templates for use in the FD, CMS, and MDS experiment-planning modules. The experimental plan editor provides a menu-driven user interface from which over 20 commands can be selected (Fig. 6). The actual number of commands varies subject to the configuration of the system hardware. The commands comprise operations such as "Fill reaction vessel with solvent," "Prepare reagent in reaction vessel," "Collect sample spectrum," "Set the reaction vessel temperature," and so forth. Each command has an associated timing formula that calculates the

A)

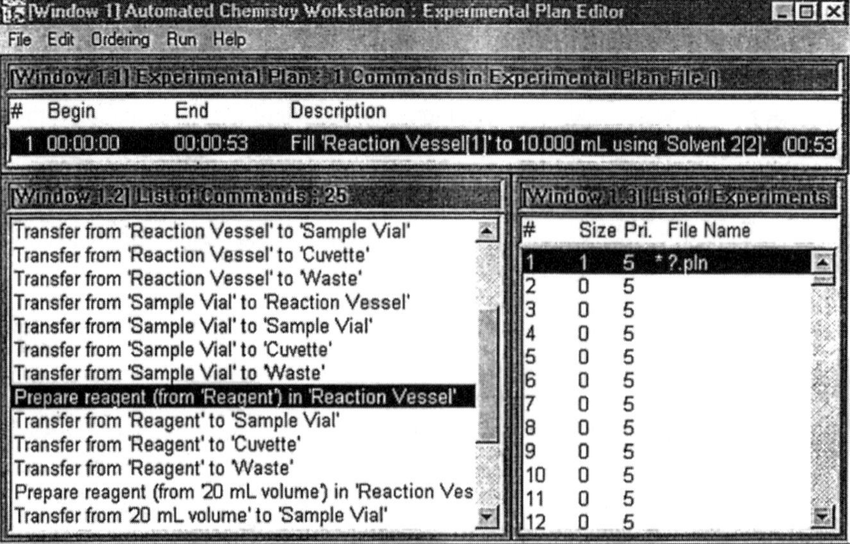

B)

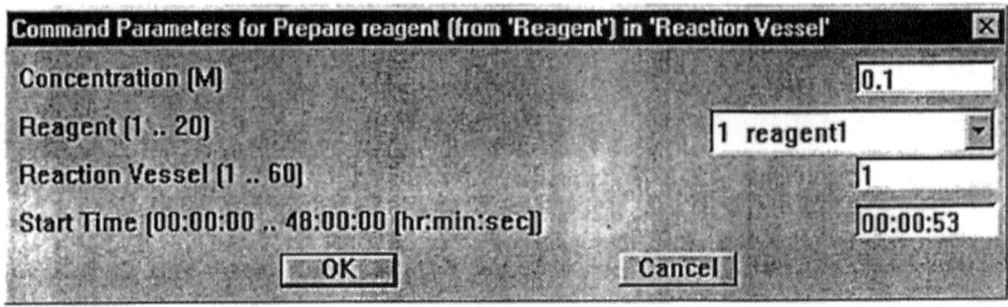

Figure 6 The experimental plan editor provides a menu-driven user interface for the composition of experimental plans (A). For example, when the user double-clicks on the command "Prepare reagent from reagent rack in Reaction Vessel" (highlighted in the List of Commands window), a dialog box prompts for parameters related to the command (B).

duration of the command (Table 3). For example, the timing formula for solvent delivery includes the time for XYZ motion of the robot and accounts for the user-specified volume of solvent to be dispensed. When commands are selected during experiment composition, the timing formulae (accessible via a function call) are engaged to compute the duration of each command.

A typical experimental plan for initiating and monitoring a reaction is shown below. The start times were selected by the chemist, but the finish times were determined automatically via the timing formulae (the timing display format is h:min:s).

#	Start	Finish	Commands
1:	00:00:00	00:00:53	Fill reaction vessel 1 to 10.00 mL using solvent 2 (0:53)
2:	00:00:53	00:01:23	Transfer 0.1 mL of reagent 1 to reaction vessel 1 to make 0.1 M (0:30)
3:	00:01:23	00:01:53	Transfer 0.1 mL of reagent 2 to reaction vessel 1 to make 0.1 M (0:30)
4:	00:01:53	00:02:43	Transfer 0.5 mL of reagent 3 to reaction vessel 1 to make 0.5 M (0:50)
5:	00:02:43	00:05:06	Analyze 0.05 mL of product by UV-Vis
6:	00:32:43	00:35:06	Analyze 0.05 mL of product by UV-Vis
7:	01:32:43	01:35:06	Analyze 0.05 mL of product by UV-Vis
8:	03:32:43	03:35:06	Analyze 0.05 mL of product by UV-Vis
9:	07:32:43	07:35:06	Analyze 0.05 mL of product by UV-Vis

In order to make the experimental plan easy to read, analysis procedures are listed here instead of the actual set of commands. Each analysis procedure is composed of several commands, depending on the desired operation. For example, the product may be analyzed by direct delivery to the cuvette, or following a workup procedure in a sample vial. In the above plan, the first UV-Vis analysis involves the combination of the following commands:

1:	00:02:43	00:03:16	Transfer 0.150 mL of reagent 3 to sample vial 1 (0:33)
2:	00:03:16	00:03:43	Transfer 0.025 mL from reaction vessel 1 to sample vial 1 (0:27)
3:	00:03:43	00:04:06	Mix 0.015 mL in sample vial 1 (0:23)
4:	00:04:06	00:04:46	Transfer 0.05 mL from sample vial 1 to cuvette (0:40)
5:	00:04:46	00:04:48	Collect sample spectrum (0:02)
6:	00:04:48	00:05:06	Clear cuvette (0:18)

Table 3 Timing Formulae for Commands

Commands	Timing formula (s)[a]	Volume (mL)	Time (sec)[b]
Fill "Reaction Vessel" with solvent	13 + 4 (solV)	5–10	33–53
Dispense solvent to "Sample Vial"	11 + 4 (solV)	1–3	15–23
Dispense solvent to "Cuvette"	19 + 4 (solV)	1–3	23–31
Dispense solvent to "Waste"	15 + 4 (solV)	1–10	15–51
Transfer from "Reaction Vessel" to "Reaction Vessel"	25.4 + 50.6 (rsV)	0.01–0.95	26–73
Transfer from "Reaction Vessel" to "Sample Vial"	25.4 + 50.6 (rsV)	0.01–0.95	26–73
Transfer from "Reaction Vessel" to "Cuvette"	37.5 + 50.6 (rsV)	0.01–0.95	38–85
Transfer from "Reaction Vessel" to "Waste"	27.4 + 50.6 (rsV)	0.01–0.95	28–75
Transfer from "Sample Vial" to "Reaction Vessel"	25.4 + 50.6 (svsV)	0.01–0.95	26–73
Transfer from "Sample Vial" to "Sample Vial"	25.4 + 50.6 (svsV)	0.01–0.95	25–73
Transfer from "Sample Vial" to "Cuvette"	37.4 + 50.6 (svsV)	0.01–0.95	38–85
Transfer from "Sample Vial" to "Waste"	26.4 + 50.6 (svsV)	0.01–0.95	27–74
Prepare reagent in "Reaction Vessel"[c]	29.4 + 50.6 (reagV)	0.01–0.95	30–77
Transfer from "Reagent" to "Sample Vial"	24.4 + 50.6 (reagV)	0.01–0.95	25–72
Transfer from "Reagent" to "Cuvette"	37.4 + 50.6 (reagV)	0.01–0.95	38–85
Transfer from "Reagent" to "Waste"	26.4 + 50.6 (reagV)	0.01–0.95	27–74
Clear Cuvette	18		18
Set temperature of "Reaction Vessel"[d]	1		1
Collect spectrum	2		2
Mix contents of "Sample Vial"	20 + 200 (mixV)	0.01–0.5	22–2:00

[a] V indicates a volume to be transferred, with the following abbreviations: solV, solvent volume; rsV, reaction sample volume; svsV, sample volume from sample vial; reagV, reagent volume; mixV, mixing volume. The constant terms are related to the robot movements. The coefficients of the volume terms reflect pumping rates; 4 (solvent syringe) corresponds to 250 µL/s, 50.6 (reagent syringe) corresponds to ~20 µL/s, and 200 (mixing contents in sample vial) corresponds to 100 µL/s.

[b] Range of duration, which is determined by the minimum and maximum values of the variable parameters.

[c] Without autodilution. If autodilution is activated, the total time for autodilution will be the sum of each operation involved, including dispense solvent to a specified container, transfer reagent to the container, and transfer the diluted reagent to a destination vial.

[d] Arbitrary number for simulation purposes.

The duration for each command is listed in parentheses. Subsequent analysis procedures use the same set of commands with a start time of the set delayed by an appropriate amount.

In this plan, the five analysis procedures are spread out in a geometrical progression (Δt = 0, 30, 60, 120, and 240 min). Each UV-Vis analysis involves transferring a sample from the reaction vessel to a vial (for oxidation, neutralization, dilution, etc.), followed by transfer to the UV-Vis absorption spectrometer where the spectrum is collected. This experimental plan is quite typical in synthetic and mechanistic chemistry: solvent is added to a reaction vessel, two reagents are added, a catalyst is added, and then samples are removed periodically for analysis. (The term reagent is used generically to encompass any species added to the reaction vessel other than solvent, including reactant and catalyst). To facilitate creation of these types of experiment plans, the chemist may copy a block of commands.

There are three aspects to copying a block of commands: (1) the number of times to copy the command, (2) the interval between the blocks of new commands, and (3) whether the new blocks should be spaced at linear or geometric times. If the timing is to be linear, the blocks of new commands are spaced at multiples of the specified timing interval. If the timing is to be geometric, the first new block is placed at the selected interval, the next interval is placed at twice the interval, and subsequent blocks are placed at double the previous interval. Editing features for deleting or copying commands enable rapid composition of experiment plans.

The plan shown above can be viewed as a timeline (Fig. 7). In the timeline it is clear that for most of the duration of the 455 min experiment, the robot remains idle as the reaction proceeds. In fact, the total duration of robot activity (14.6 min) is a fraction (i.e., utilization = 3.2%) of the experiment duration (i.e., makespan = 455 min). In order to increase the robot utilization, experimental plans can be scheduled for parallel implementation (vide infra).

Intelligent Reagent Handling

We have incorporated a number of reagent handling features in order to facilitate the experiment planning and implementation processes. These features operate in the background during the composition of experimental plans.

Reagent Database. A reagent database is used that includes information about the reagents, including their concentration and identity in the reagent racks. Commands specifying desired final reagent concentrations are converted into deliverable volumes through use of the reagent database. The conversion is performed assuming additivity of volumes of all components to be added to the reaction vessel. In the experimental plan listed above, the volume of solvent to be added will be diminished automatically in order to obtain the desired total 10

One Experimental Plan

Start	Finish	Commands
0.0	0.8	Fill reaction vessel 1 to 10.00 mL using solvent 1
0.8	1.4	Transfer 0.1 mL of reagent 1 to rxn vessel 1 to make 0.1 M
1.4	1.8	Transfer 0.1 mL of reagent 2 to rxn vessel 1 to make 0.1 M
1.8	2.7	Transfer 0.5 mL of reagent 3 to rxn vessel 1 to make 0.5 M
2.7	5.0	Analyze 0.05 mL of product by UV-Vis
32.7	35.0	Analyze 0.05 mL of product by UV-Vis
92.7	95.0	Analyze 0.05 mL of product by UV-Vis
212.7	215.0	Analyze 0.05 mL of product by UV-Vis
452.7	455.0	Analyze 0.05 mL of product by UV-Vis

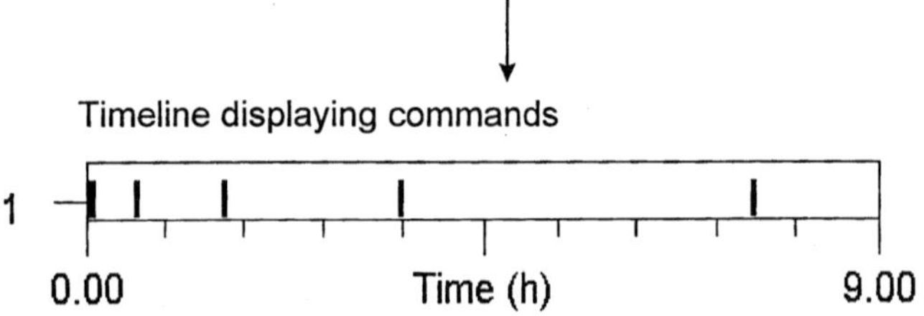

Figure 7 Example of an experimental plan (start and finish times are given in minutes). The time line describing this experimental plan (makespan 455 min) is sparsely populated by robotic operations (utilization = 3.2%).

mL reaction volume (vide infra). (Note that reagent volumes can be specified in the experimental plans by omitting use of the reagent database.) The reagent database used is a text file with the following format:

```
[Reagent Database]
reagent 1 = 10
reagent 2 = 10
reagent 2 1 = 0.1
reagent 2 2 = 0.001

. . .

compound i = concentration i

. . .
```

where compound i and concentration i represent the reagent name (used in the reagent racks) and its stock concentration, respectively. For a specific reagent, the names of all diluted samples of the same reagent must follow the rule: the "original name" (e.g., reagent 1) gives the highest stock concentration, the "original name 1" (e.g., reagent 1 1) the next highest, and so on. In the reagent database example above, the stock solutions of the reagent 2 are named as reagent

2, reagent 2 1, and reagent 2 2. Their rack positions are then defined in the same manner as all the other reagents so that the user can identify the reagents by their names rather than their actual locations.

Autoselection of Reagents. The volume range of the workstation syringe system extends from 10 to 950 μL, but this is not sufficient to span the range of final reaction concentrations often desired. A series of stock solutions of varying concentrations for a given reagent can be prepared manually by the user to cover the desired target range. When the same reagent is present at different concentrations, the reagent with the appropriate stock solution is selected automatically according to the specified lower (10 μL) limit of the syringe volume. This information is stored in a hardware configuration file. The correct vial and volume are selected automatically. If there is more than one vial with matching name and adequate concentration, the vial requiring the smallest volume above 10 μL is selected (Fig. 8). The autoselection of reagents feature guarantees enough material for all the experiments. Note that the duration of commands involving liquid transfer depends on the volume of the transfer operation. Hence the automatic selection of a larger volume for transfer will necessarily result in a lengthier operation.

Autodilution of Reagents. If a requested reagent is too concentrated for delivery (i.e., the necessary volume to be transferred is less than the minimum syringe transfer volume, commands for a 10-fold dilution are included as an automatic part of the experimental plan (with appropriate added duration for these operations). Once prepared, the new diluted solution is then available for any other experiments that require a solution of this concentration.

4.2 The Five Experiment Planning Modules

General Purpose Experiment Planning Module

The experimental plans composed in this module can vary widely, but each plan describes an open-loop experiment (i.e., includes no conditional statements). The experimental plans are composed using the experimental plan editor. A wide variety of plans can be implemented at the same time, subject only to the inherent hardware limitations. The list of experimental plans to be implemented is processed in a three-tiered scheme as shown in Fig. 9. The chemist loads and/or creates the list of experimental plans to be included in a batch for implementation. The individual experiments in a batch can be ordered by heuristics, by user-designated priorities, or by a default method. Ordering establishes the priority of the individual experiments. The use of heuristics provides a workable means of ordering the experiments without tackling the combinatorial problem of finding the ordering that gives the shortest makespan [21]. In the heuristic method, the

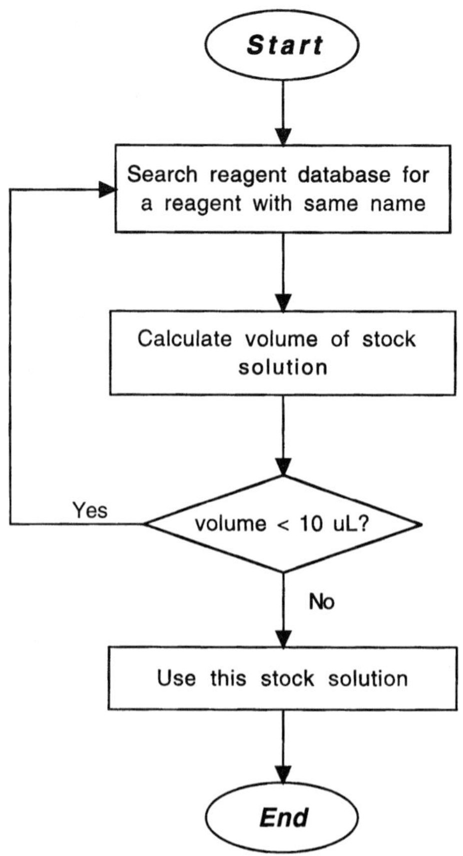

Figure 8   Flowchart for autoselection of reagent solutions having appropriate concentrations. This is one of several intelligent reagent handling features.

scientist chooses one of five heuristics to determine the priorities of a group of experiments. The five different heuristics include

1. Order the experiments with the shortest makespan last.
2. Order the experiments with the shortest makespan first.
3. Order the experiments with the shortest total command duration (active time) last.
4. Order the experiments with the shortest total command duration (active time) first.
5. Order the experiments with the most sparse (lowest utilization) experiment first.

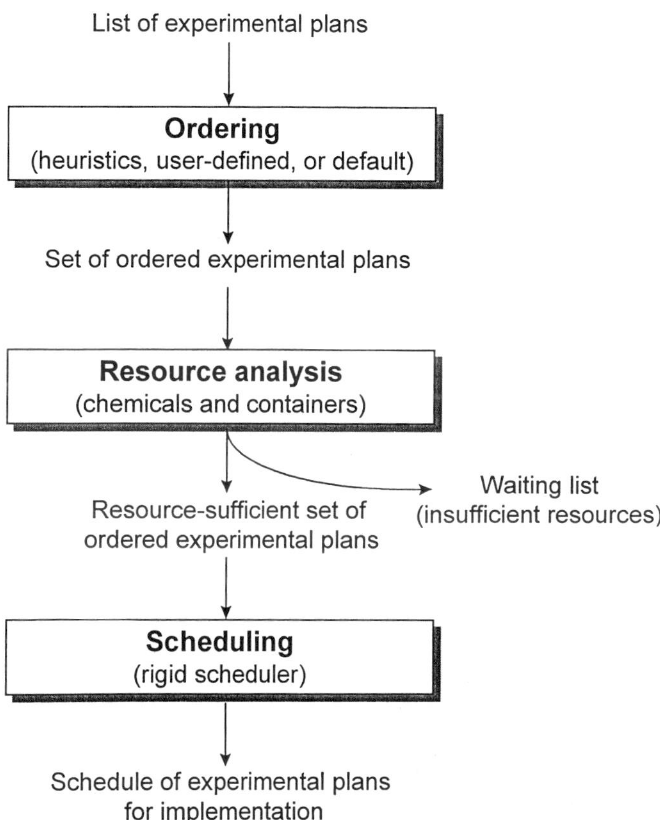

Figure 9 Processing of a list of experimental plans in the general purpose experiment planning module. Ordering establishes the priorities of the experiments, resource analysis prunes the list (in order of priority) depending on available containers and chemicals, and the rigid scheduler then offsets the start times of the individual plans in order to produce the final schedule.

The heuristic in which the longest experiments receive the highest priority is often quite attractive for experimentation [10]. In the user-designated priority method, the scientist specifies the ordering, which allows for considerable flexibility. As one example, the experiments can be ordered as desired to investigate any putative confounding effects associated with any other ordering. In the default method, the priorities are set identical to the order in which the experimental plans are loaded in the list of experimental plans (Fig. 6). After the experimental plans are ordered, causing the assignment of priorities, resource analysis is performed (vide infra). Resource analysis of the individual experiments in the batch is performed in order beginning with the highest priority experiment. Resource consumption of chemicals and containers can be calculated in advance for all of the experiments (global look-ahead) because each is an open-loop experiment. Those experiments that fail the resource test (i.e., for which resources are inadequate) are placed in a waiting list. The resulting queue of ordered experimental plans is then scheduled using the rigid scheduler. One of the great attractions of open-loop experiments is that different types of experimental plans can be scheduled in parallel (Table 2). The absence of any decision making ensures that each command can be implemented independently of all other commands, so the interleaving of experimental plans as required for parallel scheduling will not result in conflicts among experiments. Following ordering, resource analysis, and scheduling, the batch of experiments is ready for implementation.

Decision Tree Experiment Planning Module. The most versatile and complex method of describing experiments is with the decision tree planning module, which allows the chemist to describe an experiment in a C-like programming language. This language includes functions to perform robotic procedures and collect data, conditionals (if-then-else, while, for) to allow branching of the experiment or spawning of new experiments, basic math operators ($+$, $-$, $*$, $/$, log, pow), variables to store experimental results, and functions to prompt for user input. The user input can be requested in the form of dialog boxes. While the decision tree (DT) experiment planning module provides the widest possible types of experiments, the onus rests on the chemist to plan for all possible occurrences over the course of the experiments.

The DT programs are best used for specialized applications. The DT programs are not composed using the experiment plan editor but are instead composed using a related set of DT functions. Furthermore, DT programs cannot take advantage of the schedulers and are performed serially. Resource management must be done on a command-by-command basis as the program is implemented, although the chemist can arrange to perform some simulations of the DT programs to assess resource needs (Table 2). We have written a DT program to screen a list of compounds for catalytic activity [16] and have applied this in porphyrin chemistry [15]. Decision tree programs can also be used as part of objective functions in CMS and MDS experiments, and as early termination functions in FD, CMS, and MDS experiments. The DT experiment planning module, the catalyst screening program, and a complete list of DT functions are described in the accompanying paper [16].

Factorial Design Experiment Planning Module

In scientific studies of a phenomenon involving a large number of variables, one often wants to understand the effect of each variable. A comprehensive investigation of all levels of each variable or factor is achieved by examination of a regular grid spanning the factors, termed a grid search, factorial search, or factorial design. The factorial design (FD) planning module can be used to generate response surfaces, prepare indexed combinatorial libraries, or screen candidate catalysts in parallel. The FD module includes the following features:

> Automated layout of selected patterns of points in an n-dimensional search space
> Use of templates so that generic procedures can be reused without reprogramming
> Use of continuous (e.g., concentration) or discrete (e.g., chemical components) variables
> Automated early termination of individual experiments during parallel experimentation
> Opportunistic rescheduling following each decision making (early termination) event

The FD module does not contain any provisions for statistical analysis. The individual experiments of a factorial design are implemented in parallel. In general, the experiments are open-loop, since all designated points in the search space will be examined. However, the possible early termination of individual experiments can accelerate the completion of the grid search. Thus decision making is done locally on individual experiments and affects only the rate, but not the extent, of the global search (Table 2). (The early termination procedure can be omitted as appropriate for factorial design studies.) The factorial design (FD) experiment planning module is described in the accompanying paper [17].

Composite-Modified Simplex Experiment Planning Module

The Simplex algorithm is a widely used method for unconstrained nonlinear optimization [22]. The Simplex approach is attractive because the calculations are straightforward and the search space is expandable to encompass multiple dimensions. A simplex is an n-dimensional polygon with $n + 1$ vertices, where n is the number of experimental control variables. (Note that we use the capitalized ''Simplex'' to refer to the algorithm and the small case ''simplex'' to refer to the geometrical object.) Thus the simplex is triangular in a two-dimensional space, tetrahedral in a three-dimensional space, and so forth. Each vertex of a simplex has n coordinates and an associated response from the experimental system. After obtaining responses for each of the $n + 1$ vertices of an initial simplex, a new vertex is projected (reflection, expansion, and/or contraction) by moving away from the vertex in the current simplex having the worst response. The new vertex defines a set of experimental conditions whose response is to be tested. Repetition of this cycle of experimental measurement of the system response forms the basis for the most elementary Simplex algorithm.

Many modifications to the original Simplex algorithm have been made over the past 30 years. Betteridge has developed and tested a composite modified simplex (CMS) algorithm that incorporates the best features of all of these approaches [23,24]. The CMS has features to accelerate convergence, switch between coarse and fine graining, avoid oscillating on ridges in the response surface, and deal with experimental problems such as sharp peaks, boundaries, noise, and local maxima. We have implemented the CMS with only minor modifications in our automated chemistry workstation [11]. The one feature proposed by Betteridge as part of CMS that is not present involves ''suboptimal redirection.'' The nature of the Simplex algorithm in the CMS module can be altered in sophistication by turning some features on or off in software. The Simplex algorithm is well suited for optimization but does not provide flexible decision making capabilities.

A Simplex search is inherently serial. The n + 1 points of the first Simplex can be examined in parallel (in the first cycle), but each subsequent cycle involves only one experiment. An objective function is used to evaluate the experimental data as part of the search strategy. An early termination function can be used to evaluate the progress of each individual experiment. Thus decision making must be performed to guide the nature of the global search but also can involve individual experiments (Table 2). A major limitation of the CMS module is the serial nature of experimentation.

Multidirectional Search Experiment Planning Module

The multidirectional search algorithm is a powerful new pattern search method developed by Torczon [25]. Factorial designs and related grid searches can be implemented in parallel, while Simplex experiments are implemented sequentially. The multidirectional search (MDS) algorithm enables directed evolutionary searches in a parallel mode. Like FD experiments, MDS experiments are performed in parallel batches with the option for early termination of individual experiments. Similar to CMS experiments, decision making is performed after each cycle to identify the next set of experiments. Thus MDS experimentation involves global decision making and can also involve local decision making concerning individual experiments (Table 2). The MDS experiment planning module is described in the accompanying paper [18].

5 RESOURCE MANAGEMENT

Scientists plan experiments at a hierarchy of levels prior to starting laboratory work. Perhaps the lowest level of experiment planning involves securing supplies, reagents, and instruments and performing calculations concerning amounts of reagents, sample preparation procedures for analysis, and so forth. Although part of this bookkeeping is usually done in advance as part of an experimental plan, other facets often are addressed reflexively as the experiment progresses. Indeed, these resource issues are often not given much thought by scientists working manually, until a key experimental step must be performed in a timely fashion and a crucial item cannot be found, in which case the issue of resources reaches a crisis.

Resource issues become preeminent as procedures become more elaborate, or numerous commands must be performed in timed sequence, or when a large number of samples are being processed in parallel. With an automated chemistry workstation, a large number of experiments can be performed in short periods, and experimentation can proceed over long periods. However, preparing detailed plans for automated experimentation, including specifications for all consumable resources (reaction vessels, sample vials, solvents, reagents, etc.) can be extraordinarily tedious. When decisions are to be made during the course of experimentation it is impossible to predict the total resource demand; thus resource management is handled differently for open-loop and closed-loop experiments.

To free the scientist of these essential but tedious tasks, we have developed a sophisticated Resource Management module that performs bookkeeping prior to and during experimentation concerning consumables, provides a look-ahead feature for experiments in queue, and alerts the user when impending commands are in danger of depleting consumables. In addition, the resource management module enables autoselection of containers (vessels and vials) in experimental plans and provides safeguards against inadvertent overflows. Such resource management features are essential for the operation of autonomous experimentation machines.

5.1 Types of Resources

The consumable resources in the automated chemistry workstation can be grouped into two classes, chemicals and containers. The chemicals are finite in volume and include solvents for three locations (reaction solvent, UV-Vis spectrometer solvent, sample transfer syringe solvent) as well as 24 reagents. For these consumables the resource issue concerns possible depletion during experimentation. Other workstation components such as analytical instruments or robotic arms are not treated as resources using this approach (i.e., the robotic arm is the limiting resource from a scheduling perspective, but it is not a consumable and is not included in the Resource Management table).

Mixtures that are formed in containers and used during experimentation constitute a type of temporary chemical resource. These temporary mixtures in containers include reaction mixtures (in reaction vessels), reaction samples (in vials prior to analysis or for storage), and autodiluted reagents. The major resource issue is that the temporary mixtures can be too low in volume for sampling (depletion) or too large for the container size (overflow). Unlike consumables, these temporary mixtures are a consequence of experimentation. Temporary mixtures are not readily replenished in the case of depletion, and overflow can threaten the integrity of the experiment and the workstation.

5.2 Resource Management Table

A Resource Management table provides a tally of each resource. This table is initialized by the user prior to experimentation and is updated automatically each time a command is implemented; thus the table displays the total resources available in real time for experimentation. Fundamentally different approaches are employed for monitoring the raw materials (chemicals and containers) and the temporary mixtures.

The Resource Management table (Fig. 10) consists of eight windows (subject to system configuration) listing reagents, reactions, samples, and solvents (reaction solvents, spectrometer solvent, and robotic arm syringe solvent). The windows for stock chemicals (24 reagents and four reaction solvents) have columns for index, current volume, and name, which display information for their consumption. The current volume column displays the total volume that is in the container. The available volume column takes into account the inaccessible material at the bottom of the vial (that cannot be reached by the syringe). For temporary mixtures, depletion and overflow are equally important, so their windows have columns for container index, current volume, and total added volume for the 60 reaction vessels and 264 sample vials. The total added volume is the maximum volume during experiments. Each type of container has user-specified minimum and maximum values that are used as criteria for the Resource Management module to check for depletion or overflow. Initially, the Resource Management table defaults to certain values based on the information stored in related configuration files. For example, all of the reaction vessels and sample vials are assumed to start out empty, so zero is the default value for current volumes and total added volumes.

5.3 Resource Analysis Look-Ahead

The Resource Management table is accessed prior to experimentation to simulate resource availability (with user control in the GP module; automatically by the FD, CMS, and MDS modules). The simulation mode is used to look ahead and identify resource problems so that experimental plans can be altered or provisions made to intervene when resource problems are identified (Table 2). In the simulation mode the workstation hardware is effectively turned off. (The simulation mode can be used to test a DT program, where prompts are displayed at each

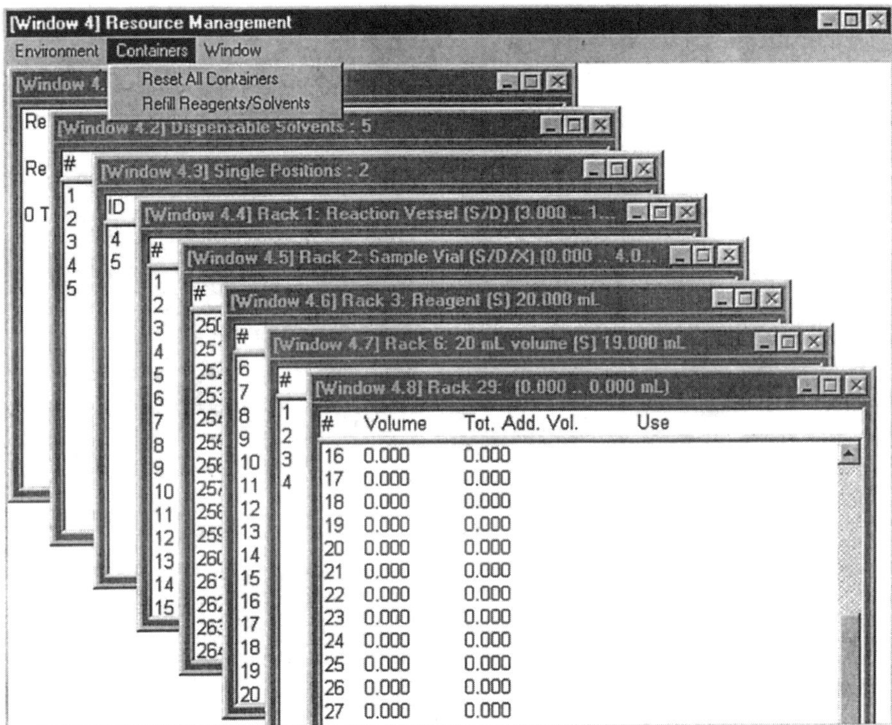

Figure 10 The Resource Management table screen display. The volume in each reagent vial can be changed via the menu. This table displays the currently available volumes and is updated each time a command is called.

point in the program when data would be acquired. The user provides responses to the prompts and in this manner can simulate experiments to test various branches of the program by providing appropriate responses [16].) As commands are passed to the hardware driver, the appropriate resource values are passed to the Resource Management table, but commands are not issued to the hardware. The following types of analysis are performed.

Autoassignment of Containers

Reaction vessels and sample vials are used for temporary storage during many experiments. The locations of specific vials/vessels are not specified in the experimental plans (GP) or templates (FD, CMS, and MDS). The Resource Management module tabulates all requests for vials as part of a look-ahead resource analysis. This module will modify the vial assignments for all experiments ensuring that those containers (reaction vessels or sample vials) used in one experiment will not be used in different experiments. Any early termination that occurs will deallocate containers that otherwise would have been used. (In DT programs the Resource Management table can be exploited to locate the next container by incorporating appropriate resource status queries. For instance, the support function "find_generic_container" leads to the identification of the next empty vessel.)

The reaction vessels and sample vials are available in an indexed order. The program then proceeds by performing the operation in the specified vessel. The ability to locate the next available clean container means that detailed provisions or counters do not have to be written into the experimental plans in order to keep track of available containers. This provides great simplicity in planning experiments since the bookkeeping features of the Resource Management table can be exploited to gain such information. This feature is ideally suited for individual experiments or searches that consume large numbers of containers.

Intelligent Safeguards Against Depletion and Overflow

Depletion of chemicals or containers is guarded against in the following manner. The look-ahead analysis causes the appropriate values for requested resources to be passed to the Resource Management table. Any insufficiency in containers is easily identified because all of the containers are indexed. The possible depletion of the contents of a container requires knowledge of the container size and the volume of liquid in the container. An automatic volume checking feature ensures that all command parameters in an experimental plan are within the physical limits of the hardware, thereby preventing inadvertent out-of-bounds values. For example, the minimum volume value of a container is tested when a command requires a transfer from that container, whereas the maximum value of a container is tested when a command requires a transfer to that container. The user sets the threshold values in a Hardware Configuration file (e.g., 5 mL is the lower volume limit for reaction vessels, 1 mL for reagent vials). If there are insufficient quantities of chemicals or containers, a warning signal is displayed.

The overflow of reaction vessels is guarded against by the assumption of volume additivity in the experimental plans, which causes the amount of added solvent to satisfy the specified total reaction volume. The possible overflow of vials is identified by the Resource Management table. Resource problems identified during the prerun simulation are tabulated and a warning is displayed. The user can then alter the experimental plans to avoid the resource problem, run a subset of resource-sufficient experiments, or go ahead and implement the experiment(s) unchanged. The latter is done with the knowledge that some experiments may fail due to insufficient resources unless the user intervenes at an appropriate time.

Total Reaction Volume Control

During composition of experimental plans (for GP, FD, CMS, and MDS experiments), the total reaction volume is set by the user at a specified volume. A look-ahead test assesses the total volume for each reagent to be added. Assuming additivity of volumes, the volume of solvent to be added is decreased appropriately so that the total requested volume of the reaction mixture is achieved.

5.4   Distinctions in Resource Management for the Five Experiment Planning Modules

Resource simulation is handled slightly differently in each of the five experiment planning modules. In open-loop experiments (GP, FD without early termination) the exact course of experimentation is known in advance, and consequently total resource consumption can be calculated in advance. The total resource consumption that is calculated for open-loop experiments in the simulation mode should be identical with that in actual implementation. Thus in open-loop experiments this resource management approach provides a global look-ahead feature (Table 2). In closed-loop experiments the exact course of experimentation is not known in advance. For CMS and MDS experiments, the resources required for each cycle can be calculated upon projection of the new points to be examined. The resources for CMS involve the one new experiment (one reaction) in the cycle, and those for MDS involve all the experiments (all reactions) in the batch to be performed in the cycle (in MDS the size of the batch can be quite large). This resource analysis is referred to as a one-cycle look-ahead analysis. For DT experiments, resource analysis is performed on a command-by-command basis upon implementation. Functions can be incorporated in the decision-tree programs to assist in resource analysis, but this requires programming and is not done automatically as part of the Resource Management module.

5.5   Resource Management During Operation

During implementation of experiments (following scheduling, vide infra), resource assessment is made on a command-by-command basis. The Resource Management table is updated with each command (via the Hardware Controller, see Fig. 5), and the results can be viewed by the user. The user can update the display concerning chemicals, but the assignment of containers cannot be interrupted (the latter is automatically carried out by the Resource Management module).

During experimentation, information flows back and forth between the particular experiment planning module that is in use and the Resource Management table. The ability to gain information from the Resource Management table during operation has advantages beyond merely identifying if adequate materials are present for a given command. During lengthy experimentation, the user can estimate when resources need to be replenished. This is especially useful in closed-loop experiments where the course of experimentation cannot be predicted at the outset. Furthermore, the incorporation of function calls in a DT program enables resource management to be performed as the program is being implemented. This topic is discussed at length in the companion paper [16].

6 SCHEDULER

Most experimental plans have lengthy intervals where the robot remains idle as the chemistry proceeds. In the example shown in Fig. 7, the total duration of robot activity (14.6 min) is a fraction (i.e., utilization = 3.2%) of the experiment duration (i.e., makespan = 455 min). In order to increase the robot utilization, experimental plans can be scheduled for parallel implementation. The experiment scheduling problem involves generating a sequence of commands that satisfies the temporal constraints specified in each experimental plan while minimizing the overall makespan of the set of experiments. The experimental scheduling problem has features distinct from those in other areas in the scheduling field [14]. The scheduler for the second-generation system treats availability of the robot as the sole resource capacity constraint [10]. This scheduler interleaves the commands of individual experimental plans by offsetting their start times and then interleaves the experimental plans without altering the relative times of the procedures within each plan. Thus this scheduler adheres to the rigid time constraints specified in the composition of individual experimental plans. An example is given for eight identical experiments (Fig. 11). The total makespan of the schedule is 523 min, and the robot activity (117 min) affords a utilization of 22%, nearly seven times faster than with serial implementation (3640 min) of the eight experiments. The Gantt chart shown in Fig. 11 resembles those that can be viewed during operation of the workstation. The Gantt chart is updated as each command is implemented, displaying the progress of the experiments.

In many chemistry applications such a high level of rigidity in the timing of operations is not required. Recently we developed a scheduler that allows specified amounts of flexibility in the start times of commands [14]. In that scheduler, each individual command was subject to variation in start time. Furthermore, a series of after-relations coupled the differences in start times between com-

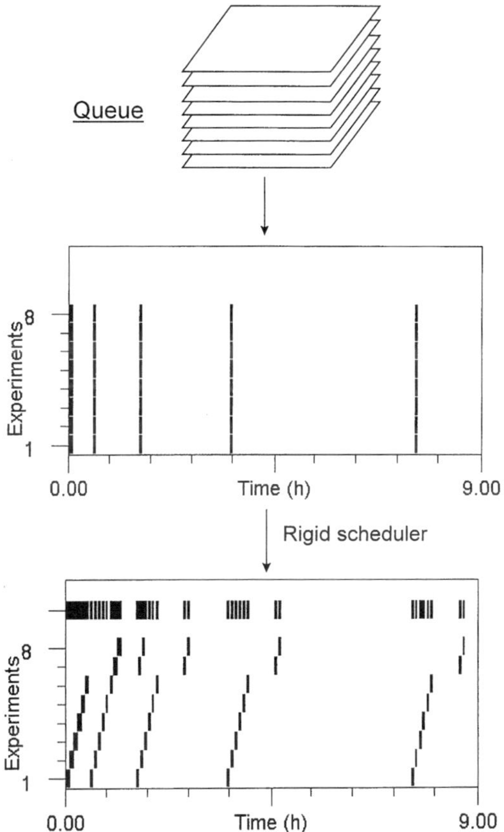

Figure 11 Scheduling of eight identical experiments. The rigid scheduler interleaves the eight experiments by offsetting their start times. At 1 h, for example, six experiments are underway. At 3 h all eight experiments are underway. Parallel implementation of experiments leads to a significant compression in duration of experimentation (makespan = 523 min; utilization = 22%) compared with serial implementation (makespan = 3640 min; utilization = 3.2%).

mands, causing the start time of a command to be dependent on that of the preceding command. With large numbers of lengthy experimental plans, the scheduling process itself took several minutes. More recently we have developed an improved flexible scheduler. The architecture of the experiment manager software has been modified so that either the rigid scheduler or the new flexible scheduler can be selected in template-related experimentation (FD, CMS, and MDS).

The following modifications have been made to implement the new flexible scheduler. (1) Many individual commands corresponding to steps of a chemical process can be grouped together. For example, the collection of a sample spectrum involves several steps including transfer of the sample to the cuvette, collection of a UV-Vis spectrum, and cleaning of the cuvette. We shall use the term metacommand to denote any collection of contiguous commands (Fig. 12). It is logical to schedule these contiguous commands as a ganged block rather than individually, given that a contiguous schedule represents the fastest possible schedule, and their grouping often reflects a coupled series of operations that underpins a chemical process. (2) In contrast to the previous flexible scheduler, where the start times of individual commands are coupled by a series of after-relations, in the new flexible scheduler such constraints on the start times of commands have been removed. We assume that the start time of each metacommand can be independently advanced or delayed. The chemist specifies the acceptable range of start times of a metacommand. Note that an individual command that is not contiguous with any other command is treated in the same manner as a metacommand. (3) A forward-matching strategy has been used to find a proper shift. This matching strategy first calculates the duration of the metacommands in the next experimental plan to be scheduled and then searches for the earliest time blocks in the already scheduled set so that the new experiment can be scheduled into those time blocks without a time-sharing violation. (Definitions: A time block is any period of robot inactivity. The already scheduled set consists of all commands from those experimental plans that have been scheduled.) The new algorithm is

1. Order all unscheduled experimental plans by some criterion in order to start scheduling. (Here the default priorities are applied, because in template-related experimentation all experiments have the same composition of commands in each experimental plan and differ only in specific parameter values.)

2. Repeat the following steps until the current experimental plan is scheduled.

    2.1 Search for an available time block in the already scheduled set where the first metacommand can fit without conflict. Perform the search by shifting the experimental plan a specified time that is equal to the duration of a command block in the already scheduled set.

    2.2 Check to see if the rest of the metacommands can be fitted into the already scheduled set. The metacommand fits if any of the following (a, b, or c) is satisfied:

    (a) If (i) and (ii) are valid, then the start time of the metacommand is set to ST_i.

        (i) $ET_i - ST_i \geq D_m$, where ST_i and ET_i denote the start

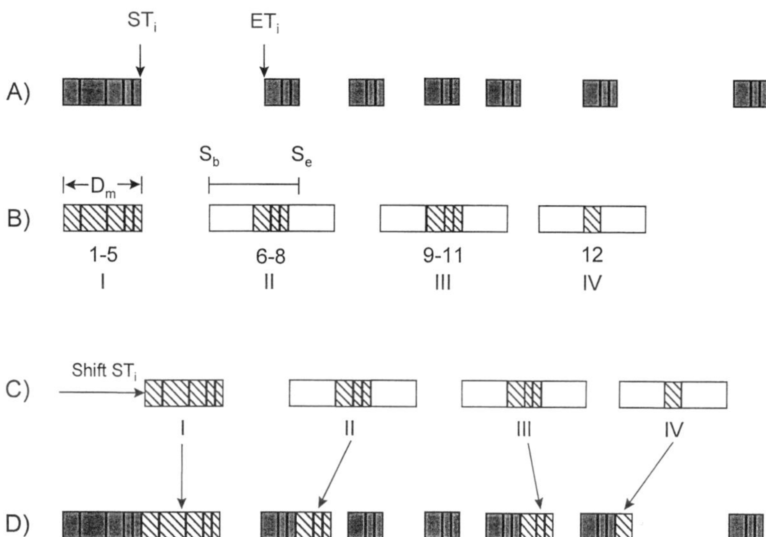

Figure 12 Operation of the flexible scheduler. (A) This time line displays the commands of all experimental plans that have been scheduled (termed the already scheduled set). ST_i and ET_i denote the start time and end time, respectively, of a time block (a period of robotic inactivity) in the already scheduled set. (B) This time line displays an experiment that has not yet been scheduled. Each of the 12 commands is displayed as a striped box. The commands can be organized as four metacommands, I, II, III, and IV (commands 1–5 are contiguous and constitute metacommand I, etc.). D_m is the duration of the metacommand. The user-defined allowable range of start times (denoted by S_b and S_e) are shown for metacommand II. An identical range pertains to the other metacommands (except the first one in the time line). The open blocks surrounding the commands (striped boxes) illustrate the possible location of the metacommand given the allowable start times. (C) Scheduling is done by shifting each metacommand in the experimental plan by an amount equal to ST_1, the start time of the first available time block. Note that without flexible start times, metacommands II, III, and IV each have conflicts with the already scheduled set (compare with row A). Each metacommand is then shifted flexibly within the range defined by S_b and S_e in order to find an open time block. Preference is given to shifting each metacommand as early as possible within the allowable start time. However, the start time can also be delayed in order to locate an open time block. If an open time block cannot be found for any of the commands, the entire experimental plan is shifted by the amount equal to the next available ST_i and the process is repeated. In this case, metacommands II and IV are advanced and metacommand III is delayed. (D) Once scheduled, the commands are implemented at the times shown in this completed schedule.

time and end time of the ith time block in the already scheduled set, respectively, and D_m is the duration of the metacommand being scheduled.

    (ii) $S_b \leq ST_i$ and $S_e \geq ST_i$, where S_b and S_e bracket the range of possible start times (beginning, end) of a metacommand.

(b)   If (iii), (iv), and (v) are valid, then the start time of the metacommand is set to S_b.

    (iii) $ET_i - ST_i \geq D_m$.

    (iv) $S_b \geq ST_i$

    (v) $S_b + D_m \leq ET_i$.

(c)   If (vi) and (vii) are valid, then the start time of the metacommand is set to S_b.

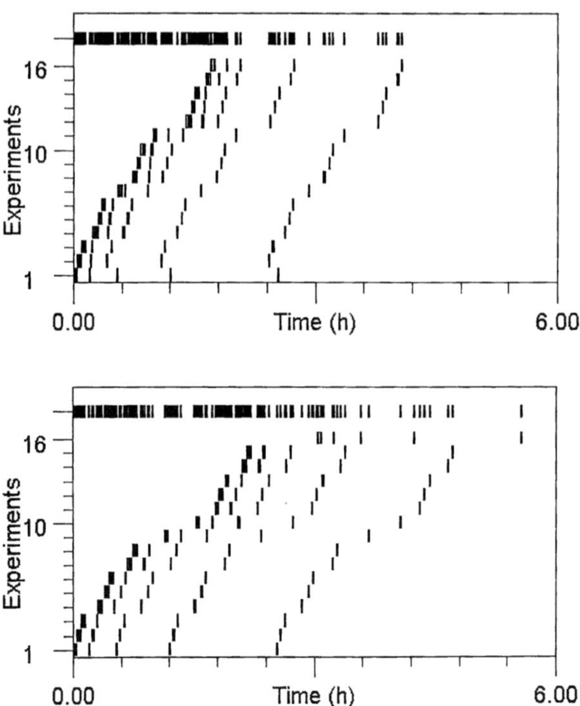

Figure 13 For 16 experiments, the flexible scheduler (top) gives a 27% increase in throughput compared with the rigid scheduler (bottom). The flexibility of timing for all metacommands (except the first one in each experimental plan) is $(-10, 10)$ in units of min.

> (vi) ST_i is the makespan of already scheduled commands.
>
> (vii) $S_b \geq ST_i$

Implementation of the flexible scheduler requires the user (1) to check the flexible scheduler flag and (2) to define the flexible ranges for the metacommands (the metacommands are grouped automatically based on the start time and duration of individual commands in the template). The flexible ranges are defined absolutely in minutes. Figure 13 shows the results of scheduling 16 experiments with the flexible scheduler and the rigid scheduler. The flexible scheduler gives a 27% increase in throughput compared with the rigid scheduler on this test set. Currently, the flexible scheduler can only be used in template-related experimentation, e.g., with FD and MDS experiments (and also for the first cycle of CMS experiments). As more sophisticated schedulers are developed, these also can be incorporated into the existing software.

7   OUTLOOK

An experiment planner composed of five modules has been developed that facilitates composition of plans for many types of chemistry experiments. The automated chemistry workstation hardware is two to three times faster than our previous system. The flexible scheduler affords an additional improvement in throughput. Provisions for automated decision making based on experimental data are readily incorporated into various experimental plans. In addition, provisions for intelligent reagent handling and resource management have been implemented that enable the workstation to function as an autonomous experimentation machine. Still, there are many limitations in our third-generation workstation that must be addressed in the next generation system. These include

1.   At present we cannot perform conditional branching (except for early termination of individual experiments) during parallel experimentation. To accomplish this we must join a conflict resolution approach with the scheduler and experimental planners.

2.   The hardware components are restricted to rather clean chemistry. More sophisticated components, particularly a more versatile reaction station, are required to broaden the scope of available chemistry. Modular hardware units are attractive for compatibility with different types of chemistries.

3.   The software is amenable to interaction with a workstation equipped with a variety of analytical instruments. However, at present the analytical capabilities are restricted to UV-Vis absorption spectroscopy. Additional instruments need to be added in order to examine a wider spectrum of chemical reactions.

Some form of mass spectrometry is needed as a more universal detector than absorption spectroscopy.

4. At present, no feedback is available concerning the integrity of workstation operation. A series of self-validation routines, implemented periodically and in response to aberrant behavior, are needed to ensure the proper functioning of the workstation. At a minimum, the results from such self-validation routines should provide a record of workstation integrity and can be used to shut down the system when malfunctions occur.

5. The workstation is capable of high throughput, with fast robot speed and software capable of scheduling >100 reactions (~15 commands each). The major bottleneck to prolonged operation at present is the limited number of reaction vessels (60) and sample vials (264). Inevitably, it is the expenditure of these containers that limits the duration of workstation operation. An end effector for replacing containers is needed to realize the aspiration of continuous long-term autonomous experimentation rather than batch operation as is practiced now. Utilization is generally higher during continuous operation than during batch operation [14].

6. Scheduling and resource management are now done separately. In GP experimentation, for example, batches of experiments are ordered based on heuristics dealing with makespan or robotic utilization, assessed for resources (chemicals and containers), and scheduled based on availability of the robotic arm. As new analytical instruments are added, the workstation will more closely resemble a workcell. The availability of analytical instruments can assume as much importance as that of the robotic arm. New scheduling and resource management procedures will need to take into account the workcell environment.

7. Reagent handling and solvent handling tasks are now performed in individual experimental plans as the schedule is processed during implementation. Syringe washing steps are performed following each manipulation. Intelligent reaction setup procedures would preview all experiments in queue and make additions of common reagents, invoking washing procedures only when appropriate to avoid contamination.

8. The five experiment planning modules now offer the opportunity for wide-ranging experimentation. However, at present it is not possible to perform any of these simultaneously, nor is it possible automatically to perform a breadth-first survey followed by an in-depth search. The latter would require data sharing and evaluation features that span the various experiment planning modules.

9. The workstation at present is knowledgeless. No chemistry data are available that would prevent, for example, setting a reaction temperature above that of the boiling point of a solvent. Rudimentary chemistry data need to be incorporated into an expanded reagent/solvent database that can be accessed for safety purposes as well as for intelligent experiment planning.

10. The results from experiments are now stored electronically. However, unless specified in an experiment planning module, data from prior experiments are not generally available for automatic evaluation of ongoing experiments, planning of new experiments, or unnecessary repetition of an experiment. Electronic data sharing in the form of an intelligent workstation notebook is needed to achieve this objective.

A major objective of this work has been to achieve the capability for parallel and adaptive experimentation. The advantages of parallel experimentation are apparent, and workstations of various types have been constructed that enable various degrees of parallelism. Adaptive automation systems constitute a central theme in robotics research. However, few adaptive systems for scientific experimentation have been constructed. The ability to perform decision making during parallel experimentation should afford both flexibility and high throughput, making the workstation a more useful tool in scientific experimentation.

8 SUMMARY

We have designed and constructed a third-generation chemistry workstation and accompanying software for automated experimentation in relatively clean domains of chemistry. The workstation is designed for flexible, high-throughput, microscale chemistry and includes a reaction station with 60 reaction vessels. The experiment manager software includes a large number of features to facilitate experiment planning and implementation. The significant advances in software from the second-generation system are as follows. (1) Five experiment planning modules have been incorporated to match different types of scientific experimentation, including open-loop and closed-loop experiments. The five modules enable plans to be composed for general purpose (GP), decision-tree (DT), factorial design (FD), composite-modified Simplex (CMS), and multidirectional search (MDS) experiments. (2) Parallel experimentation and adaptive experimentation have been blended so that some decision making can be done using objective functions or early termination functions while multiple experiments are in progress. Some planning modules (CMS, MDS, and DT) can cause new experiments to be spawned based on the chemistry data obtained. (3) Decision making can involve the local course of individual experiments and/or the global course of sets of experiments. (4) Resource management features provide the capability to look ahead at experiments in queue, assess resource (chemicals, containers) availability, and designate suitability for implementation. (5) Intelligent reagent handling features support the autodilution of reagents, autoselection of appropriate stock solutions, autoassignment of containers, and control over reaction volumes. (6) Opportunistic rescheduling features have been provided so that pending experiments can be advanced in queue whenever possible, leading to compressed experimental time frames. (7) Templates composed for FD, CMS, and MDS experiments can be stored and reused with modifications as required for specific experiments. (8) The scheduling module contains both a rigid scheduler and a flexible scheduler. (9) The software is modular and runs under Windows 95. This workstation is especially designed for fundamental investigations of chemical reactions and optimization of experimental reaction conditions.

ACKNOWLEDGMENTS

We are deeply grateful to Takeda Chemical Industries, Ltd. (Osaka, Japan), for funding the development of this third-generation automated chemistry workstation. The hardware was developed jointly with Scitec, Inc. (Lausanne, Switzerland).

REFERENCES

i.   M. Harre, U. Tilstam, H. Weinmann. Breaking the new bottleneck: automated synthesis in chemical process research and development. Org. Proc. Res. Dev. 3:304–318, 1999.

ii. See Ref. 15.

iii. D. G. Cork, T. Sugawara, J. S. Lindsey, L. A. Corkan, H. Du. Further development of a versatile microscale automated workstation for parallel adaptive experimentation. Lab. Robotics Autom. 11:217–223, 1999.

iv. H. Du, L. A. Corkan, K. Yang, P. Y. Kuo, J. S. Lindsey. An automated microscale chemistry workstation capable of parallel, adaptive experimentation. Chemom. Intell. Lab. Syst. 48:181–203, 1999.

1.   T. Sugawara, D. G. Cork. Past and present development of automated synthesis apparatus for pharmaceutical chemistry at Takeda Chemical Industries. Lab. Robotics Autom. 8:221–230, 1996.

2.   R. A. Rivero, M. N. Greco, B. E. Maryanoff. Equipment for the high-throughput organic synthesis of chemical libraries. In: Combinatorial Chemistry: A Short Course (A. Czarnik, S. H. DeWitt, eds.). American Chemical Society, Washington, DC, 1997, pp. 281–307.

3.   S. H. DeWitt, A. W. Czarnik. Automated synthesis and combinatorial chemistry. Curr. Opin. Biotechnol. 6:640–645, 1995.

4.   J. H. Hardin, F. R. Smietana. Automating combinatorial chemistry: a primer on benchtop robotic systems. Mol. Diversity 1:270–274, 1995.

5.   J. S. Lindsey. Automated approaches toward reaction optimization. In: Combinatorial Chemistry: A Short Course (A. Czarnik, S. H. DeWitt, eds.). American Chemical Society, Washington, DC, 1997, pp. 309–326.

6. J. S. Lindsey. A retrospective on the automation of laboratory synthetic chemistry. Chemom. Intell. Lab. Syst. 17:15–45, 1992.

7. J. S. Lindsey, L. A. Corkan, D. Erb, G. J. Powers. Robotic workstation for microscale synthetic chemistry: on-line absorption spectroscopy, quantitative automated thin layer chromatography, and multiple reactions in parallel. Rev. Sci. Instr. 59:940–950, 1988.

8. A. Corkan, J. S. Lindsey. Design concepts for synthetic chemistry workstations. In: J. R. Strimaitis, J. P. Helfrich (eds.). Advances in Laboratory Automation Robotics, Vol. 6. Zymark, Hopkinton, MA, 1990, pp. 477–497.

9. L. A. Corkan, E. Haynes, S. Kline, J. S. Lindsey. Robotic thin layer chromatography instrument for synthetic chemistry. In: Ali M. Emran (ed.). New Trends in Radio-pharmaceutical Synthesis, Quality Assurance and Regulatory Control. Plenum Press, New York, 1991, pp. 355–370.

10. L. A. Corkan, J. S. Lindsey. Experiment manager software for an automated chemistry workstation, including a scheduler for parallel experimentation. Chemom. Intell. Lab. Syst. 17:47–74, 1992.

11. J.-C. Plouvier, L. A. Corkan, J. S. Lindsey. Experiment planner for strategic experimentation with an automated chemistry workstation. Chemom. Intell. Lab. Syst. 17:75–94, 1992.

12. L. A. Corkan, J.-C. Plouvier, J. S. Lindsey. Application of an automated chemistry workstation to problems in synthetic chemistry. Chemom. Intell. Lab. Syst. 17:95–105, 1992.

13. J. S. Lindsey, L. A. Corkan. Toward high-performance parallel experimentation machines. Use of a scheduler as a quantitative computer-aided design tool for evaluating workstation performance. Chemom. Intell. Lab. Syst. 21:139–150, 1993.

14. R. Aarts, J. S. Lindsey, L. A. Corkan, S. Smith. Flexible protocols improve parallel experimentation throughput. Clin. Chem. 45:1004–1010, 1995.

15. R. W. Wagner, F. Li, H. Du, J. S. Lindsey. Investigation of reaction conditions using an automated microscale multi-reactor workstation: the synthesis of *meso*-tetramesitylporphyrin. Org. Proc. Res. Dev. 3:28–37, 1999.

16. H. Du, W. Shen, P. Y. Kuo, J. S. Lindsey. Decision-tree programs for an adaptive automated chemistry workstation: application to catalyst screening experiments. Chemom. Intell. Lab. Syst. 48:205–217, 1999.

17. P. Y. Kuo, H. Du, L. A. Corkan, K. Yang, J. S. Lindsey. A planning module for performing grid search, factorial design, and related combinatorial studies on an automated chemistry workstation. Chemom. Intell. Lab. Syst. 48:219–234, 1999.

18. H. Du, S. Jindal, J. S. Lindsey. Implementation of the multidirectional search algorithm on an automated chemistry workstation. A parallel yet adaptive approach for reaction optimization. Chemom. Intell. Lab. Syst. 48:235–256, 1999.

19. R. W. Wagner, H. Du, D. G. Cork, K. Yang, L. A. Corkan, F. Li, J. S. Lindsey. Automated chemistry workstation operations guide and help manual. North Carolina State University, Department of Chemistry Technical Report 98-1.

20. M. Baudin. Manufacturing Systems Analysis with Application to Production Scheduling. Prentice-Hall, New Jersey, 1990.

21. T. E. Morton, D. W. Pentico. Heuristic Scheduling Systems. John Wiley, New York, 1993.

22. D. A. Pierre. Optimization Theory with Applications. Dover, New York, 1986, pp. 193–263.
23. D. Betteridge, A. P. Wade, A. G. Howard. Reflections on the modified simplex-II. Talanta 32:723–734, 1985.
24. D. Betteridge, A. P. Wade, A. G. Howard. Reflections on the modified simplex-I, Talanta 32:709–722, 1985.
25. V. Torczon. Multi-directional search: a direct search algorithm for parallel machines. Ph.D. thesis, Rice University, 1989. Available as TR90-7, Department of Mathematical Sciences, Rice University, Houston, TX, 77251-1892.

4
Automated Purification Systems

Michele R. Stabile-Harris and April Ciampoli
ArQule, Woburn, Massachusetts

1 INTRODUCTION

A number of improvements in purification techniques have enabled chemists to perform simultaneous production of thousands of compounds for lead discovery. While liquid–liquid extraction has been a staple for all chemists, use of this standard technique has only allowed for the purification of a few compounds at a time. Now, with the aid of fluid handlers, hundreds of samples can be freed from impurities in a matter of minutes. The use of solid-phase resin scavengers also aids in the removal of excess reagent that may have been present to force a reaction to completion. A third purification technique involves the application of solid-supported reagents. This method delivers ease of purification at the completion of a reaction step with a simple filtration. The following pages illustrate several practical, effective purification protocols that have been used extensively in our laboratories. There are a number of useful publications and web pages that present other procedures for resin strategies that will not be covered in this chapter [1].

2 LIQUID–LIQUID EXTRACTION

ArQule's liquid–liquid extraction (LLE) workstation is designed to perform the process of separating undesired products, inorganic salts, and unreacted starting materials from the desired product by their distribution between two immiscible liquid phases. Isolation of the desired product is performed by one of the following procedures:

1. The liquid phase containing the desired product is transferred from the original reaction vial and dispensed to a clean vial.
2. The liquid phase containing the undesired components is transferred from the original reaction vial to waste.

The liquid–liquid extraction operation usually encompasses processes such as liquid handling, solution agitation, and solution separation. These operations are designed specifically to operate independently from one another. This document will only describe the development of the liquid handling operation.

2.1 Low-Throughput Automated Approach

Initially, the liquid handling portion of LLE was carried out either manually and/ or on a Gilson 215 Liquid Handler, which is a single-probe instrument. Performing the extractions manually became quite tedious and error prone when one had to process numerous samples at a time. Utilizing the Gilson to perform the liquid handling process was the first approach taken to automate LLE. This was feasible at the time because the required Gilson was inexpensive and relatively easy to program. All the required hardware including needles, syringes, and decks are commercially available from Gilson. Limitations of our first-generation Gilson system became evident after the production of a few libraries. The limitations included these:

1. The deck capacity only accommodated ten of our custom 24-well reaction blocks, which resulted in only 240 compounds per deck run.
2. Only one transfer could be carried out at a time because of the single-probe condition, which resulted in a bottleneck when purifying thousands of compounds.
3. The user would have to modify the program to adjust the Z-height at which to send the probe to allow for transferring of either the top or the bottom layer.
4. Manual changing of the system solvent each time requires a different solvent. For example, the user had to remove the lines from the solvent container to switch between the extraction and wash solvents.

Our initial use of the Gilson gave us the experience and insight to design a much higher throughput system that encompassed the hardware and software requirements for LLE.

2.2 High-Throughput Automated Approach

The liquid handling portion of LLE was then implemented on a TECAN Genesis 200/4 unit. This fluid handler consists of a 2.0 meter deck and four independent pipetting tips. The unit was purchased without the standard diluters that are available with TECAN instruments. Instead, the pipetting tips are plumbed to two customized Cavro XL 3000 Modular Digital Syringe Pump assemblies, which are also available from Tecan. Each assembly consists of two 10 mL syringes, two six-port valves, and an I/O board for serial communication.

One of our system requirements was the ability to dispense multiple solvents. Thus the design incorporated the use of six port valves. The ports are plumbed to the six most commonly used solvents, including ethyl acetate, methanol, DMSO, chloroform, water, and methylene chloride.

An aluminum deck was designed in-house to accept 48 of our custom reaction blocks. Teflon-coated stainless steel needles were also designed in-house for the Tecan unit. The needles are designed with a blunt end to allow for the pipetting of any undissolved particles that may not settle in the interface between the two immiscible liquid phases.

Custom LLE software was also developed in-house to drive the Tecan robot and the Cavro pump assemblies. The LLE software was designed to perform the following liquid handling functions:

1.   Dispense liquid to a vial
2.   Aspirate a layer of liquid to waste
3.   Transfer a layer of liquid to a collection vial

These three functions are distinct modules within the application. These modules are very practical for the liquid handling function of LLE. The flexibility is also needed either to remove the waste layer or to transfer the product layer. Some processes could require multiple wash steps. For example, a basic wash and a water wash could be needed. In this example, the waste layer that contains the basic layer is aspirated to waste, and then the water is dispensed to the original vial to extract any remaining impurities that are still contained in the organic layer. The organic layer is then transferred to a clean vial for isolation of final product.

These modules (dispensing and transferring) are separated from one another because of other processes that could occur during liquid–liquid extraction and increase the throughput of a production facility. For example, once the two immiscible liquid phases are dispensed to the vials, the vials must be processed through solution agitation and solution separation. The Tecan could be utilized for a different library during this time. Depending on the size of the library, the compounds could be separated into batches where one batch could be running on the Tecan while another one was running through solution agitation, separation, and evaporation.

The LLE application allows users to create and save a specific method by choosing one of the three predefined processes and supplying the required parameters for that process.

The "Dispense to a vial" module requires the specification of the following parameters: number of reaction blocks, solvent selector valve number [1–6], and the volume to dispense. This module also can be used for an application other than LLE, such as reconstitution of samples after evaporation.

The "Aspirate to waste" module and the "Transfer to a collection vial" module require specification by the user of the following parameters: number of reaction blocks, total volume within the vial, removal volume, removal layer (top or bottom), wash solvent selector valve number [1–6], and the wash solvent volume. If the bottom layer is specified as the removal layer, the Tecan will move the tips to the bottom of the vessel and aspirate the specified volume. If the top layer is specified as the removal volume, the LLE software calculates where to aspirate the specified removal volume based on the current volume, the removal volume, and the geometry of the vessel. The wash solvent parameters correspond to the solvent that is used to wash the pipetting tips of the Tecan between each transfer.

The "Dispense to a vial" and "Aspirate to waste" modules can each process 48 (24-well) reaction blocks. The "Transfer to a collection vial" module can only process 24 reaction blocks, because the remaining 24 deck positions are used for the collection blocks. Currently, the system throughput is 1,200 compounds at one time with two extraction processes per compound. The next section describes a library preparation with an example of an extractive workup using the Tecan LLE system.

2.3   Removal of Succinimide from an Isoxazoline Library

Automated liquid–liquid extraction was used to purify a library of isoxazolines [1h]. The by-product from this reaction was succinimide, generated from the in-situ preparation of nitrile oxides. The synthesis, shown in Scheme 1, and the workup were designed as follows.

Oxime solutions (65 µL, 1 M in DMF) were dispensed into 2 dram vials. N-Chlorosuccinimide (NCS) (65 mL, 1 M in DMF) was added to the reactions,

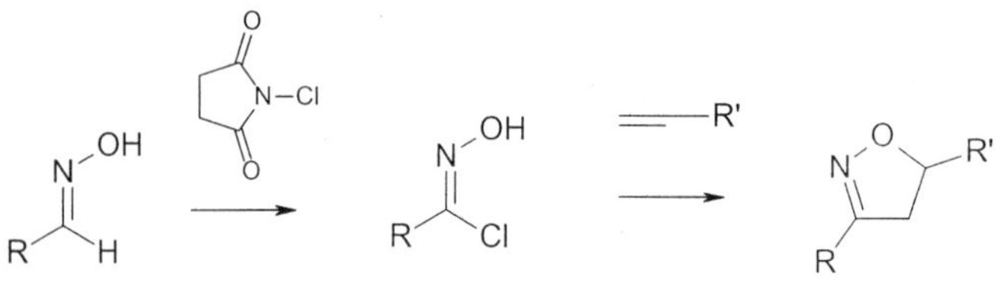

Scheme 1   Synthesis of isoxazolines

and the vials were capped and agitated for 3 h at room temperature. After removing the caps, triethylamine (65 µL, 1 M in DMF) was added with agitation, and a solution of the alkene (65 µL, 1 M in DMF) was dispensed. The vials were capped again and allowed to shake overnight at room temperature. Workup: DMF was evaporated using a Savant. Chloroform (2 mL) and water (1 mL) were added using the Tecan system described in the previous section. The vials were vortexed to ensure complete mixing of the water and chloroform and then centrifuged to remove any emulsions. After removing the top water layer, additional water (1 mL) was added, and the extraction procedure was repeated. However, instead of removing the water layer, the chloroform was removed, transferred to a clean vial, and evaporated to give the final product. Overall purity of the 1760 compound library was 93% after workup.

3 SOLID PHASE EXTRACTION USING RESIN SCAVENGERS

With the advent and success of high-throughput synthesis, techniques for the purification of combinatorial products quickly evolved with the use of solid-supported reagents. Kaldor and coworkers published the first report describing the use of solid-supported reagents for the purification of compound libraries [2]. In general, this purification method consists of the addition of a resin at the completion of a reaction to bind unreacted or excess reagent and filtration to remove the resin and bound by-products. The concept of selective binding is based on standard chemical reactivity and involves either covalent or ionic interactions. For instance, a pyridine-type resin will remove acid components in an ionic fashion (Fig. 1), while an isocyanate bound reagent will covalently bind amines (Fig. 2). Multiple resins can be used throughout the reaction sequence, and the ease of purification is highlighted by the fact that the purification steps involve (1) dispensing the resin to the reaction, (2) agitation and/or heating of the mixture, and (3) filtration of the resin and evaporation of the reaction solvent to afford the pure product. A compilation of resin scavengers and commercial sources is listed in Table 1.

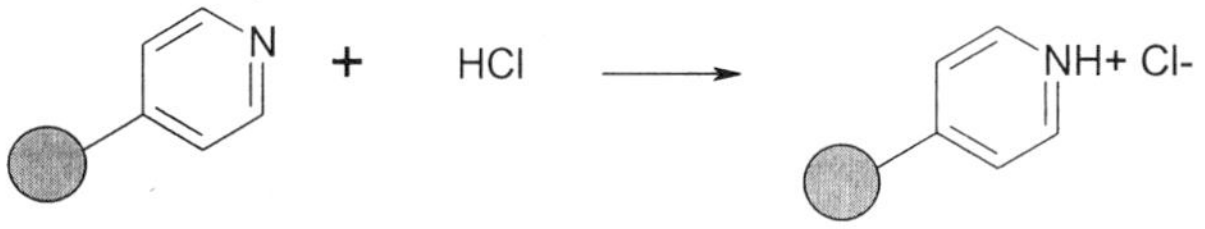

Figure 1 Reagent scavenging using ionic interactions.

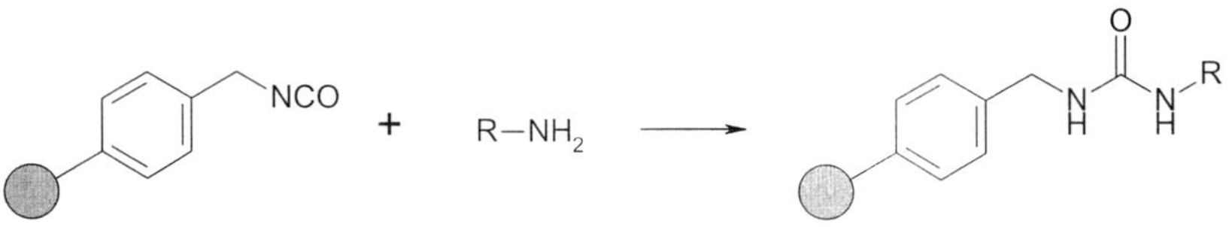

Figure 2   Reagent scavenging using covalent interactions.

3.1   Filtration of Resins

In order to have practical use with a large array of compounds, the solid-supported reagents described in the next paragraphs were used in conjunction with a commercially available apparatus from Tomtec. This apparatus allows for quick filtration and separation of resin bound by-products from numerous samples at one time with a 96-well solid-phase extraction vacuum box [3].

3.2   Application of Polyvinyl Pyridine (PVP) as an Acid Scavenger

A simple example of automated purification that eliminates the need for a water workup is with amide formation (Scheme 2). The procedure is as follows. PVP resin (2 equivalents) was dispensed into glass vials in 80-well format. Amine solutions in THF (500 µL, 0.1 M) were dispensed using a Tecan fluid handler followed by THF solutions of acid chlorides (100 µL, 0.5 M). The vials were capped and heated with shaking at 60°C. To recover the product, the THF was evaporated and replaced with DMSO (800 µL). After shaking for 1 h at 60°C in capped vials, the Tomtec was used to transfer the slurry and a subsequent 800 µL DMSO wash into commercially available filter plates. The solutions were filtered into spatially addressed 96-well plates with a 2 mL volume capacity for storage.

3.3   Use of Amberjet™ Strongly Basic Resin to Scavenge Residual Boronic Acids

A second use of ionic interactions for the removal of impurities is highlighted in this section. The synthesis of amino sugars has been pioneered by Petasis [4]. In automated array format, using a slight excess of boronic acid forces the reaction to completion with most substrates. However, because boronic acids are known to be protease inhibitors [5], the residual amount of the acid must be completely removed at the end of the reaction. While application of liquid–liquid extraction could separate the starting material from the product, the final compounds have a water-soluble sugar moiety attached resulting in migration of the product to the water layer and an undesirable low mass recovery. Use of a resin scavenger in this case is the ideal solution to remove any by-products. The procedure and workup conditions for the formation of the amino sugars (Scheme 3) is outlined in the next paragraph.

Sugar solutions (0.25 M in water, 200 µL) were dispensed into glass vials using a Tecan fluid handler. The water was evaporated in a Savant at medium heat over 3 h. Amine solutions (0.25 M in methanol, 200 µL) were then added. Boronic acid solutions (0.25 M in methanol, 210 µL) were dispensed into the reaction mixture. Vials were capped and heated with agitation for 3 h at 70°C. Reaction vials were uncapped and 50 to 100 mg of Amberjet™ resin was added; then the vials were agitated for 1 to 2 h. After uncapping the vials, 600 µL of solution was transferred from the vials to a filter plate and filtered into a 2 mL microtiter plate. The resin was rinsed two times with 400 µL of methanol and agitated for 30 s to ensure that the desired product was completely recovered from the resin. Microtiter plates were placed in a laboratory hood overnight for initial evaporation of the solvent. When the plate had less than 500 µL of remaining methanol, the plates could be placed in a Savant and the remaining solvent evaporated without worries of solvent "bumping." Alternatively, prepacked plates can be purchased from Polyfiltronics with the resin of choice, and the solution can be transferred directly to the filter plates, covered, agitated, and rinsed. However, the chemist must take care when using methanol, since it tends to leak during the agitation process. The results of the library speak for the purification technique in that the 2270 compounds were synthesized with 88% average purity.

3.4 Isocyanate Resin for the Removal of Amines

The isocyanate resin is just one of the scavengers that removes nucleophiles in a covalent manner from reaction mixtures. Isocyanate is the reagent of choice for the removal of amines because the reaction of the nucleophile with resin occurs without liberation of by-products. Excess amine was used in the following example (Scheme 4) to force the reaction to completion with a variety of nucleophiles. The optimized procedure for this method is as follows. Step 1: A solution of the α-amino-γ-butyrolactone hydrobromide (400 µL, 0.125 M in 2:2:1 isopropanol:acetonitrile:triethylamine) was dispensed into glass vials in 80-well format. A solution of acid chloride (200 µL, 0.25 M in acetonitrile) was added, and the vials were sealed and gently vortexed. The solvents were removed under reduced pressure in a Savant. Step 2: Amine solutions (400 µL, 0.25 M in dioxane) were dispensed into the reaction mixture from Step 1. The vials were capped and heated to 90°C for 48 h. The solvents were removed under reduced pressure, and 600 µL of THF was added to each vial. Isocyanate resin (~ 65 mg) was added in an automated fashion to each well. The resin slurry was agitated at room

Table 1 Polymer Supported Reagent Scavengers

Entry	Chemical name	Structure	Use	Reference	Commercial source
1	(PS-Thiophenol) 3-(3-mercaptohenyl)-propanamidomethylpolystyrene	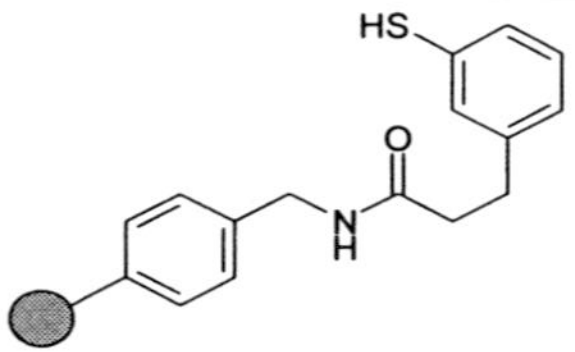 HS	Scavenges alkylating agents (alkyl halides)	1a	Argonaut
2	(PS-trisamine) tris-(2-aminoethyl)amine polystyrene	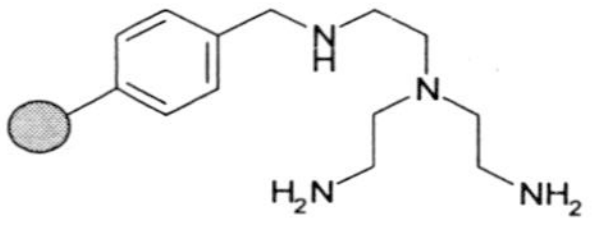 H_2N, NH_2	Scavenges acid chlorides, sulfonyl chlorides, isocyanates, electrophiles	11, 12, 53	Argonaut; Aldrich; Advanced Chem Tech
3	(PS-TsNNH2) Polystyrene sulfonyl hydrazide	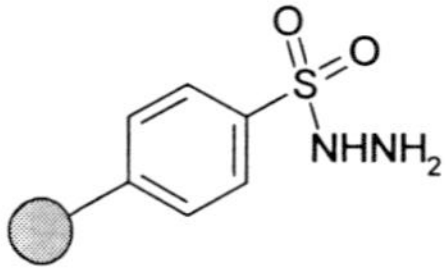 $NHNH_2$	Scavenges aldehydes and ketones	54, 55	Argonaut
4	(PS-TsCl) Polystyrene benzenesulfonyl chloride	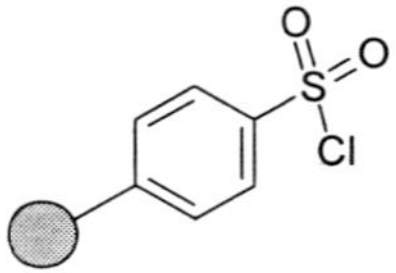 Cl	Scavenges nucleophiles, amines, hydrazines, alcohols, organometallics	1d	Argonaut
5	MP-Carbonate macroporous triethylammonium methylpolystyrene carbonate	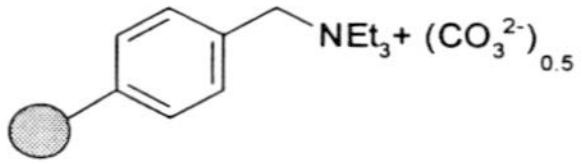 $NEt_3{}^+ (CO_3{}^{2-})_{0.5}$	Scavenges carboxylic acids, acidic phenols	1d	Argonaut; Fluka
6	(PVP) Poly(4-vinylpyridine)	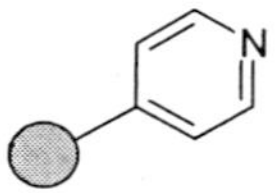 N	Scavenges acids	43	Aldrich
7	(PS-benzaldehyde) Polystyrene carboxaldehyde	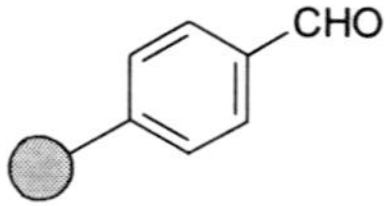 CHO	Scavenges nucleophiles, amines, hydrazines, Meldrum's acid, organometallics	56, 57	Advanced Chem Tech; Argonaut

#	Name	Structure	Application	Ref.	Supplier
8	(PS-isocyanate) Polystyrene methylisocyanate	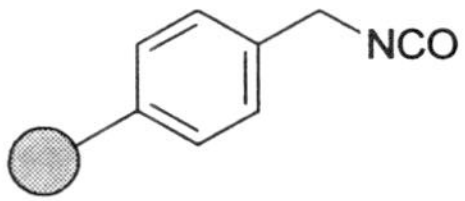	Scavenges nucleophiles, amines, hydrazines	1a, 11, 56, 58, 59	Aldrich; Argonaut
9	(PS-triphenylphosphine) Diphenylphosphino-polystyrene	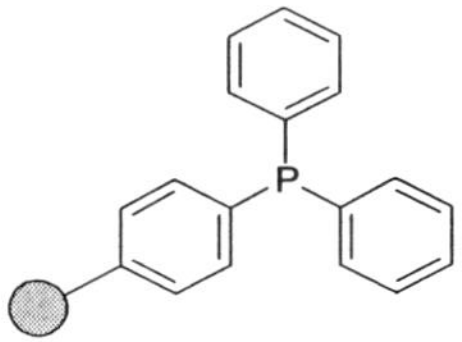	Scavenges alkyl halides	34, 39	Argonaut; Novabiochem
10	(PS-NH2) Aminomethylated polystyrene	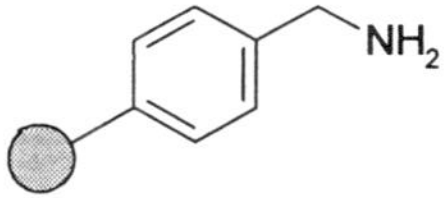	Scavenges carboxylic acids, sulfonyl halides, isocyanates	28	Aldrich; Novabiochem
11	N-(2-Mercaptoethyl)aminomethyl polystyrene	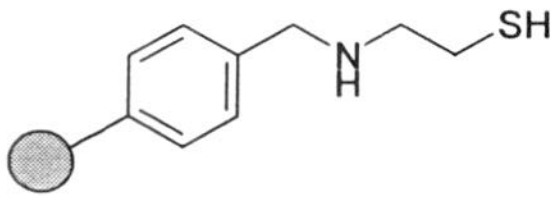	Scavenges alkyl halides	19	Novabiochem
12	Methylthiourea polystyrene	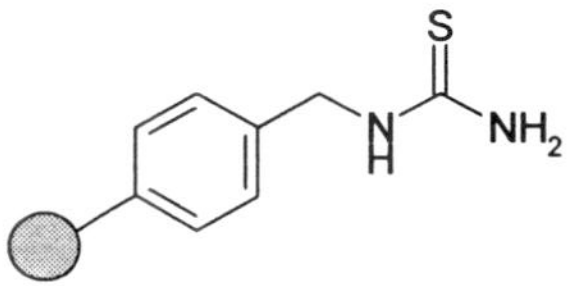	Scavenges alkyl halides and bromo ketones	60	Novabiochem
13	3-[4-(Tritylmercapto)phenyl] propionyl AM resin	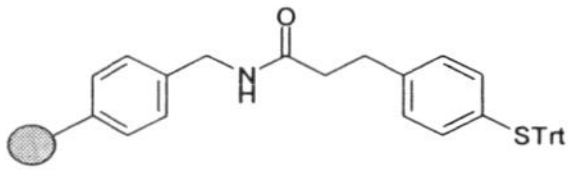	Scavenges alkyl halides, mesylates, tosylates and α, β-unsaturated carbonyl compounds	1f	Novabiochem
14	N-Methylisatoic anhydride polystyrene	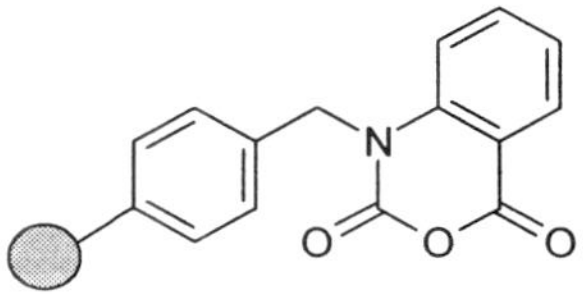	Scavenges nucleopilic amines and hydrazines	61	Novabiochem

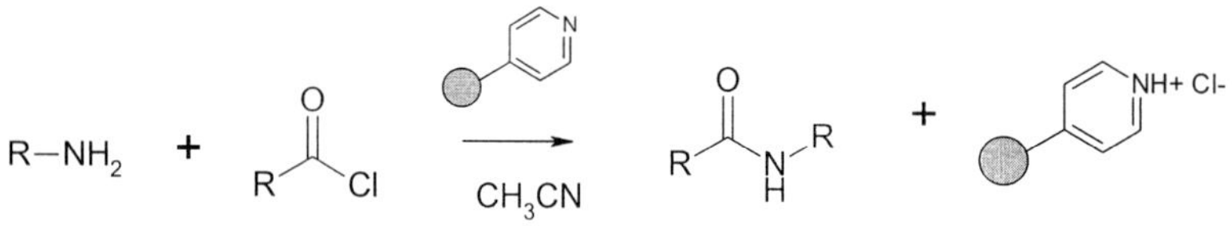

Scheme 2   Polyvinylpyridine (PVP) used as an acid scavenger.

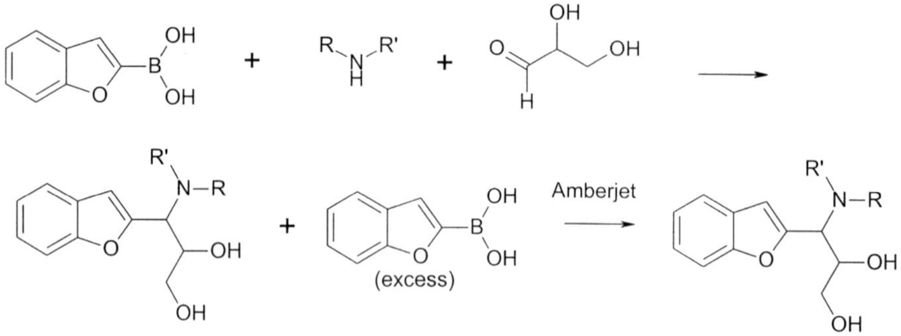

Scheme 3   Removal of excess boronic acid with Amberjet resin.

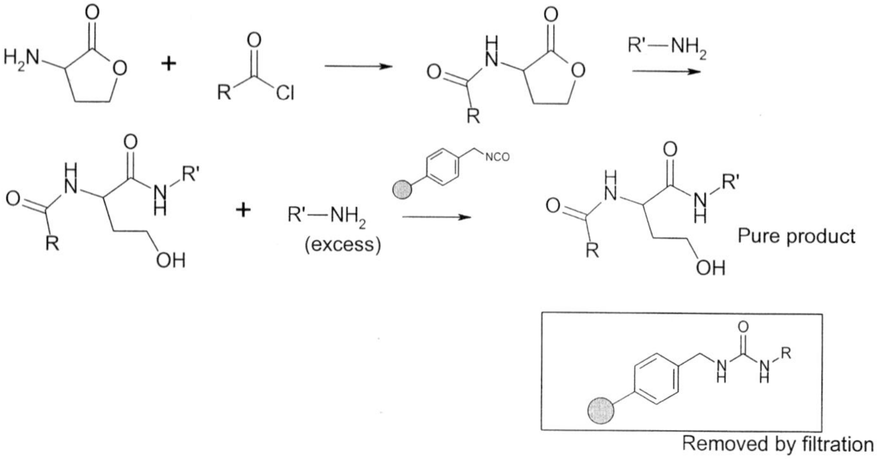

Scheme 4   Isocyanate resin removing excess amine.

temperature for 4 h. The THF was evaporated, and DMSO was added to solubilize products. Resin was removed via filtration to afford a 5,928 compound array with a final average purity of 93%. In some cases, the desired product can react with the by-product scavenging resin. With this library, the primary alcohol in the product did not react with the resin bound isocyanate.

4 SOLID-SUPPORTED REAGENTS

Chemical reagents that promote or catalyze chemical reactions commonly produce by-products that are removed at the end of a standard synthetic step via water wash. In some cases, as in amide coupling reactions, the coupling agents are difficult to remove completely with simple extractions. A water wash and the possibility of product loss in the water layer are avoided with the use of solid-supported reagents. Use of reagents on solid support allows a reaction to take place followed by a filtration to afford a clean product. A number of commercially available resin bound reagents are available in addition to some that have literature preparations. This list can be found in Table 2.

4.1 Use of Polymer-Supported Borohydride in a Reductive Amination Library

A number of methods have been developed for reductive amination reactions [6]. These methods range from evaporating a toluene/water azeotrope to form the imine in the initial stages of the reaction to adjusting the pH at the end of the reaction during an extractive workup to afford a pure product. The following procedure outlines an automated method for the preparation of secondary amines from primary amines, aldehydes, and polymer bound borohydride (Scheme 5) performed in our laboratories. The products are obtained in an overall average 90% yield and 90% purity. The procedure is as follows. Molecular sieves (3 Å, ~110 mg) are dispensed into 2 dram vials. Vials are capped with pierceable teflon caps and purged with argon. Using a septa-piercing method on the Tecan, aldehydes (110 µL, 0.5 M in anhydrous DMF) are dispensed to each vial followed by primary amines (500 µL, 0.1 M in anhydrous methanol). The reaction mixtures were agitated for 24 h at room temperature. Sets of reaction vials were uncapped and the borohydride resin (~35 mg) was dispensed to each vial. Vials were capped again and the reactions were agitated for a further 24 h at room temperature. To isolate the product, vials were uncapped and 1 mL of MeOH was dispensed into each reaction using the Tecan. Solutions were vortexed to ensure that the final compound was in solution, and 1.4 mL of reaction mixture was transferred into clean reaction vials. An additional 1 mL of MeOH was added to the original vials to rinse the resin. After vortexing, the top solution was again transferred to the above collection vials, leaving the resin and molecular sieves behind at the bottom

Table 2 Polymer Supported Reagents

Entry	Resin type	Structure	Use	Reference	Commercial source
1	(PS-DMAP)N-(methylpolystyrene)-4-(methylamino)pyridine	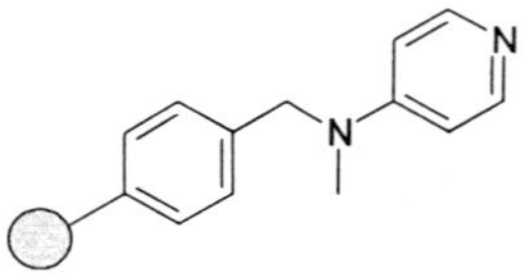	Catalyst for acylation reactions	8c	Argonaut
2	(PS-DIEA) N, N-(Diisopropyl) aminomethyl-polystyrene)	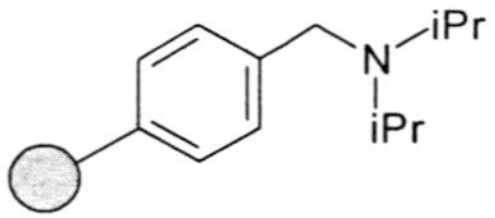	Tertiary amine base	9, 10	Argonaut
3	(PS-NMM) 3-(Morpholino)-propylpolystryene-sulfonamide	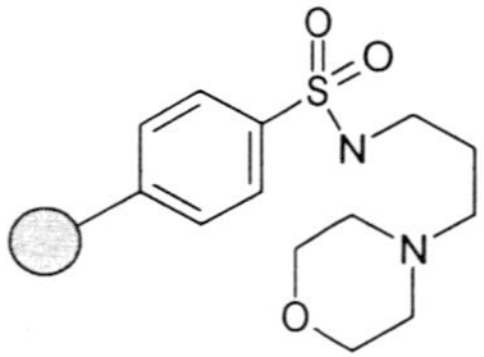	Tertiary amine base	9	Argonaut
4	Morpholinomethyl styrene	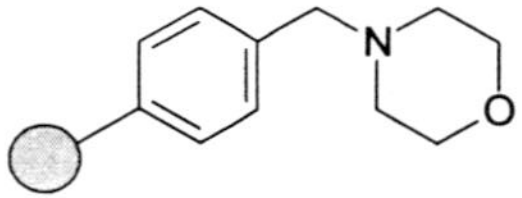	Tertiary amine base	9, 11, 12	Aldrich; Novabiochem
5	Piperidinomethyl polystyrene	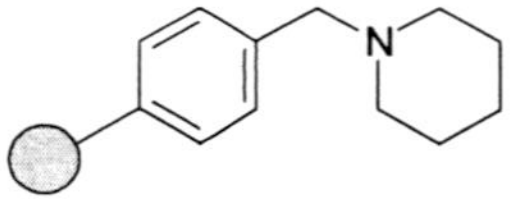	Tertiary amine base	9	Aldrich; Novabiochem

		Use		Supplier
6	(PS-TsCl) Polystyrene sulfonyl chloride	Intermediate in synthesis of tertiary amines	13, 14, 15, 16	Argonaut
7	(MP-carbonate) Macroporous triethylammonium methyl-polystyrene carbonate	General base for neutralization, scavenging carboxylic acids, acidic phenols	1d	Fluka; Argonaut
8	(PS-carbodiimide) N-cyclo-hexyl-carbodiimide-N-propyloxymethyl polystyrene	Coupling agent for amide and ester synthesis	17, 18, 19, 20	Argonaut
9	(PS-HoBt) 1-Hydroxy-benzotriazole-6-sulfonamidomethyl polystyrene	Coupling agent for acids and amines	21, 22, 23, 24, 25	Argonaut
10	PS-EDC	Coupling agent for acids and amines	19	None
11	PS-DCC	Coupling agent for acids and amines	26, 27	Novabiochem
12	(MP-TsOH) Macroporous polystyrene sulfonic acid	Removal of primary, secondary, tertiary amines	28, 29, 30, 31, 32, 33	Argonaut

Table 2 Continued

Entry	Resin type	Structure	Use	Reference	Commercial source
13	(PS-TPP) Diphenylphosphino-polystyrene	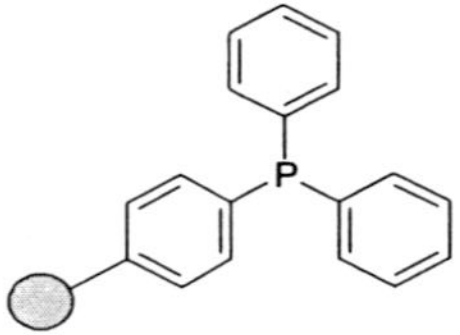	Wittig, Mitsunobu and ozonolysis reactions	34, 35, 36, 37, 38, 39, 40, 41, 42	Argonaut; Novabiochem
14	(PS-PVP) Poly(4-vinylpyri-dine)	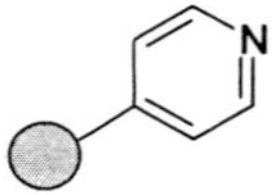	General base	43	Aldrich
15	Polystyrene diphenyl phosphine dichloride	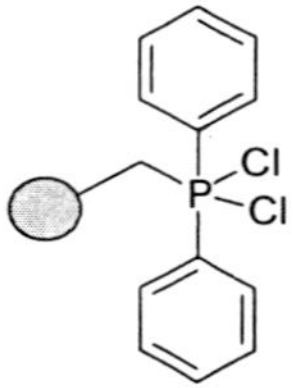	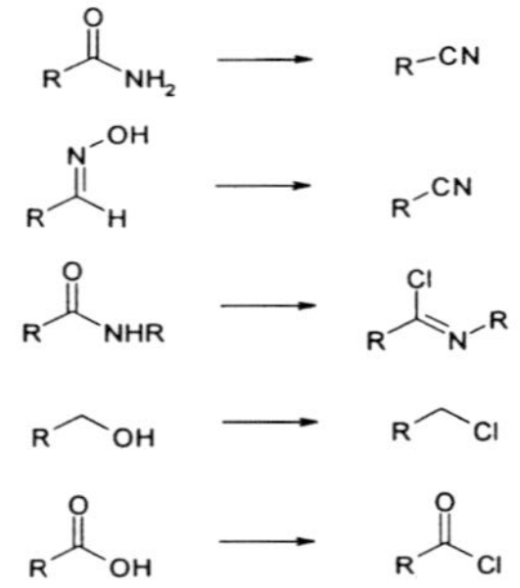 $R-CONH_2 \rightarrow R-CN$; $R-CH=N-OH \rightarrow R-CN$; $R-CONHR \rightarrow R-C(Cl)=N-R$; $R-OH \rightarrow R-Cl$; $R-COOH \rightarrow R-COCl$; $R-NH-CO-NH-R \rightarrow R-N=N-R$	1a	Can be prepared in situ
16	Polystyrene diphenyl phosphine dibromide	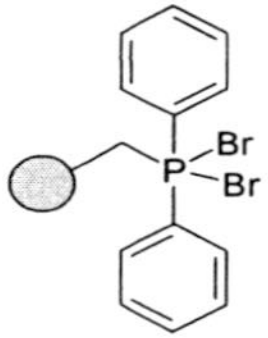		1a	Can be prepared in situ

No.	Name	Structure	Application	Ref.	Source
17	Polystyrene diphenylphosphine diiodide		$R\text{-}COOH \rightarrow R\text{-}COOMe$	1a	Can be prepared in situ
18	(PV-PCC) Poly(4-vinylpyridinium chlorochromate)	$ClCrO_3^-$	Oxidizing agent	44, 45	Fluka
19	PV-PDC	$Cr_2O_7^{2-}$	Oxidizing agent	1a	Fluka
20	Silica supported periodate	$NMe_3^+\ IO_4^-$	Oxidative cleavage of diols conversion of sulfides to sulfoxides	1a	None
21	PVP-osmium	OsO_4	Catalytic hydroxylation	1a	Fluka
22	Amberlyst A-26 Br_3^- form	Br_3^-	Brominating agent	46, 47, 48	Fluka

Table 2 Continued

Entry	Resin type	Structure	Use	Reference	Commercial source
23	PVP-dibromide	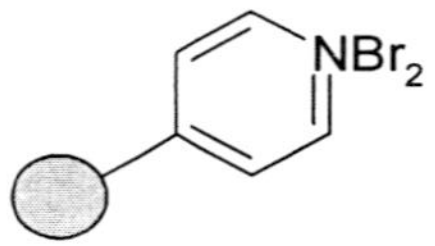	Brominating agent	1a	Fluka
24	Amberlyst A26 F- form	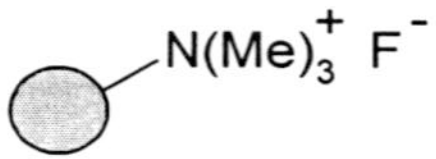	Fluorinating agent	1a	Aldrich
25		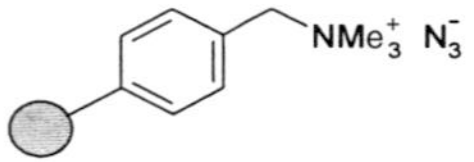	$R{-}X \rightarrow R{-}N_3$	1a	None
26	PS-phthalimide	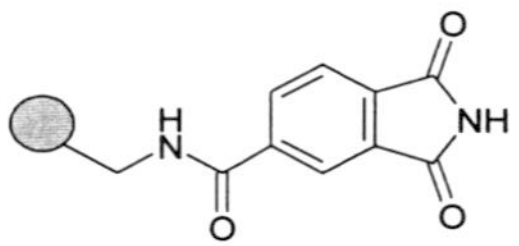	Mitsunobu reactions	1a	None
27	Polymer supported diol (2,3-dihydroxy-1-propylthiomethyl)polystyrene	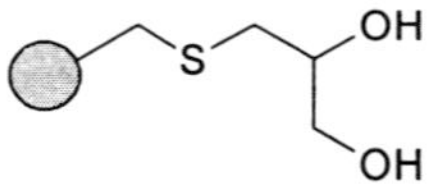	Polymer supported protecting group for aldehydes and ketones	1a	Aldrich

28	Amberlyst A-26 BH_4^- form	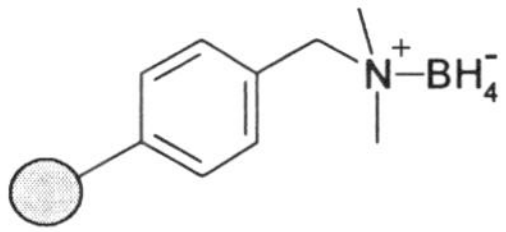	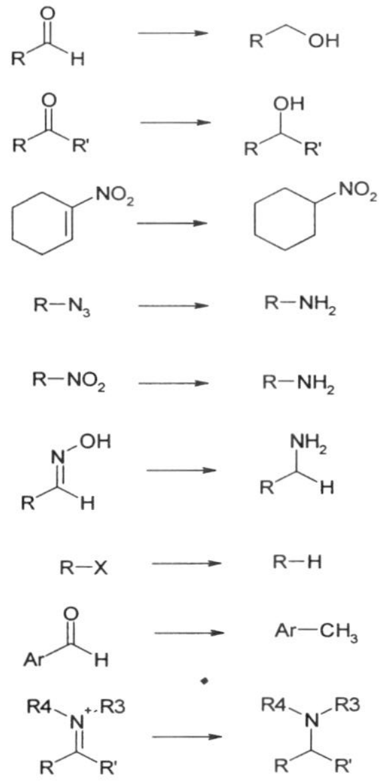	49	Aldrich Fluka

Table 2 Continued

Entry	Resin type	Structure	Use	Reference	Commercial source
29	6-(Methylsulfinyl)-hexanoylmethyl polystyrene		Sulfoxide reagent for Swern oxidation	50	Novabiochem
30	Piperidine-4-carboxylic acid polyamine resin HL		Removal of Fmoc protecting groups	51	Novabiochem
31	N,N′-Di-Boc-isothiouronium-methyl polystyrene		Preparation of guanidines	52	Novabiochem

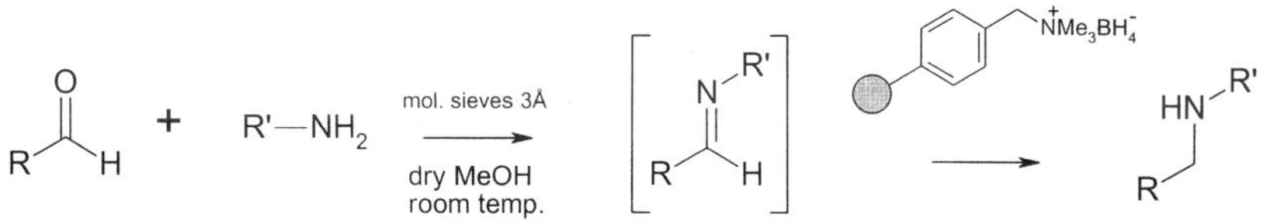

Scheme 5 Polymer supported borohydride used in reductive amination.

of the reaction vessel. The solutions were then evaporated using a Savant to afford the desired secondary amine in high yield and high purity.

4.2 Lanthanide Catalyst for Use in a Diels–Alder Array

In addition to polymer-supported reagents and polymer-supported scavengers, catalysts bound to solid phase are also of interest. Polymer-supported catalysts are convenient to use because of the ease of removal at the end of a reaction by filtration. Two groups have reported the use of polymer bound $PdCl_2$, palladium phosphine and (R)-BINAP attached to polystyrene [7]. Scandium and ytterbium as well as a number of other lanthanides have been prepared on ion exchange resins [8]. Such resins have utility in aldol reactions, Diels–Alder transformations, and epoxide openings. In this example, a three-component Diels–Alder library was prepared from aldehydes, anilines, and cyclopentadiene (see Scheme 6). Lanthanide resin (30–40 mg) was dispensed into glass vials in 80-well format using a solid dispenser. Aniline (200 µL, 0.25 M, 1/l ethanol:toluene) and aldehyde solutions (200 µL, 0.25 M, 1:1 ethanol:toluene) were dispensed into the vials using a Tecan. The solvents were then removed using a Savant and freshly distilled cyclopentadiene (200 µL, in acetonitrile 15 parts cyclopentadiene:185 parts acetonitrile) was added. The vials were capped and the reaction mixture was agitated overnight at room temperature. Removal of the resin was achieved

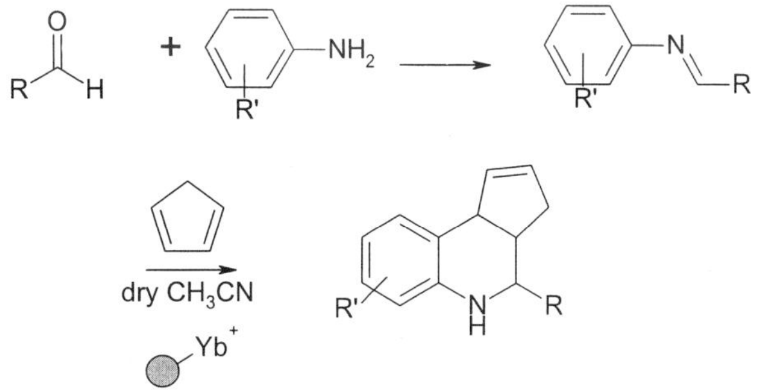

Scheme 6 Diels–Alder library generated via solid-supported lanthanide catalyst.

using the Tomtec vacuum filtration apparatus, and the solvents were evaporated to give the desired products. This library of 3,600 compounds had an average purity of 84%.

5 FUTURE PROSPECTS REGARDING AUTOMATED PURIFICATION IN COMBINATORIAL CHEMISTRY

Solution-phase combinational chemistry currently relies on liquid–liquid extraction, solid-phase resin scavengers, or solid-supported reagents to achieve the desired purity of target compounds from a reaction pathway. These techniques, in conjunction with ample amounts of careful process chemistry, have contributed to the production of thousands of compounds for biological screening. In order for a compound to perform well in a screen, the desired purity is generally greater than 90%. While the methods discussed in the previous section can sometimes achieve compounds with such high purity, this is not the norm for arrays of over 3,000 compounds.

Thus, in the future, process chemistry will be relied upon to produce the desired target as the major component of a reaction mixture and liquid–liquid extraction will be replaced by high-throughput purification using HPLC. Purifying each compound will allow a reduction in the time from idea generation to compound production to testing. A large portion of the time in the lead discovery pipeline is spent in process chemistry, trying to discover a method to produce a pure compound. Less time should be spent at this stage trying to discover a method to synthesize compounds that are 90% pure and alter the process to accept a purity level of at least 50%. Resin scavengers and liquid–liquid extraction will not be relied upon for complete purification of the final samples, since by-products or excesses of reagents will be removed using a high-throughput purification method. Thus compounds will be obtained with the desired purity, and assays can be run in a timely fashion while reducing the rate of false positives. However, until the time when high-throughput purification can compete with the speed, cost-effectiveness, and consistency of LLE and polymer-supported reagents/ scavengers, more resin bound compounds need to be developed to keep up with high-throughput combinatorial synthesis.

ACKNOWLEDGMENTS

The authors would like to thank Justin Caserta, Joseph Castellino, Mary A. Korpusik, Emily Marler, Christine Martel, Jennifer Mills, David Paisner, Jennifer Dyer Summerfield, John Walsh, Libing Yu, and Yuefen Zhou for their valuable contributions to this work.

REFERENCES

1. (a) Drewry, D. H., Coe, D. M., Poon, S. Med. Res. Rev. 1999, 97. (b) Booth, R. J., Hodges, J. C. Acc. Chem. Res. 1999, 32:18. (c) Flynn, D. L., Devraj, R. V., Parlow, J. J. Curr. Opin. Drug Disc. Dev. 1998, 1:41. (d) www.argotech.com. (e) www.sigma-aldrich.com. (f) www.nova.ch. (g) www.peptide.com. (h) Weidner-Wells, M. A., Fraga-Spano, S. A., Turchi, I. J. J. Org. Chem. 1998, 63:6319.
2. Kaldor, S. W., Siegel, M. G, Fritz, J. E., Dressman, B. A., Hahn, P. J. Tetrahedron Lett. 1996, 37:7193.
3. www.tomtec.com/Pages/spevacbx.html.
4. Petasis, N. A., Akritopoulou, I. Tetrahedron Lett. 1993, 34:583.
5. (a) Wolfenden, R. Acc. Chem. Res. 1972, 5:10. (b) O'Connor Westerik, J., Wolfenden, R. J. Biol. Chem. 1972, 247:8195. (c) Lienhard, G. E. Science, 1973, 180:149. (d) Westmark, P. R., Kelly, J. P, Smith, B. D. J. Am. Chem. Soc. 1993, 115:3416.
6. Magid-Abdel, A. F., Carson, K. G., Harris, B. D., Maryanoff, C. A., Shah, R. D. J. Org. Chem. 1996, 3849.
7. (a) Villemin, D., Goussu, D. Heterocycles, 1989, 29:1255. (b) Bayston, D. J., Fraser, J. L.; Ashton, M. R., Baxter, A. D. Polywka, M. E. C., Moses, E. J. Org. Chem. 1998, 63:3137.
8. (a) Yu, L, Chen, D., Li, J., Wang, P. G. J. Org. Chem. 1997, 62:3575. (b) Kobayashi, S., Nagayama, S. J. Am. Chem. Soc. 1996, 118:8977 (c) Shai, Y, Jacobson, K. A., Patchornik, A. J. Am. Chem. Soc. 1985, 107:4249.
9. Booth, R. J., Hodges, J. C. J. Am. Chem. Soc. 1997, 119:4882.
10. Hulme, C., Peng, J., Louridas, B., Menard, P., Krolikowski, P., Kuman, N. V. Tetrahedron Lett. 1998, 39:8047.
11. Creswell, M. W., Bolton, G. L., Hodges, J. C., Meppen, M. Tetrahedron 1998, 54:3983.
12. Blackburn, C., Guan, B., Fleming, P., Shiosaki, K, Tsai, S. Tetrahedron Lett. 1998, 39:3635.
13. Rueter, J. K., Nortey, S. O., Baxter, E. W., Leo, G. C., Reitz, A. B. Tetrahedron Lett. 1998, 39:975.
14. Baxter, E. W., Reuter, J. K., Nortey, S. O., Reitz, A. B. Tetrahedron Lett. 1998, 39:979.
15. Zhong, H. M., Greco, M. N., Maryanoff, B. E. J. Org. Chem. 1997, 62:9326.
16. Hunt, J. A., Roush, W. R. J. Am. Chem. Soc. 1996, 118:9998.
17. Parlow, J. J., Mischke, D. A., Woodard, S. S. J. Org. Chem. 1997, 62:5908.
18. Weinshenker, N. M., Shen, C. M. Tetrahedron Lett. 1972, 13:3281.
19. Desai, M. C., Stramiello, L. S. M. Tetrahedron Lett. 1993, 34:7685.
20. Adamczyk, M., Fishpaugh, J. R., Mattingly, P. G. Tetrahedron Lett. 1995, 36:8345.
21. Pop, I. E., Deprez, B. P., Tartar, A. L. J. Org. Chem. 1997, 62:2594.
22. Huang, W., Kalivretenos, A. G. Tetrahedron Lett. 1995, 36:9113.
23. Dendrinos, K., Jeong, J, Huang, W, Kalivretenos, A. G. Chem. Commun. 1998, 499.
24. Dendrinos, K. G., Kalivretenos, A. G. Tetrahedron Lett. 1998, 39:1321.
25. Dendrinos, K. G., Kalivretenos, A. G. J. Chem. Soc. Perkin Trans 1 1998, 1463.
26. Weinshenker, G. M. Org. Synth, Coll. Vol. VI 1988, 951.
27. Sturino, C. F., Labelle, M. Tetrahedron Lett. 1998, 39:5891.

28. Flynn, D. L, Crich, J. Z., Devraj, R. V., Hockerman, S. L., Parlow, J. J., South, M. S., Woodard, S. J. Am. Chem. Soc. 1997, 119:4874.

29. Gayo, L. M., Suto, M. J. Tetrahedron Lett. 1997, 38:513.

30. Siegel, M. G., Hahn, P. J., Dressman, B. A., Fritz, J. E., Grunwell, J. R., Kaldor, S. W. Tetrahedron Lett. 1997, 38:3357.

31. Shuker, A. J., Siegel, M. G., Matthews, D. P., Weigel, L. O. Tetrahedron Lett. 1997, 38:6149.

32. Parlow, J. J., Flynn, D. L. Tetrahedron, 1998, 54:4013.

33. Suto, M. J., Gayo-Fung, L. M., Palanki, M. S. S., Sullivan, R. Tetrahedron 1998, 54:4141.

34. Bernard, M., Ford, W. T. J. Org. Chem. 1983, 48:326.

35. Relles, H. M., Schluenz, R. W. J. Am. Chem. Soc. 1974, 96:6469.

36. Regen, S. L., Lee, D. P. J. Org. Chem. 1975, 40:1669.

37. Landi, J. J. Jr., Brinkman, H. R. Synthesis 1992, 1093.

38. Bolli, M. H., Ley, S. V. J. Chem. Soc., Perkin Trans. 1 1998, 2243.

39. Hughes, I. Tetrahedron Lett. 1996, 37:7595.

40. Stanetty, P., Kremslehner, M. Tetrahedron Lett. 1998, 39:811.

41. Jang, S.-B. Tetrahedron Lett. 1997, 38:1793.

42. Fenger, I., LeDrian, C. Tetrahedron Lett. 1998, 39:4287.

43. Yoshida, J., Hashimoto, J., Kawabata, N. Bull. Chem. Soc. Jpn. 1981, 54:309.

44. Frechet, J. M. J., Warnock, J., Farrall, M. J. J. Org. Chem. 1978, 43:2618.

45. Bergbreiter, D. E., Chandran, R. J. Am. Chem. Soc. 1985, 107:4792.

46. Cacchi, S., Caglioti, L. Synthesis 1979, 64.

47. Bongini, A., Cainelli, G., Contento, M., Manescalchi, F. Synthesis 1980, 143.

48. Smith, K., James, D. M., Matthews, I., Bye, M. R. J. Chem. Soc. Perkin Trans. 1 1992, 1877.

49. (a) Sande, A. R., Jagadale, M. H., Mane, R. B., Salunkhe, M. M. Tetrahedron Lett. 1984, 25:3501. (b) Dumartin, G., Pourcel, M., Delmond, D, Donnard, O., Pereyre, M. Tetrahedron Lett. 1998, 39:4663.

50. Liu, Y., Vederas, J. C. J. Org. Chem. 1996, 61:7856.

51. Carpino, L. A., Mansour, E. M. E., Cheng, C. H, Williams, J. R., MacDonald, R., Knaczyk, J., Carman, M. J. Org. Chem. 1983, 48:661.

52. Dodd, D. S., Wallace, O. B. Tetrahedron Lett. 1998, 39:5701.

53. Katoh, M., Sodeoka, M. Bioorg. & Med. Chem. Lett. 1999, 9:881.

54. Emerson, D. W., Emerson, R. R., Joshi, S. C., Sorensen, E. M., Turek, J. E. J. Org. Chem. 1979, 44:4634.

55. Kamogawa, H., Kanzawa, A., Kodoya, M., Naito, T., Nanasawa, M. Bull. Chem. Soc. Jpn. 1983, 56:762.

56. Kaldor, S. W., Siegel, M. G., Fritz, J. E., Dressman, B. A., Hahn, P. J. Tetrahedron Lett. 1996, 37:7193.

57. Frechet, J. M., Schuerch, C. J. Am. Chem. Soc. 1971, 93:492.

58. Rebek, J., Brown, D., Zimmerman, S. J. Am. Chem. Soc. 1975, 97:4407.

59. Kaldor, S. W., Fritz, J. E., Tang, J., McKinney, E. R. Bioorg. Med. Chem. Lett. 1996, 6:3041.

60. Warmus, J. S., Ryder, T. R., Hodges, J. C., Kennedy, R. M., Brady, K. D. Biorg. Med. Chem. Lett. 1998, 8:2309.

61. Coppola, G. M. Tetrahedron Lett. 1998, 39:8233.

5
Parallel Purification

Janice A. Ramieri
Biotage, Inc., Charlottesville, Virginia

1 INTRODUCTION

Although automated parallel synthesis has improved the rate of production of novel organic compounds, compound purification is emerging as the next bottleneck. By purifying samples in parallel rather than sequentially, you can enhance the productivity of both lead discovery and lead optimization efforts.

In 1997, Biotage, Inc., a Dyax Corp. Company introduced the Parallex™ HPLC system which includes four separate HPLC columns, each with its own pump head, flow stream, detection, and fraction collection. The system uses deep-well microtiter plates for both injection and collection; separates fractions based on time or on slope/threshold information from two UV traces; and tracks all sample information through the system in Microsoft Access™ using bar-coded tags on the sample plates. This system offered a solution for combinatorial chemistry.

Pharmaceutical companies are now pushing the medicinal chemistry groups to increase synthetic productivity. These groups do not produce as many compounds as combinatiorial groups but are increasing their productivity daily. We identified a demand for a new parallel purification system to meet the needs of medicinal chemistry groups. In January 2000, Biotage introduced the Parallex Flex, a scalable parallel system that offers from one to four channels of HPLC purification. This system not only offers a solution for combinatorial chemistry but also offers a solution for medicinal chemistry groups as well.

Designing a parallel purification system for use by multiple medicinal chemists who may have little training in preparative HPLC, required a number of design decisions that were not typical of traditional single-channel preparative HPLC systems. The chapter reviews the following important areas of design:

1.1 Preparative vs. Analytical Chromatography:

The fundamental difference between analytical and preparative chromatography is whether to analyze the sample or to purify the sample. The objective of analytical chromatography is to acquire data about the sample. At the end of an analytical run, the instrument sends the sample to waste and the user has important data about that sample. Since preparative chromatography is for purification purposes, collecting the sample isolated from impurities is the objective of the run. Usually larger quantities of the sample need purifying therefore; to load as large an amount as possible without losing resolution (overloading) is the usual practice. When scaling up from analytical to preparative chromatography, the user should consider the following variables: sample load concentration, sample dissolution solvent composition, sample dissolution volume, and flow rate.

1.2 Software

Biotage designed the Parallex Flex software as an open access system for use by multiple medicinal chemists. Our design goal was to make the user interface as simple as possible. The Flex requires a trained administrator, who is responsible for system maintenance and training. The administrator can restrict inexperienced users to running preset methods, while more experienced users may define new methods. Walk-up Wizard software allows users to start a run with only minimal training. All information is tracked in Microsoft Access, and simple graphical software brings chromatograms, plate or vial maps, and textual information together in an intuitive fashion, both for current samples and fractions and for historical data runs months ago. Chemists can review the information from their desks as well as from the instrument.

1.3 Hardware

There are four important design criteria for the hardware subsystems within the Parallex Flex: *Solvent delivery*: The solvent delivery subsystem insures constant volume delivery to each channel, irrespective of variations in back pressure. Each channel operates with independent flow rates and timing, so that the operator can run different methods on different channels. *Loader*: When injecting synthetic organic compounds onto a reverse phase column, one area of concem is precipitation of the sample in the sample loop at the starting HPLC solvent. We have designed the loader module to separate the sample from the solvent upon injection. We also load each sample loop during the previous samples gradient, increasing overall throughput. *Detector*: The detector offers two UV wavelengths for one to four HPLC channels in a single compact package, sharing a common deuterium source and optics. *Fraction Collection*: The collector is designed for precise fractionation, with the divert valve positioned immediately above the collector vessels. It is capable of fractionation by slope (for optimization of purity, >95% is easily achieved) or by threshold (for optimization of yield) on either or both UV wavelengths.

1.4 Integration with Characterization and Quantitation

Purification is only part of the problem. In order to develop structure/activity relationships, chemists must characterize and quantitate the samples as well. The Parallex Flex fits into the High Throughput Organic Chemistry (HTOC) [1] process specification. HTOC tracks data through purification, characterization by mass spectrometry, and quantitation, automating all data transfer. The Microsoft Access database on the Parallex Flex serves as a central repository to track this information. The Parallex Pilot, a separate Biotage product, allows chemists to reformat the fractions, returning them to the original sample plate format.

1.5 Cost per Compound

For large numbers of compounds, cost per year quickly becomes an important issue. The Parallex Flex has been designed to minimize five cost factors: the amount of solvent used, the amount of solvent collected, the amount of solvent evaporated, utilization of the mass spectrometer, and labor costs. This section will summarize the areas of design and show how each affects the cost per compound. It is possible routinely to purify, characterize, and quantify 25 mg^s of compounds at high throughput for less than \$15 per compound, using the HTOC process.

2 PREPARATIVE vs. ANALYTICAL CHROMATOGRAPHY

When scaling up from analytical to preparative chromatography, consider the following variables: column size, sorbent selection, sample load, the sample dissolution solvent composition, sample dissolution volume, and flow rate. To accurately illustrate scale-up, we will set the dimension of an analytical column as 4.6 mm in diameter $\times$ 150 mm in length; the dimension of the preparative column is 20 mm in diameter $\times$ 50 mm in length. The following sections illustrate practices for the drug discovery laboratory. Each section offers solutions for this specific environment. Other solutions or recommendations may be applicable for other environments.

2.1 Sample Loading Concentration

The amount of sample to load on any column (analytical or preparative) is proportional to the amount of silica in the column and the difficulty of the separations. The amount of silica is directly proportional to the volume of the empty columns. In this case, the volume of the prep column (V_p) and the volume of the analytical column (V_a) can be calculated as follows.

$$V_p = \pi r^2 l \qquad V_p = \pi(1)^2 5$$
$$V_p = 15.71 \text{ mL}$$
$$V_a = \pi r^2 l \qquad V_a = \pi(0.23)^2 15$$
$$V_a = 2.49 \text{ mL}$$
$$R = \frac{V_p}{V_a}$$
$$R = \frac{15.71}{2.49}$$
$$R = 6.3$$

The ratio (R) of the two column volumes is 6.3:1. Therefore, to calculate the load on the preparative column, if the load on the analytical column is 0.32 mg,

$$L_p = L_a \times 6.3$$
$$L_p = 0.32 \text{ mg} \times 6.3$$
$$L_p = 2 \text{ mg}$$

When overloading, a user can typically load up to 40 mg of crude sample onto a 20 × 50 mm column and achieve sufficient purity for most separations. To test a 20 mg overload of the preparative column on an analytical column, one would do the following:

$$L_a = \frac{L_p}{6.3}$$
$$L_a = \frac{20 \text{ mg}}{6.3}$$
$$L_a = 3.17 \text{ mg}$$

2.2 Sample Dissolution Volume

You will achieve the best chromatographic separation when you dissolve the sample in as small a volume as possible. For an analytical injection, this is not difficult. The concentrations typically loaded onto these columns are quite small.

For preparative injection, it is sometimes harder to load the necessary concentration in a directly scaled volume. For example if you inject the 0.32 mg in a 10 µL volume on the analytical column, your injection volume (VL_p) onto the preparative column should be 63 µL.

$$VL_p = VL_a * 6.3$$
$$VL_p = 10 \text{ µL} * 6.3$$
$$VL_p = 63 \text{ µL}$$

You may be able to dissolve the directly scaled load of 2 mg into a direct-scaled volume 63 µL, but to dissolve the overloaded concentration of 20 mg into 63 µL is difficult. As a rule, your injection volume should not exceed 10% of the volume of the column.

$$VL_p \leq 0.1 \ V_p \qquad VL_p \leq 0.1 * (15.71)$$
$$VL_p \leq 1.57 \text{ mL}$$

$$VL_a \leq 0.1 \ V_a \qquad VL_a \leq 0.1 * (2.49)$$
$$VL_a \leq 0.25 \text{ mL}$$

2.3 Sample Dissolution

The choice of dissolution solvent is important in all chromatographic separations. The ideal situation is to dissolve the sample in the same composition as the starting gradient. The composition of the starting gradient of a reverse phase separation is always low organic, typically 10–20%. Usually synthetic samples will not be soluble in mostly aqueous conditions. Therefore the user must use some amount of organic solvent to dissolve the sample. How much organic solvent depends on how much sample preparation time the user is willing to invest. For a large number of samples, it is easier to have a universal dissolution solvent. Universal dissolution solvents are typically very strong organic solvents, i.e., DMSO or 100% MeCN. When using these strong solvents, the user should keep the volume as small as possible. As discussed in Sec. 2.2, the injection volume should be less than 10% of the column volume.

2.4 Flow Rate

When scaling up the flow rate you want the linear velocity through each column to be equal. We can scale the flow rate in proportion to the cross-sectional areas (CA) of the columns.

$$CA_p = \pi r_p^2$$
$$CA_a = \pi r_a^2$$

$$R = \frac{CA_p}{CA_a}$$

$$R = \frac{r_p^2}{r_a^2}$$

$$R = \frac{(1)^2}{(0.23)^2}$$

$$R = 18.9$$

If your analytical flow rate is 1 mL/min, the cross-sectional area of the two columns varies in a ratio of 18.9:1, so you should set the flow rate at 18.9 mL/min for this preparative column.

2.5 Gradient

You should run the same gradient (and equilibration and wash steps) on the preparative system that you have developed on the analytical system. To run the same conditions, you will have to change the times of the steps to compensate for the differences in volumes of each system. For example, if the analytical system has a volume of 0.1 mL from the pump outlet to the column inlet, then at 1 mL/min it will take 6 seconds before the column sees the gradient change at the inlet. For the Parallex the volume from the gradient valve on the inlet of the pump to the column inlet is 15 mL. Therefore, at a flow rate of 25 mL/min, it will take 36 seconds for a change in composition to hit the inlet of the column.

2.6 Sorbent Selection

The advantages of columns packed with 10 μ silica instead of 5 μ are:

> They are easier to pack and therefore less expensive.
> They require less special care and therefore foul less.
> They are more reliable.
> They operate at lower pressures and therefore you can increase the flow rate during equilibration and wash.

The disadvantages of columns packed with 10 μ silica instead of 5 μ are

> There is a slight decrease in resolution. Although for this type of application, the decrease in resolution does not usually affect the ability to achieve over 95% purity.

2.7 Column Size

Typically, the selection of columns for the Parallex is either a 20 × 50 mm column or a 20 × 100 mm column for reverse phase separations. These two columns can accommodate loading capacity from 1 mg up to 100 mg of crude sample. The shorter columns are quicker to reequilibrate and wash, but they cannot handle the higher injection volumes of straight organic. For very high throughput applications you pick the shortest columns and concentrate the sample in as small a volume as possible. You can then reduce the cycle time to approximately 6 minutes. For the 20 × 100 mm columns, the equilibration and wash time have to be at least 1 minute long, which is a bit more than one column volume at 25 mL/min. It is best to put more than one column volume for both equilibration and wash. To accomplish this for the longer columns, you can either extend the time during wash or increase the flow rate for that segment. Most manufacturers typically recommend a maximum flow rate of 60 mL/min through the 20 × 100 mm columns. It is not advisable to operate the wash or equilibration at that high a flow rate, since maximum flow rates can shorten the life of a column, but increasing the flow rate to 35 mL/min during the wash and equilibration and increasing the time to 1.5 minutes for the longer columns is acceptable.

3 PARALLEX FLEX

The Parallex Flex system, designed by Biotage, a Dyax Corp. Company, is the first high-resolution workstation designed to *scale* to meet the purification throughput requirements of both combinatorial and medicinal chemists (Fig. 1).

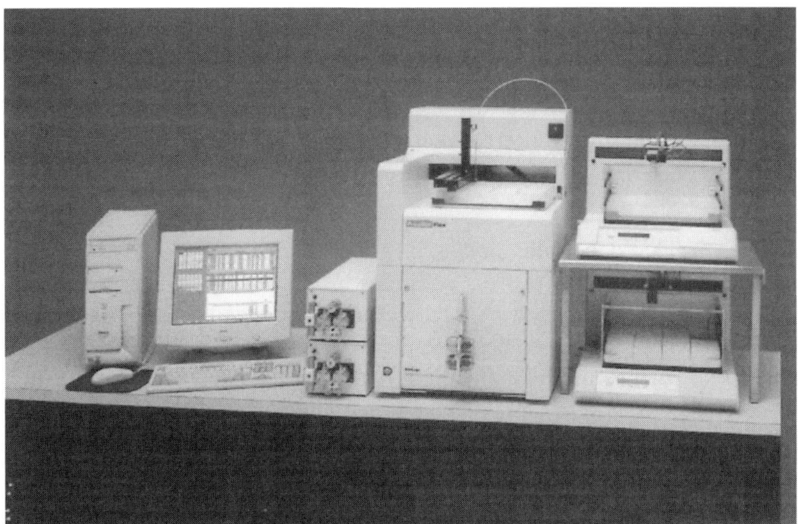

Figure 1 Picture of a two-channel Parallex Flex system.

The Parallex Flex is an *open access* high-throughput purification tool based on bar-coded deep-well plates for both sample input and fraction collection. The system has the *flexibility* to incorporate other forms for sample input and fraction output, i.e., test tubes, vials etc. We utilize a Microsoft Access database to track data and ease the search for historical data and have designed the information pathway to interface and track sample plates from synthesis through the Parallex Flex and on to the mass spectrometer. This section will focus on the five major hardware areas.

3.1 Parallex Flex Hardware

Once you have determined the scale-up chromatography variables, operating a system with multiple columns in parallel takes some specific design criteria. Throughout this section, we will review specific design criteria and the impact they have on the chromatographic variables.

3.1.1 Solvent Delivery

When deciding on a pumping system to deliver the elution solvent to the column, we considered many variables. The pump module is a two-headed design to reduce the pulsation on the column. The module incorporates a pressure transducer, a proportioning valve, and a mixer for accurate gradient control. Each pump module delivers flow at the rate of 1 mL/min to 50 mL/min against a pressure up to 4,000 psi. Each channel has its own pump module, Fig. 2, for independent control. The software controls a switching valve to form a binary gradient between inlet A and inlet B. Since each pump has its own valve and mixer, the user has the ability to operate a different gradient for each channel. The valve also has an inlet C. The operator can use this inlet to wash the column with a solvent different from the elution solvent.

3.1.2 Loader

Biotage has considered the solubility issues that arise in combinatorial or medicinal chemistry. Since reverse phase separations usually start with a composition of mostly water, this could cause most compounds to precipitate during the load cycle. We have designed the loader to sandwich the sample in between slugs of a sample wash solvent that is compatible with the dissolved compound. This prevents the precipitation of compounds in the tubing of the loader. The loader module (Fig. 3) consists of a liquid handler capable of holding up to eight microtiter plates or racks on a removable tray. We accomplished injection automatically via a syringe pump, a needle, a six-port injection valve, and a loop. Multiple samples are loaded into the loops using just the one syringe pump and needle.

Figure 2 Two-channel pump module.

We define loading a sample as taking the sample from the well/vial to the loop. The loading of each loop is accomplished while the previous sample run on that channel is being processed, so that there is no time lost in the load step.

We designed the loader to pull the sample into the loop. To load the sample into a 2 mL loop, we shall consider the following.

The sample sandwich volume is 2 mL minus the volume of sample loading divided by 2. For example, if the sample size is 1 mL, then the sandwich is 0.5 mL on the front side of the sample and 0.5 mL on the back side of the sample. An additional 1 mL is picked up after the sample and sandwich to make up for the volume of the needle, tubing to the loop. We store this volume in the database in the Volumes table. Therefore if the tubing needs to be replaced we can adjust for any changes. The sample wash solvent should be close to the composition of the solvent in which the sample is dissolved. Since we load our loops with sample wash solvent on each side of the sample, you must also consider the

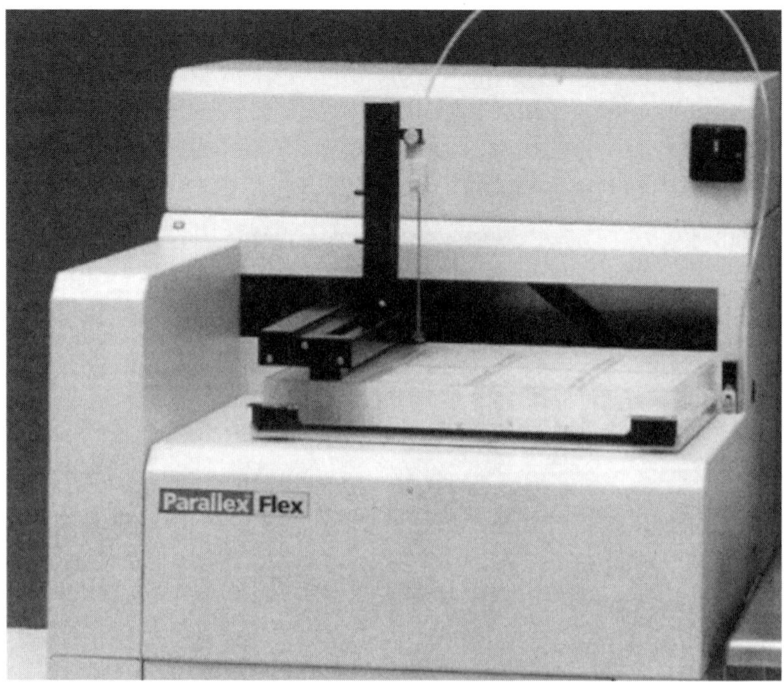

Figure 3   Loader/injector module.

composition of the dissolution solvent. A 100% organic solvent will affect the separation, if the volume of sample and sandwich is greater then 10% of the column volume.

Since we have a 2 mL loop, even if you pick up 63 µL of sample, the loop will be filled up to 2 mL with your sample wash solvent, which most likely is 100% organic solvent. A 2 mL injection is 21% of the total volume of the preparative column. 10 µL is only 0.6% of the volume of the 4.6 × 150 mm analytical column.

There are two solutions to this situation. One is to exchange the 2 mL loop for a smaller one closer to the volume to be injected. The other is to use a second wash solvent of less organic solvent. This gives you the ability to use only part of the 2 mL loop. Thus, the sample can still be sandwiched in a strong organic solvent, but the rest of the loop is filled with equilabration solvent.

3.1.3 Detector

Biotage designed the detector to achieve two main goals. First, to make it scalable from 1 to 4 channels. The second, to make sure you do not miss the compounds of interest. Two wavelengths (typically 280 and 254 nm) are monitored and displayed on the computer. The UV detector is unique and an integral part of the Flex system.

Scale: It would add significant cost to add an entire detector with each channel. Therefore we designed a flow cell (Fig. 4) that can handle up to four streams of flow, the single-channel version can be field-upgraded to include additional channels. All channels are monitored simultaneously while keeping the fluid isolated.

No compound loss: We designed the detector (Fig. 5) with a 1.5 mm path length. This offers the user more sensitivity for flow sample concentration. Although high concentration compounds may go off scale, we monitor at two wavelengths so that we can use one of the wavelengths to desensitize the detector. We also designed the detector with filters that have a ± 15 nm bandwidth. Compare this to current analytical detectors on the market, which are variable wavelength with a ± 2 nm bandwidth. A tight bandwidth is paramount for analyzing

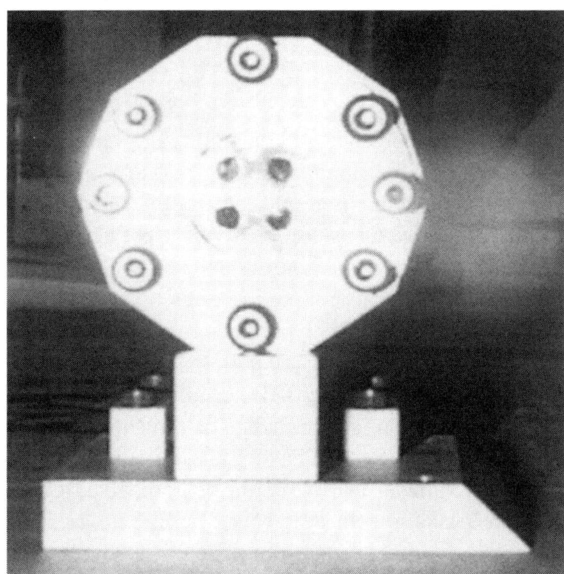

Figure 4 Four-channel flow cell.

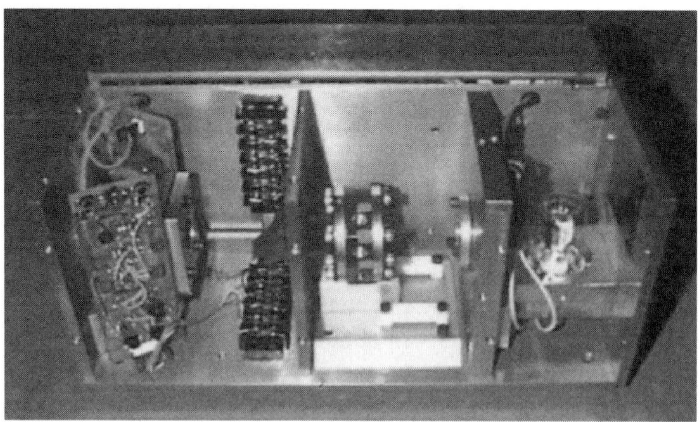

Figure 5   Multichannel UV detector.

data. It offers accuracy and reproducibility. We have chosen the filters to make sure we do not miss any compounds. Since the chemist may not know the λmax of the compound, having a broader bandwidth will offer us a better opportunity to detect it as it passes through the detector.

3.1.4   Fraction Collection

Biotage has designed the software with sophisticated collection algorithms. The system has one collector per channel (Fig. 6). The Parallex Flex Control software controls each collector and triggers collection based on time, volume, slope and/ or threshold. Biotage mounted the three-way waste/collect valve just before the wells to divert flow to waste when the instrument is not in collection mode. The user has the ability to program the software to collect fractions optimizing for yield or purity. The slope field is used when higher purity is the desired objective. During a run of unresolved compounds, the mixture portion of the compound is diverted to waste and the more pure portions are collected in wells. In Fig. 8 the two peaks are separated with a higher purity for each. The threshold field is used when higher yield is the desired objective. We have built into our threshold collection algorithm a detection for valleys. During a run of unresolved compounds, the mixture portion of the peaks is split at the valley. In that case, the user has two wells, each with a slightly lower purity, but none of the compound has been sent to waste, as seen in Fig. 7. Peaks 4 and 5 are separated into two wells. Each peak is slightly less pure, but the system did not send compound to waste.

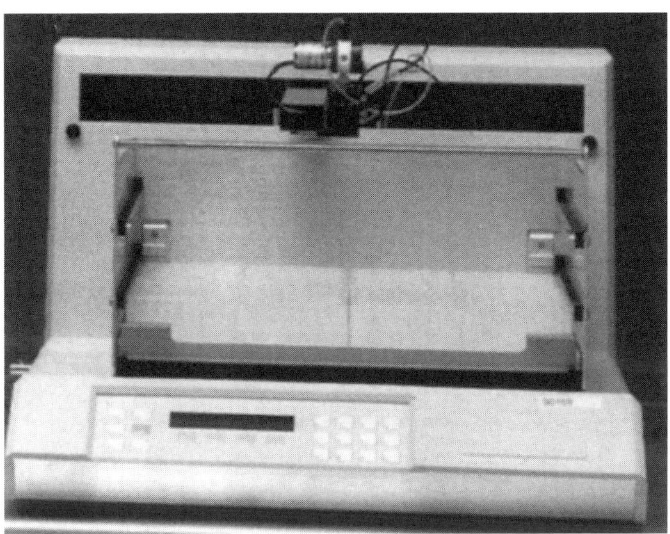

Figure 6   Fraction collector.

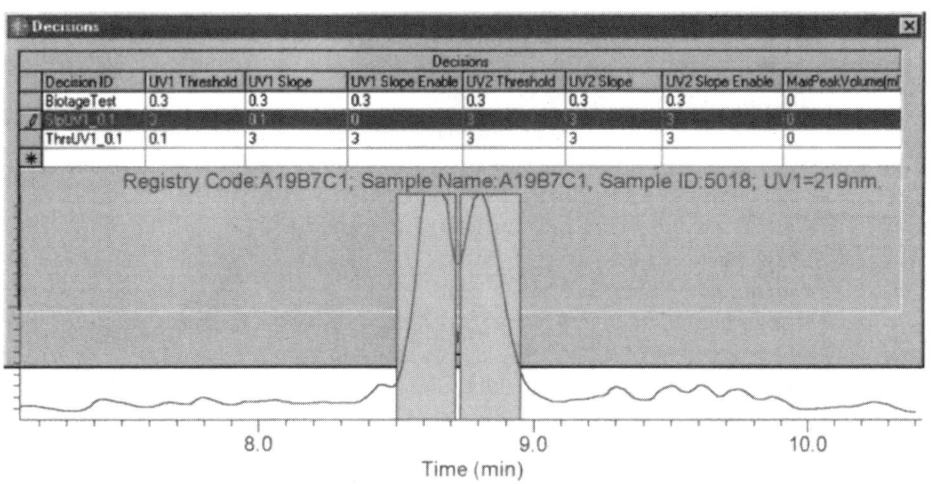

	Decision ID	UV1 Threshold	UV1 Slope	UV1 Slope Enable	UV2 Threshold	UV2 Slope	UV2 Slope Enable	MaxPeakVolume(ml
	BiotageTest	0.3	0.3	0.3	0.3	0.3	0.3	0
	SlpUV1_0.1	3	0.1	0	3	3	3	0
	ThrsUV1_0.1	0.1	3	3	3	3	3	0
*								

Figure 8   Example of collection by slope.

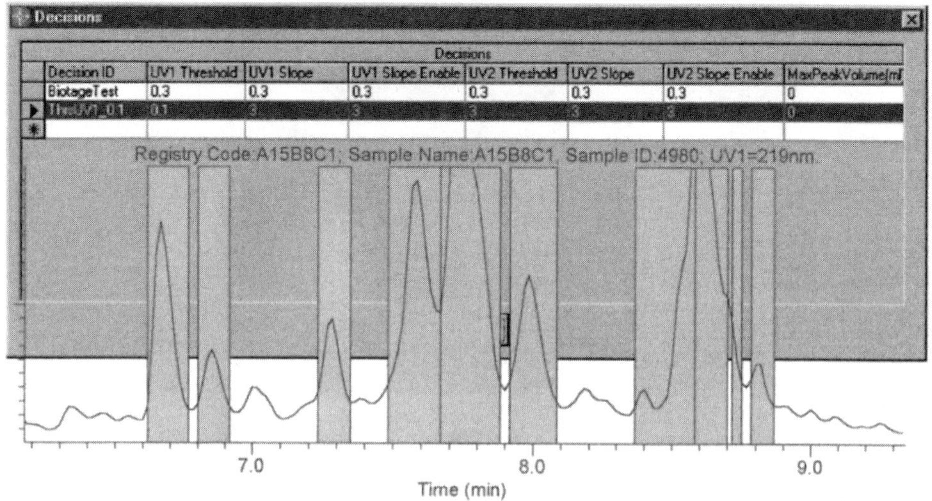

Figure 7 Example of collection by threshold.

3.2 Parallex Flex Software

The Flex Software runs under Windows NT and includes the following modules.

Microsoft Access: Database for storing and tracking all sample and fraction data.

Microsoft Excel: Used for Pilot reports

System Status: A program that displays the real-time status of all channels.

Walkup Wizard: User interface for loading and running the instrument.

Flex Explorer: User interface for viewing data and reporting chromatograms, and database information, plate maps.

Flex Control: Control of all hardware, i.e., injection, collection, and chromatographic hardware components (loader, injection valves, pumps, syringe pump, collectors, waste divert valves). This system displays real-time chromatograms.

Configure: An interface that allows the system administrator to define objects (users, columns, trays, etc.) and to configure the instrument (change columns, UV filters, solvents, etc.).

Pilot Wizard: User interface for the Parallex Pilot, a program that combines fractions to return to the original plate format.

Chromerge: Subsystem developed by LC Resources, Inc., used for viewing of Chromatogram files.

Customer Software Extensions: The database specification, so customers may add tables and queries to incorporate links to synthesis or analysis, or may link the Flex database to other databases.

HTOC Export: Biotage has taken the lead in defining the HTOC specification, which provides import and export capability using CSV files.

The software package Parallex Pilot can be added to the Parallex Flex to reconstitute fractions. In this section we will discuss the six key modules (Microsoft Access, System Status, Walk-up Wizard, Flex Explorer, Configure, Flex Control) and the Parallex Pilot software.

3.2.1. Microsoft Access

This program is part of the Microsoft Office 2000 Suite (included with the Flex) and is used to store and track all fraction and sample information. The Access schema includes information on samples, fractions, methods, plates, trays, batches of samples, etc. During the Walkup Wizard, the user imports all information on samples, loads microtiter plates onto trays, defines methods, configures the instrument—all of which is stored in Access. When the instrument is fully loaded. Flex Control starts automatically and reads the information from Access, fractionates the samples, and updates the Access database with fractionation information. The user then views the results in the Flex Explorer, selecting data using Access queries and reporting using Access reports.

3.2.2 System Status

This program displays the status of each channel, showing the method running, the operator, the number of sample and fraction wells that have been filled, etc.

3.2.3 Walk-up Wizard

This program takes the user through a systematic process of deciding on a method, identifying the sample plate, loading the samples and collector vessels, and starting the instrument (Fig. 9).

There are four operator security levels:

Level 0: Users at this level can use the Drop Off Sample function but cannot define a new method as part of this operation.

Level 1: Users at this level have all Level 0 privileges and can use Examine Data, Run Reports.

Level 2: Users at this level have all Level 1 privileges and can define new methods while in the Drop Off Sample operation.

Level 3: Users at this level have all Level 2 privileges and can use Configure

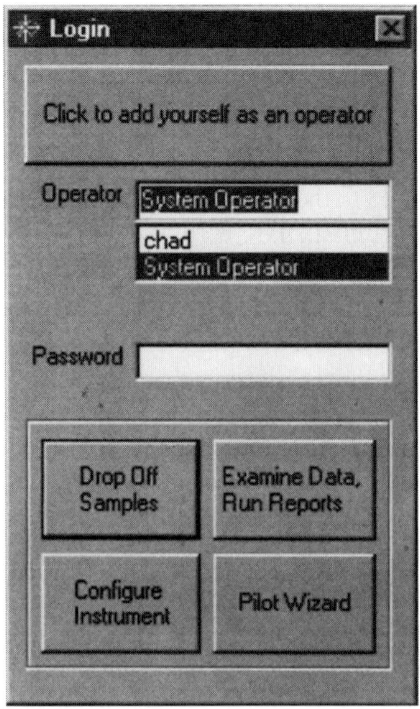

Figure 9 Login screen (Level 3).

Instrument (Fig. 9). To be assigned to this level requires a special procedure.

Once a menu of options is entered using the Configure screen, any operator (referred to as chemist in the text) will be able to perform a separation on the Flex system.

To perform a separation, the chemists will log on and indicate they want to Drop Off Samples. This will open the first Walk-up Wizard screen. The Walk-up Wizard will lead the chemists through a systematic sequence to choose the specific conditions of their separation, to label and locate the sample, fraction wells, and trays and to initiate the separation. For a Level 1 user (lowest security), the first screen looks as in Fig. 10.

3.2.4 Flex Explorer

This program is the primary user interface for reviewing data and reporting. Using the Explorer, (Fig. 11) the user views results, exports data to other instruments or computers, and generates reports. The Explorer is isolated from actual instrument operation: it writes data to Microsoft Access and reads results from Microsoft Access. As such, it may be run under Windows 95 or Windows 98 as well as

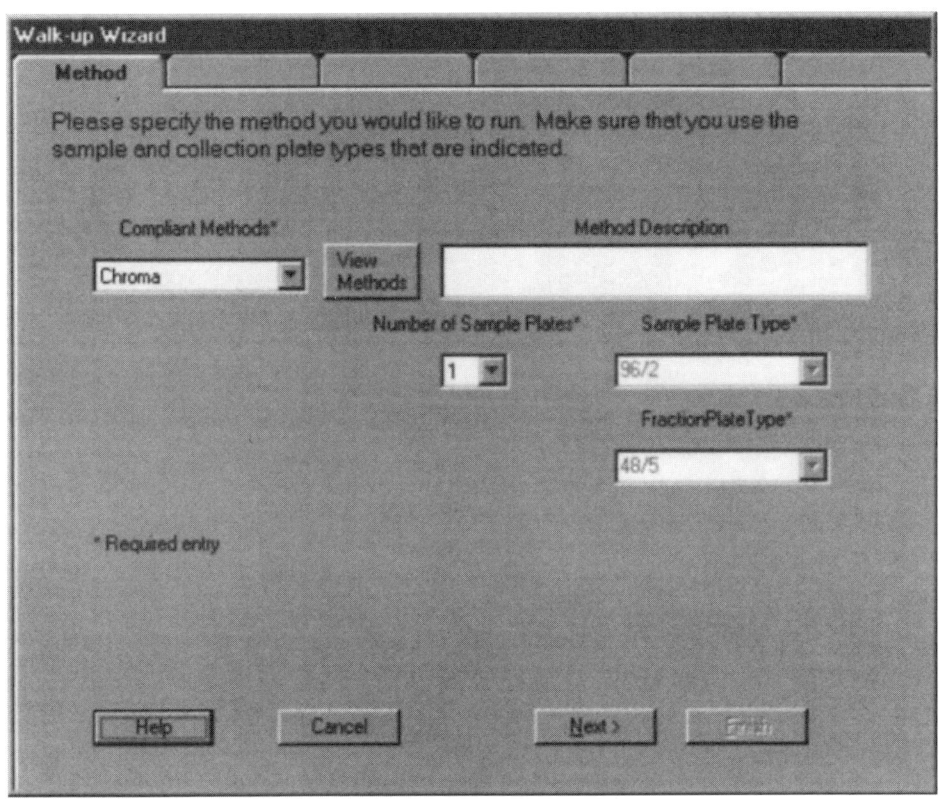

Figure 10 Level one Method Screen (first of the Walk-up Wizard).

under Windows NT, and it may be run locally on a user's desktop computer. This separation of the control process from the user interface provides added stability to the Flex.

The review of chromatograms and data is simplified in the Flex Explorer portion of the software. It provides a dynamic display of any selected sample or fraction and allows several display options. A visual presentation of the associated chromatograms appears below the descriptions of the sample and fraction well information.

3.2.5 Configure

The System Operator or Level 3 operator uses this set of screens to set up choices of hardware, methods, and solvents to be used in separations. They log into the software and choose the Configure Instrument selection.

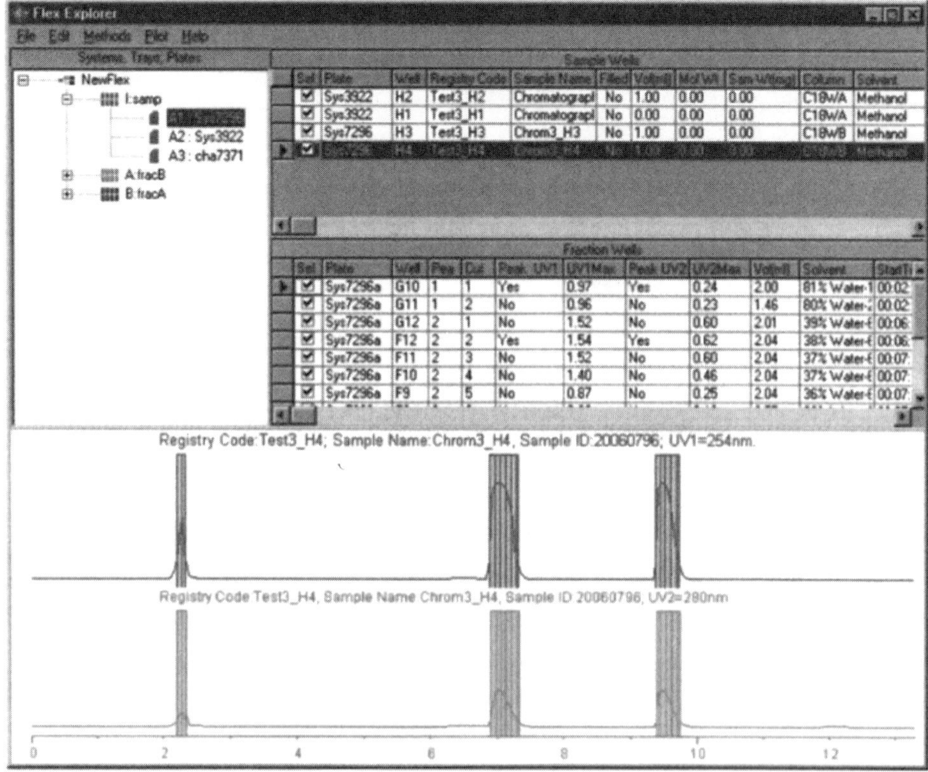

Figure 11   Flex Explorer screen.

Configuring the instrument includes registering hardware and operators, loading specific parameters to be used in separations, enabling channels, and entering methods and decisions for separations. Most of this operation does not have to be repeated, unless new hardware (i.e., columns) or methods are chosen later.

3.2.6   Flex Control

This program is responsible for overall operation of the HPLC instrumentation. It reads information from the Microsoft Access database (as set up by Walk-up Wizard), operates the loader, controls the gradient, acquires data from detector, digitally filters the data and converts to absorbance, makes fractionation decisions as defined by the method chosen, operates the collectors, and updates the Access database with information describing the fractionation and collection process. Flex control is automatically started by the Walk-up Wizard.

3.2.7   HTOC Export

Biotage has taken the lead in proposing a specification called HTOC (High Throughput Organic Chemistry) that provides a straightforward method of passing data and material among synthesis, purification, characterization, quantitation, and combination/evaporation systems. The goal is to provide purified, characterized, quantified samples in the same plate format as the original synthesis format. The Access database in the Flex is the heart of the HTOC process. Flex Walk-up wizard allows specification of which fractions are to be exported to the MS for analysis at the end of each batch.

Flex Explorer will then automatically read the response from the mass spectrometer as it is generated with the understanding that a default file name be specified. From Fig. 12 you can see the ability to turn it on or off by checking the Enable HTOC Export. This actually runs a query in the database. If the user does not happen to click this enable box, they are still capable of exporting the data. The user can also build their own queries and execute them from Flex Explorer module.

3.2.8   Flex Pilot Software

Pilot Wizard: The Pilot Wizard allows the user to return the samples to the original plate format utilizing the data from characterization and quantitation and to combine the fractions. At this point samples have been purified, characterized, and quantified and are ready for bioassay and structure–activity analysis.

Flex Pilot Operation: The Pilot Select controls a Gilson 215 liquid handler to move fractions from the HTOC fraction tray (shown in Fig. 13 top right-set of eight plates) to empty transfer plates (shown in Fig. 13 top left set of four plates). The user then takes the transfer plates to an evaporator to remove the solvent from the fractions. The transfer plates are then returned to the 215 liquid handler. The Pilot Combine dissolves the fractions into smaller volumes and then transfers those fractions to the final plate combining fractions of the same peak into one well. The Pilot is a separately purchased software option. The Flex Pilot drives operation through two programs, Select.exe and Combine.exe, which control Gilson 215 liquid handlers.

3.3   Integration with Characterization and Quantitation

Purification is only part of the overall problem that faces drug discovery today. At the completion of synthesis it is not even certain that the reactions have succeeded. Compounds are almost certainly impure, and the compound mass is un-

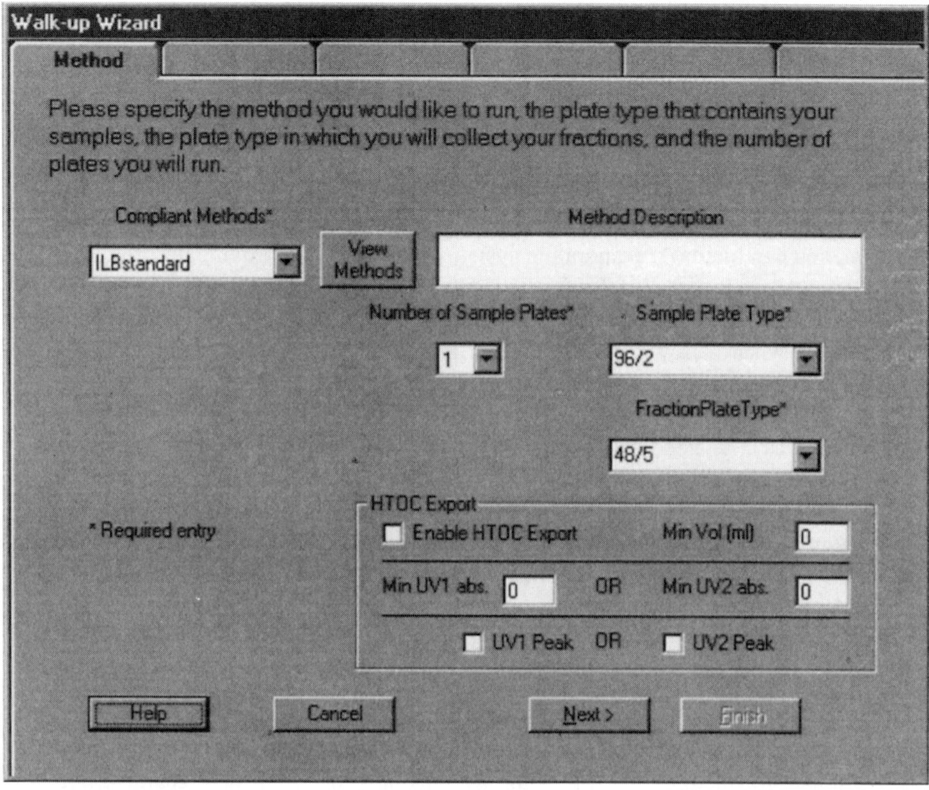

Figure 12   Walk-up Wizard with HTOC Export.

confirmed and undetermined, which precludes use of the samples for determining structure–activity relationships (SAR). Ideally, a scientist would like to go from newly synthesized compounds (Fig. 14, top Left Plate) to compounds ready for assay and SAR, as easily and cost effectively as possible (Fig. 14). The High Throughput Organic Chemistry (HTOC) [1] process achieves the assay plate with a cost under $15.

Why utilize the HTOC process at all? Why not just synthesize and assay? When a hit was found one could go back and deconvolute the information. The question can be answered simply by cost. Initially, many companies planned to synthesize and to screen pools containing several compounds at once. While these efforts at pooled synthesis continue, analysis of the results of pooled screens has not been as successful as was first hoped. False positive and negative assay results are quite common with pools, and identifying the particular component within the

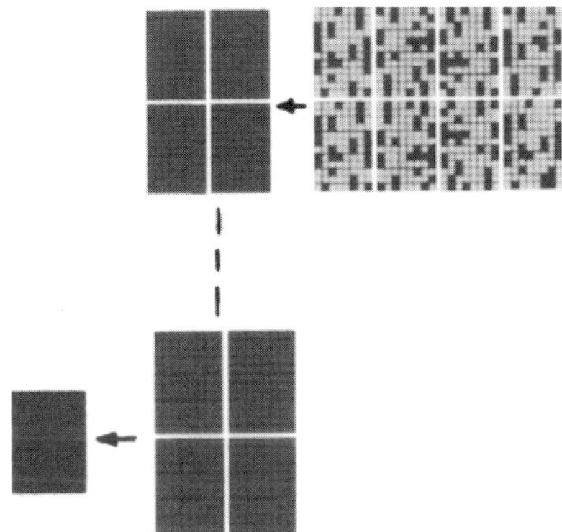

Figure 13 Pilot Selection and Combination station.

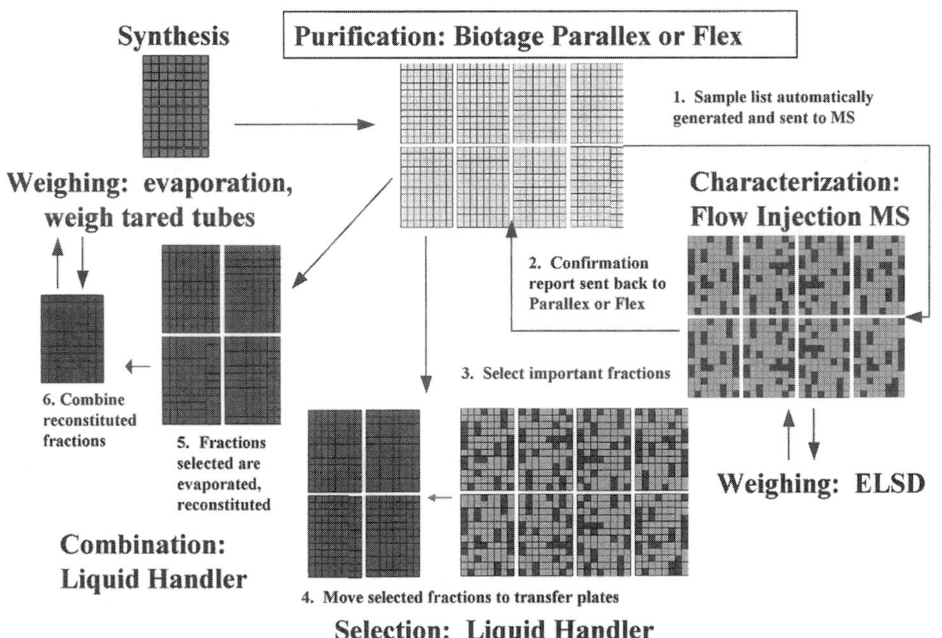

Figure 14 HTOC process.

pool that was responsible for a positive result is often difficult (the deconvolution problem). The past several years has seen a definite swing in most company strategies toward automated synthesis of single compounds, running many reactions in parallel in separate reaction vessels.

4   COST PER COMPOUND

Your company has adequate assay ability but limited synthesis capability. You wish to buy compounds for assay work. You have your choice between Option A, a compound straight out of synthesis for $25 and Option B, a compound with data and guaranteed to be above 85% pure for $40. Which one would you buy? Most people would buy Option B. So now the question becomes, can you complete the HTOC process for $15?

We attempted to determine synthesis cost by splitting it into the components that make up the entire process. The synthesis system is almost certainly the most expensive part of the overall HTOC process, requiring the most careful chemical planning and validation. All activities within the synthesis system will incur substantial costs. Companies consider the following for high throughput synthesis:

> Capital equipment costs
>> Service contracts on capital equipment
>> Maintenance, floor space, and utility costs for capital equipment
>> Reaction validation costs
>> Costs of resins, reagents, supplies, cleavage reagent, solvents with added purchasing and handling overhead
>> Labor costs for loading and operation of synthesizers, including labor overhead
>> Data tracking costs
>> Disposal costs of waste materials
>> Costs of evaporation of TFA and solvents
> Library planning costs

Since Biotage does not have expertise in synthesis, we asked for this information from large pharmaceutical companies to smaller contract companies. We found that the cost was either unknown or would not be divulged. We needed to come up with a way to estimate the synthesis cost. The equation below gives us this ability. Estimates of the cost per compound in this system are difficult to obtain. Many companies do not fully track costs on this basis, and others regard the cost figures as confidential information. We have consulted with different drug discovery groups throughout the pharmaceutical industry about the estimated cost, and this appears to be a valid formula:

Compounds/12 hours	411	
Compounds/year	106,971	
Equipment cost	$500,000	Parallex, MS, 2-Gilsons 215's, ELS detector & Genevac
Amortization period (yrs)	3	
Maintenance factor	18.0%	Service contracts
Overhead factor	20.0%	Utilities, floor space
Equipment cost/compound	$3.33	
Operator time (man-years)	1.5	One operator, 12 hours/day
Operator cost/year	$150,000	Salary, benefits, overhead -- no equipment overhead
Labor cost/compound	$2.10	
Injections/guard column	96	
Guard column cost	$40	
Injections/column	1000	A conservative estimate
Column cost	$600	
Compound weight (mg)	20	
Cycle time (min)	7	
Flow rate (ml/min)	20	
Water use/compound (ml)	70	
Acetonitrile use/compound (ml)	70	
Water cost/ml	$0.00	
Acetonitrile cost/ml	$0.02	
Peak width (min)	0.5	
Solvent disposal cost/ml	$0.02	
Solvent evaporation cost/ml	$0.20	
compound plate cost/compound	$0.26	Based on single use
Fraction plate cost/compound	$0.06	Based on single use
Column cost/compound	$1.02	
Solvent cost/compound	$1.19	
Solvent disposal cost/compound	$1.40	
Solvent evaporation cost/compound	$4.00	
Variable cost/compound	$8.16	
Total cost/compound	$13.60	

Figure 15 Cost per compound analysis.

$$\text{Cost per compound} = \text{\# of compounds Man-years} * \$250{,}000 + \text{Variable cost}$$

$250,000/year is a conservative estimate of the loaded cost per person, including salary, benefits, purchasing, management overhead, lab overhead—includes equipment costs. Variable costs include costs of chemicals, resins, consumables, waste disposal, storage—probably $5–10 for 10 mg, and scales almost linearly with increasing weight.

Most companies are spending $15–100 per compound for dirty, unidentified compounds of unknown quantity. Calculating the cost of purification to final plate is much easier. Figure 15 shows the cost per compound for 100,000 compounds per year.

5 SUMMARY

Automated parallel synthesis is moving pharmaceutical companies further ahead in both lead discovery and lead optimization compound production. As chemistry advances and problems become more complex, one universal system is not enough. Individual instruments must integrate into the larger bioinformatic structures of the individual company. In this chapter we have discussed the details of a system that offers not only a solution to higher purification throughput but also a solution to compound tracking, data management, reformatting of fractions, and data integration.

REFERENCE

1. High Throughput Organic Chemistry (HTOC) process specification published in American Laboratory, February 1999.

6
Future Prospects for Automation in Chemistry Discovery

Brian H. Warrington
GlaxoSmithKline, Harlow, England

1 THE FUTURE NEED FOR AUTOMATION

1.1 Chemical Productivity

What is clear is the need for higher productivity in discovery chemistry. As well as being safe, new pharmaceuticals are expected to address unmet medical needs, modify disease, or otherwise show substantial socioeconomic benefit. The drug industry also faces cost reduction pressures from governments and health insurers and the upcoming expiry of patents that will for example expose more than US\$30 billion of US sales to generic competition. Pharmaceutical companies cannot therefore rely on new drugs in existing categories to maintain revenues but must develop breakthrough drugs that have no substitutes. To sustain current growth rates, this will require the launch of one or two new blockbusters a year in new therapeutic areas, or roughly three to four times the current rate [1,2]. This need to be first to market with breakthrough drugs has set pharmaceutical companies in a race, assisted particularly by genomic prediction, to identify new disease targets including those for which we have no experience in designing ligands. Because of the urgency, empirical high-throughput screening to find developable leads is normally deployed ahead of characterization of target protein structure or identification of a natural ligand so that an early start can be given to chemical optimization, albeit with an SAR often based on structure alone.

Although views differ on the ideal size of a screening collection, the number of screening candidates available to most pharmaceutical companies in the early 1990s—sometimes only tens of thousands—was almost certainly too small to provide sufficient diversity to furnish leads for all targets screened. To increase the probability of finding a usable lead, campaigns were initiated to increase the size and diversity of corporate compound collections as rapidly as possible. To achieve increases in compound numbers measured in orders of magnitude without a commensurate increase in the chemist workforce, traditional methods were abandoned in favor of new methodologies designed to increase chemist productivity such as simultaneous or parallel synthesis.

1.2 Simultaneous Synthesis

Using solid-phase chemistry techniques, simultaneous (''mix and split'') synthesis of mixture libraries is so highly efficient both in respect of the number of reactions required and the number of vessels to contain them that a single chemist could quickly synthesize a large library without need for automation. Consequently, for many companies this technique provided an early entry to high-throughput chemistry. However, the realization of the product as a mixture had severe resource implications in other areas. For a library screened as a mixture by bulk cleavage of the resin, further chemical resource was required either for identification of active components by deconvolutive resynthesis of sections of the library as single compounds or for orthogonal tagging during synthesis and postscreen tag reading. The alternative strategy of screening single beads was technically demanding in that it requires large uniform beads and incurs substantial extra time and reagent overheads during screening because of the high levels of oversampling required for an acceptably high level of library representation.

Overall, the screening of mixtures by either method still gives at best only semiquantitative results and is prone to false positives and negatives. In addition, every apparently viable hit from the screen must be followed up and validated via a series of essentially manual serial processes. To ensure the presence of most library members and avoid a disruptive level of unproductive hit identification cycles due to impurities, synthesis must be preceded by exhaustive iterative validation work and followed by rigorous quality control. The many months spent on library validation become worthwhile only if a very large number of diverse compounds can then be realized. In reality, validation will usually show that the formation of a large number of satisfactory products under a single set of conditions can be achieved only with a restricted set of the potential reagents—simultaneous reactions require chemical similarity rather than diversity! With other constraints such as the limited scope of solid phase chemistry and the restrictions imposed on reagent choice to assist mass spectral identification of actives, it can be seen that although mix and split libraries can be very large, they tend to have narrow diversity and a limited repertoire. As a result, much of the industry has turned away from them in favor of targeted libraries of discrete compounds [3].

1.3 Parallel Synthesis

An alternative approach is parallel synthesis, i.e., single compounds made in multiples using a common process, using either solid-phase or solution-phase synthesis. The obvious benefits for the screener are less work and more robust results. For the chemist, faster quantitative screening results lead to earlier SAR. A further advantage is that there is usually sufficient retained unbound compound to explore activity further without resynthesis. There are also some technical benefits: validation times can be much shorter as products from imperfect reactions can be removed or rescued by purification and thus do not invalidate the whole library, and the technique is not dependent on solid-phase chemistry or coding. This means that a very much broader range of chemistry can be employed than for simultaneous synthesis, and different reaction conditions can be used for different sections of the array. Overall, it is therefore possible to create a wider range of diversity within the library. The main issue for parallel synthesis is that the very large number of operations and vessels necessary for the rapid synthesis and purification of large libraries lies beyond the reach of normal manual chemistry.

1.4 How Big Should a Compound Collection Be?

Despite substantially enhanced corporate compound collections, high-throughput screening still has only limited success, particularly against adventurous novel targets. Failure means that not only is a potential drug lost but the biological target is not validated for its therapeutic potential. Worse from a commercial point of view is failure due simply to the lack of sufficient diversity in the corporate compound collection, because the abandoned target might subsequently be successfully screened by a slower competitor with a different compound collection.

To be able to define the number and type of compounds required for successful screening would imply a perfect understanding of terms such as ''diversity,'' ''bioactivity,'' and ''developability,'' but this is still far from the case. However, from the foregoing it can be seen that of the high-throughput strategies, the superior scope and flexibility of parallel synthesis is more likely to deliver diversity than the restrictions of simultaneous synthesis. This thesis appears to be supported by what is known of current results. Organizations that have used predominantly simultaneous synthesis to amass enormous mixture collections have published little evidence that success in finding leads that can be progressed is simply proportional to the number of screening candidates available. In fact, if this were the case, their pipelines should by now be noticeably exploding with development compounds! In contrast, parallel synthesis appears to provide a richer source of leads and an effective methodology for lead optimization. For example, Bristol-Myers Squibb report that since 1995 the use of parallel synthesis methodology has resulted in a three- to fourfold increase in new drug candidates, 50% fewer staff requirements per drug candidate, and a 40% reduction in lead optimization time [4]. However, it must be remembered that corporate collections also contain a third set of compounds—those accumulated over many years by iterative synthesis. Anecdotal reports suggest that this historic set, despite the limited number of compounds it contains, shows the highest density of developable leads, even for novel targets.

The number of ways the common biological elements can be selectively fitted into the dimensions of a small molecule active site is nearly infinite [5] and is probably limited by the number of atoms available rather than the possible variations in position and connectivity. In reality, the upper limit will always be the maximum the current technology will allow. Clearly, these very large numbers represent a ''perfect'' drug ligand and probably ignore the effects of bioisosterism and assume no role for intelligent lead optimization. A more important figure to define is the minimum number of compounds required to guarantee the identification of an optimizable lead. The lowest figure suggested appears to be about 100,000 for a diverse set of peptides acting at olfactory receptors [6,7]. However, this may be a special case based on ''diversity'' in a severely restricted class of compounds taken as the only possible ligands for a large repertoire of related receptors. A more realistic guess for corporate collections might be based on past performance. For hand-crafted compounds, the historic probability was that one compound in 1000 made in discovery reached the market and only three of ten products that made it to market gave a positive return on investment [2]. On the assumption that a pharmaceutical company will need to be four times better and lack of success was for want of the right compound, simplistic proportionality suggests that a 12-fold enhancement of the collection to a figure in the low millions might have served current needs. However, when further factors are taken into consideration, such as the need to find leads for novel targets and the clustering of diversity that results from high-throughput methods, this number could be much higher.

1.5 What Should a Compound Collection Contain?

Clearly, apart from being bioactive, biocompatible, and bioavailable, an enhanced collection should provide new diversity. Some clues as to how this can be done already exist. For example, Chiron's [8] comparison of compound collections estimated the different fragments in one or more of the compounds thus: all possible nucleic acids, 200; carbohydrates, 400; peptides, 500; 7-transmembrane receptor ligands, 600; peptoid polymers, 600; benzodiazepines, 800; carbamate polymers, 1200; isoquinolines, 1200, diketopiperazines, 1300; 10 best-selling drugs, 4300, 100 best-selling drugs, 5800. Thus, although a combinatorial library may be claimed to be internally diverse, it is deficient compared to typical drug collections. Success in screening may be related less to the internal diversity of individual clusters ("libraries") and more to the number of different "clusters of diversity" provided by the libraries forming a collection. Thus a historic database may contain many thousands of notional clusters each of broad diversity accumulated from a large variety of past chemical programmes. For material made by high-throughput technologies, the relatively shorter validation times required for parallel synthesis allows the synthesis of a large number of small broad library clusters within the same time frame as that required for the simultaneous synthesis of only a few large but narrow libraries. Support for positive relationship between the pharmacophoric coverage of a collection and the number of biohomogeneous clusters it contains can also be seen from Abbott's comparison of its 1992 and 1995 compound collections [9].

1.6   The Long-Term Future of Chemical Automation

If the aim is to produce a general lead-seeking collection composed of a few million high-quality singles, automation becomes essential on the grounds of practicality and economics. Figure 1 compares the logistics and time frames for the production of a library of 1000 compounds by different methods. Only by automation can the increased number of operations required for the parallel methodology be achieved within a similar or even shorter time frame than for simultaneous synthesis using similar human resources. Against the annual cost of a chemist with overheads in the region of US\$250,000–350,000, a machine costing a similar amount that can improve a bench chemist's productivity from about 50 compounds p.a. by handcrafting to tens of thousands is money well spent, provided they are of the right type. Paradoxically, to minimize screening cycle time and avoid the logical inconsistency and wastage implied in selective screening of overrepresented sections of lead discovery collections, this argues for the lowest possible number of compounds achieving the highest density of leads for all targets. In turn, this means a commitment to synthesize highly specified diverse and developable structures rather than just those that are amenable to easy synthesis. High performance automation is therefore likely to be required.

In lead optimization, to date the use of automation has tended to be sporadic. Immediate priorities often dictate that scarce intermediates and headcount be more crucially applied to pushing a synthesis forward another step, rather than expanding the diversity at the current stage. In addition, from the point of view of a biologist using a low-throughput in-vivo assay, high-throughput chemistry during late-stage optimization is overproduction. However, with the advent of high-performance automation, the relevant chemistry can be accessed more often, and when it is in use, chemist productivity rates equivalent to about 1000 com-

Simultaneous Synthesis

Presynthetic Effort

> Confirm and optimize chemistry for each reaction in respect of each variable/tag. Confirm R1, R2 (and tags as required) present and pure by bulk and single bead cleavage by LC/MS of precursor libraries.
> TOTAL TIME = 6 months

Library Synthesis Operations

> 30 reactions in 30 reaction vessels
> 200 LC-MS (bulk and single bead) runs
> TOTAL TIME = 2 months

Postsynthetic Effort

Bulk-cleave and screen sub libraries. Resynthesize active pools as singles. Screen singles against target.	**or**	Over-array beads. Screen for actives. Identify actives by tags or LC/MS. Resynthesise actives. Confirm actives against target.

> TOTAL TIME = 6 months

Parallel Synthesis

Presynthetic Effort

> Confirm chemistry for each step.
> TOTAL TIME = 0.25 month

Library Synthesis Operations

> 5000 weighing/filling operations
> 1110 reactions in at least that number of vessels
> 1000 cleavages or l/l extractions or SPE runs
> 1000 evaporations
> 1000 LC-MS runs 1000 hplc runs producing 15000 fractions
> 1000+ LC-MS runs to pool fractions
> 2000 decisions
> 1000 evaporations
> 1000 registrations Not to mention replatings, taring, bottling, labeling, data transfer, etc. etc.
> TOTAL TIME = 2 months

Postsynthetic Effort

> Screen singles against target
> TOTAL TIME = 0.1 month

Figure 1 Comparison of the logistics and time frames for synthesis and screening of a 1000 component ($10 \times 10 \times 10$) library. Typical times (in person-months) and logistics are given for the production of a $10 \times 10 \times 10$ library of 1000 compounds by simultaneous synthesis and parallel synthesis with purification. This is a cluster size approachable by either technique that is likely to engender a collection with good diversity and that can be efficiently screened. Due to the largely serial pre- and postsynthetic overheads involved in synthesis and screening of the simultaneously synthesized library, the parallel process is preferable in terms of effort, and compounds will reach screening sooner, but extensive automation of synthesis and purification is essential. Manually (at 50 compounds per annum) the library would require 20 person-years.

pounds per annum can be reached, and cost per compound is very much less than for iterative chemistry. As the great majority of medicinal chemists work in the area of lead optimization, the use of automation, even if only for a few short runs, can easily double or triple workforce output. With this economic driver and a recent history of increasingly rapid loss of market exclusivity [2] (Fig. 2), the use of automation to increase compound count to identify the best leads and patent scope will probably increase further.

Overall, the future of automated chemistry appears assured for many years to come. Beyond setting up an adequate lead discovery collection, the aim will then be to increase numbers to provide an opportunity for finding more advanced or developable leads and thus reducing optimization cycle times and losses in development. In addition, as alternative approaches such as gene therapy are still a long way off and will probably be applicable to only a selection of disease states, there remain a large number of biological targets to be tackled chemically. It has been estimated that the human genome codes about 100,000 proteins, of which a quarter may have therapeutic potential [10] while current therapies probably address less than 500 [1]. Thus the flood of "orphan" targets, for which chemical leads must be found, is unlikely to abate in the foreseeable future, and

Drug Introduction	Follower Drug	Years of Exclusivity
Inderal – 1968	Lopressor – 1978	10
Tagamet – 1977	Zantac – 1983	6
Caoten – 1980	Vasotec – 1985	5
Prozac – 1988	Zoloft – 1992	4
Diflucan – 1990	Sporanox – 1992	2
Recombinate – 1992	Kogenate – 1993	1
Invirase – 1995	Norvir – 3/1996	0.25

Figure 2 Increase in rapidity of loss of market exclusivity. (From Harris, C., et al. Analyst Report. Life Sciences, Bringing Industrial Scale Science to the Clinical and Research Laboratory. Warburg Dillon Read, New York, 24 Jan. 2000.)

the role of discovery chemistry will become more critical, particularly when the aim to treat also polymorphic patient populations is also considered.

2 FURTHER DEVELOPMENT AND IMPLEMENTATION OF CURRENT HARDWARE

2.1 Status of Current Hardware

Chemistry automation usually requires relatively short production runs with a varied repertoire of process options that can require dry or inert containment. In addition, the physical properties of individual operands can vary greatly, and they can be reactive and aggressive. The development of equipment designed specifically to automate small-molecule organic synthesis has therefore taken longer to appear than that for biological assays, where the tendency has been toward long runs of uniform, repetitive aqueous liquid handling operations. Because of the additional challenges it has been predominantly small adventurous companies that have taken on the challenges and risks of de novo design of chemical automation, and what has evolved is a range of partial solutions addressing different niches of the synthesis and purification process. For reviews of currently available equipment and what can be done with it, the reader is referred to the other chapters of this book and to an excellent review by Hird [11].

The critical requirement when selecting equipment is fitness for purpose, which is usually defined by performance, throughput, and sample size. Although most current items of equipment achieve good performance against one of these criteria at reasonable cost, high performance in all areas cannot be achieved without complexity, reliability, and price becoming serious issues. As with manual chemistry, the chemist will have to make a skilful choice from the range of tools available to achieve the task. For many potential purchasers, the easy option has been to enter an evaluation stasis where any appraisal is quickly invalidated by new developments—ad infinitum! Those who have taken the risk of implementing automation have found that a more important factor in success than the choice of the correct equipment has been the development of a process for high-throughput chemistry and a commitment to make it work. Because this requires the interfacing of disparate pieces of equipment, recent years have seen increased collaborations between users and suppliers. Hopefully this trend will continue toward some standardization of interfaces.

2.2 Small Organic Molecule Synthesisers

The generic synthesis of small organic molecules is the most complex stage in the automated compound production process and has presented the most problems for the robotics engineer. The seemingly contradictory objective of containing and manipulating reactive chemicals in an inert environment challenges the limits of engineering materials and methods. Manual chemistry has shown that glass is the best material for containing chemistry but the most difficult from an engineering point of view. The focus of effort in this field will probably remain on identifying equipment that will withstand attack by reagents and at the same time prevent contamination of the reaction by either the machine or the ambient environment. Most current automation solutions provide methodologies for achieving production rates that are higher than those that can be achieved through manual iterative chemistry. However, none provides a general automated panacea for all chemical synthesis, and ultimately all fall short of man's versatility. Progress towards machines that are more competent and robust will depend largely on the development of more sophisticated materials and fabrication techniques.

For many tasks, the performance limitations of a synthesizer do not prevent its use. The job required either lies within the ability of the instrument or can be made to fit by reconfiguring the chemistry. This has usually been achieved by the use of a solid phase to support either the substrate or the reagents so that automated actions can be simplified, usually to easily performed nonstoichiometric liquid transfer routines. However, it must be noted that reaction reconfigurations require time and manual effort and might require extra steps to be added to the process resulting in reduced overall productivity. In addition, unless a purification process is also available, extensive reaction optimization will be required to reduce impurities. Although these time and effort penalties were accepted in the early days of chemistry automation, increasingly the demand is for a direct transfer of unoptimised manual chemistry to a more versatile machine coupled to a high-throughput purification process.

2.3 Simple Synthesizers

The performance limitations of some synthesizers are inherent in their design and cannot be improved except by building a different machine! For a few pieces of equipment these performance limitations will be terminal. For most, there will be recognition by users (and more importantly manufacturers) that they fulfil a need within a particular market segment in much the same way as do peptide and oligonucleotide synthesizers. Thus the way forward will probably be to tune the synthesizer and possibly the workup process for the greatest ease, speed, and reliability in performing its limited range of chemistry. In practice, the use of such machines will probably be limited to high-throughput lead generation, where there might be sufficient flexibility in the specification of the compounds required, and there can be sufficient return in terms of numbers to compensate for any time spent in reconfiguring the chemistry.

2.4 Advanced Synthesizers

A significant challenge is to automate without optimization the chemistry exactly specified by a biological SAR or some other design process. To gain acceptance the synthesizer has to offer a walk-up facility competitive with manual chemistry in terms of ease and effort and a big enough productivity and serendipity advantage to overcome the inertial barriers to use. Matched throughput purification facilities would be a necessary adjunct to unoptimized synthesis.

Currently, the most comprehensive range of synthesis services and operational modes are those offered by the Myriad Core System (Fig. 3) and its variants [12]. Key features contributing to high performance are a multitasking platform (a) that allows simultaneous processing of different stages of the procedure; a complex vessel that can be sealed and transferred to interact with sophisticated robotics in different modules (b); and a positive action delivery tip (c) able to deal with a wide range of liquids and slurries. As all functions of the machine are automated, including intermodule transfers, unattended operation is possible. The machine is produced in a high-throughput version (Myriad Core System), targeted at lead optimization, and a smaller 24-vessel system (Myriad Personal Synthesiser) for lead optimization. By implementing functionally equivalent hardware and software, the design allows for transfer of the synthesis script between machines, thus allowing high-throughput expansion on the Core System of libraries previously optimized on the Myriad Personal Synthesiser and the setting up of a common procedure bank for both machines.

Potentially, the modular high-performance high-throughput concept epitomized by the Myriad Core System could be extended by the addition of new on-line automated modules to the hardware and the software. Ultimately this could provide an almost unrestricted synthesizer, but the size of the platform is already large, and the size of an extended machine would severely restrict its widespread use for reasons of impact on infrastructure and cost. Because improvements in high-throughput screening have led to a reduction in the size of the sample required for storage, it is more likely that the future will hold more compact (and probably faster?) machines but with similar high performance and throughput. In addition, thought will be given to aligning synthesizer hardware to the strategies used in high-throughput chemistry. For example, to maximize chemical efficiency, synthesis campaigns are often conducted in a ''split-split'' manner [13] (Fig. 4). Currently, this is tackled by using a cascade of equipment, but a more convenient and efficient solution would be to achieve this within the same synthesizer, via a range of vessel sizes. In addition, during lead optimization, which is

---

Figure 3 Components of the Myriad Core System. (a) Modular platform. (b) Sample vessel actions. (c) Positive displacement tip.

(a)

Processing Module
Incubator Module
Input/output Module
Solvent processing area
Reaction Vessel tray conveyor
Ambient Processing Area
Reagent Drawers
Subambient Incubator
Super ambient Incubators
Tray Load area
Tray Storage area
= Tray of 12 Reaction vessels

(b)

KEY
Solvent
Gas
Side Chamber
Main Chamber
Glass Frit
Capillary
Cap Closed
Add Solvent
Gas Bubble
Gas Filter and Aspirate to Waste
Gas Bubble
Add Liquid Reagent
Take Slurry (Resin) Sample
Gas Filter
Take Filtrate Sample

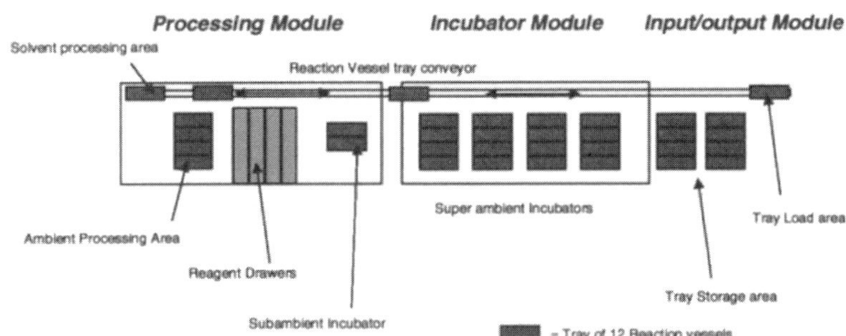

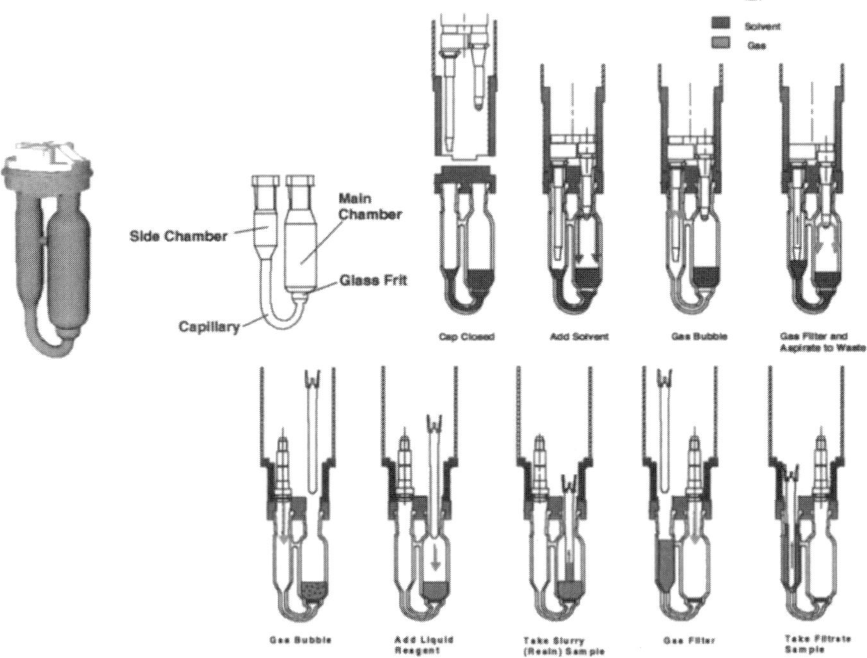

(c)

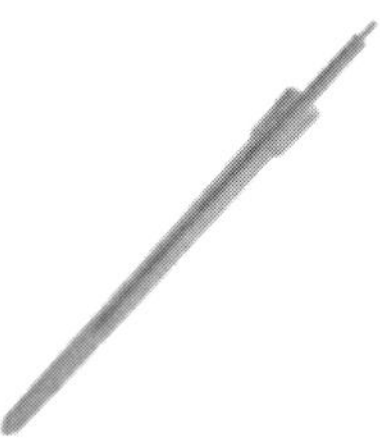

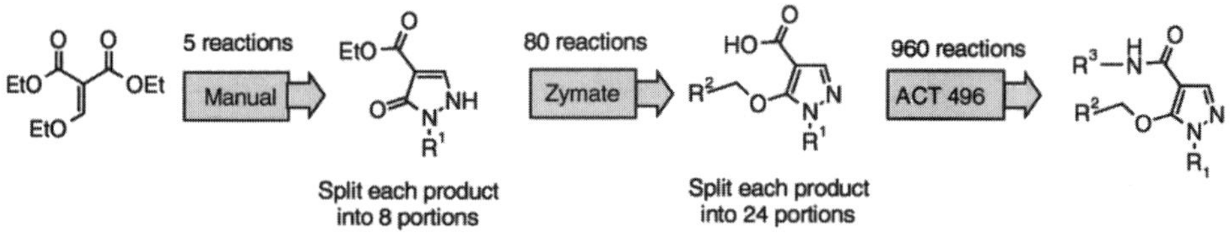

Figure 4   Split-Split Pyrazole library synthesis. (From Brooking, P., Doran, A., Grimsey, P., Hird, N. W., Maclachlan, W. S., Vimal, M. Split/split: a multiple synthesizer approach to efficient automated parallel synthesis. Tet. Lett. 1999, 40, 1405–1408.)

often carried out at the "personal" synthesizer level, the amount of material required for secondary assays is often substantial. Therefore it would be useful to have a choice of vessel sizes within the same synthesizer.

2.5   Synthesizers Beyond Human Range

Historically automation of mass production was instituted as a means of reducing manufacturing cost. More recently it has also been used to achieve tasks beyond human range. One area has been that of maintaining quality and consistency over long runs of repeated procedures and in multiplexing. Here, the machine excels in an area where man is vulnerable—reliable repetition and keeping track. It may be possible to push the multiplexing abilities of chemical automation to further advantage in the preparation of chemical diversity.

There has been a strong tendency to use automation to replicate methodology for parallel synthesis, but with greater productivity. However, this task is set to remain within man's multitasking and multiplexing abilities, and although there may be many vessels in use, apart from the introduction of different diversity, there is usually only one basic repeated procedure. Thus the analogues synthesised are related, but in theory at least, on a fully implemented modular platform there is no reason why the vessels should contain related materials or be subjected to the same processes, and more chemical ground could be covered in one run. Of course, this would probably require greater complexity at the hardware level. For example, synthesizers would probably have to have a variety of concurrently usable incubation facilities, and this could require the implementation of more individually controllable energy sources such as microwaves. The practicality and advantages of this approach to heating has recently been demonstrated in "Personal Chemistry" [14]. The greater challenge would be to write organizational and error-checking software to compensate for man's poor ability to schedule and track the complex and fast-moving events of such a synthesizer.

Almost certainly, this would require the incorporation of analytical monitors to allow reflexive control of individual reactions.

2.6 Purification Equipment

Reaction Workup

Depending on the chemistry, the stage of purification usually termed workup and quality criteria set for products workup alone may suffice to isolate an acceptable product. For solid-phase chemistry, reaction workup amounts to no more than washing the beads to remove impurities. For solution chemistry, the process is more complex. Often it will involve liquid–liquid extraction, and some synthesizers or liquid handlers can be configured to perform an extraction by use of dead reckoning to set the probe height to remove the layer above or below a precalculated level based on the volumes known to have been used.

Apart from the constraints of having to track solvent usage, efficient purification relies on rapid and good separation of the immiscible layers or, with lower recovery, an accurate prediction of a probe position where an uncontaminated layer can be found. ALLEX (Automated Liquid–Liquid Extractor), recently launched by Mettler Toledo Myriad [12], is a specialized module with multiple and back-extraction capability in which the mixed solvents flow through a narrow tube and capacitance effects are used to detect dynamically the phase boundaries between "plugs" of immiscible solvents and perform the separation. As a result separations can be faster ($\sim$60 per h), no volume estimates are necessary, emulsions can be detected, and interfacial rag layers can be separated and dealt with. The impact of this highly specialized tool on workup and purification protocols has yet to be evaluated, but what is important to note is that its conception is part of a growing appreciation that "chemistry" is not a single-step process requiring a single "synthesizer" solution.

Another approach is specific removal of known contaminants using solid phase extraction (SPE). Several instruments are well suited to automating the addition of reagents to and pressurizing SPE cartridges, such as the Hamilton Microlab 2200 [15] and the Zinsser Lissy [16]. With the expected growth in automated solution phase chemistry and expansion of the range of SPE reagents it is expected that more dedicated instruments will appear. In contrast, an aspect of workup that has received little attention is that of crystallization. It is difficult to judge whether this is a result of the difficulties of automating the process or whether there is little perceived need. However, with increasing use of parallel methodology in reaction optimization and the possibility that it may be found that gums do not represent a robust long-term storage form for samples nor necessarily the purest forms, this situation may change.

Chromatography

The purification methods described in the preceding section are somewhat dependent on the happenstance of impurities being readily and precisely separable from the target product, and selective removal requires some knowledge of the underlying chemistry. In contrast, chromatography assumes little other than approximate logP of the product. Using solvent systems targeted to separate compounds in the "developable" logP range, reverse phase preparative scale HPLC has therefore been a popular technique for high-throughput purification of libraries of single entities. The basic technology is mature, and effort has been directed mainly to achieving high-throughput purification of lead generation libraries. A good example of how this is achieved can be seen in the Biotage Parallex [17] where the hardware consists of four columns running in parallel with UV-triggered fractionation. Samples undergo an overall multiwell plate to multiwell plate transfer and with small fast columns and effective sample tracking software a throughput in excess of 300 samples per day in the 10–30 mg range can be achieved. An issue in high-throughput purification by HPLC is the handling of the enormous number of fractions generated. To aid critical fraction identification and reduce pooling, several groups [18] have described the use of MS-directed fraction collection, although the systems used could be complex. Quantification is also a problem, and further enhancement of the sensitivity of evaporative light scattering (ELS) detectors is an important goal, as these devices do not require a chromophore and show a more linear mass response than UV.

For lead optimization, other multicolumn machines, such as the Flashmaster [19], the Sympur [20], and the Quad 3 [17] for the parallel purification of smaller numbers of larger samples encountered have also appeared. However an interesting recent development is the open access Biotage Flex [17] which can be flexibly configured with one to four variably sized and independently operable columns and is thus usable in lead generation or lead optimization.

2.7 Quality Control Equipment

LC-MS

Electrospray ionization produces quasi molecular ions $(M + X)^+$ [where $X =$ H, NH_4, Na, and K] and few fragment ions, thus providing an excellent confirmation of molecular weight. As a result, LC–MS remains the most common method for confirming the identity and purity of high-throughput chemistry products, mainly because of the speed, as quantification can be poor. As with purification, this is an area where the effort has been directed to configure an essentially mature technique to a high-throughput mode. This has been achieved mainly through automated sample handling and the provision of sample and result tracking software. Machines such as the Micromass OpenLynx Diversity [21] or Finnigan aQa [22] are now commonplace and can provide almost continuous LC-MS analysis from a microtiter plate input reaching about 500 samples per day. Based on an input of expected molecular weights for products and preset criteria for UV peak areas, samples can be automatically evaluated on a pass/fail basis against QC standards. However, as the number of samples continues to increase, driven greatly by the increased use of high-throughput purification that produces multiple samples per compound, the key aim has been to achieve a more economical, productive, and integrated solution than simply deploying large multiples of machines. For example, multiway multiplex API interfaces can increase mass spectrum acquisition rates substantially. Thus by using a Gilson 215/889 liquid handling/multiple injector system in combination with an LCT orthogonal ToF mass spectrometer fitted with a novel eight-way multiplex API interface (Fig. 5), Micromass have achieved an effective eight-channel "MUX" instrument [23]. The parallelism within this instrument reduces the time required to process a 96-well plate from 300 min (for serial injections in a Micromass Platform II Diversity) to about 40 min. In theory at least, this principle can be extended to higher multiplicity, but it is likely that higher throughput will come from nanoelectrospray technology currently being explored by Advanced BioAnalytical Services [24] and PE Biosystems [25]. Very small quantities of fluid held in 100 nL reservoirs in a silicon chip are pressurized and sprayed from miniature (10 μm, i.d.) electrospray nozzles etched through the bottom of each well as recessed openings on the other side of the chip. The chips are designed to conform to 96-, 384-, or 1536-well formats, and the delivery probe is stepped from well to well to deliver the samples directly to a ToF mass spectrometer.

NMR

Potentially, NMR has much to offer in terms of both product identification and quantification. High-throughput processing of samples can now also be achieved. For example, the Bruker Efficient Sample Transfer (BEST) system [26] utilizes a flow probe and an automated sample injection directly from a microtiter plate using a Gilson XL-215 to give a throughput of 15 samples per h and excellent proton spectra from samples of 500 μg. Optional double probes have two concentric cells: the inner cell is used to hold a lock reference sample, thereby eliminating the need for deuterated solvents. Varian's versatile automated sample transport (VAST) [27] system is similarly configured [28]. However, automated triage of samples against quality control standards is less easily achieved with NMR than with LC-MS. Modern software can calculate the predicted spectrum of a pure product with sufficient accuracy to allow the presence of the product to be confirmed. Parallel synthesis of discrete compounds with common features presents a particularly favorable case. However, as it is not possible to predict the spectra of impurities in an impure product, quantitative assessments of purity can

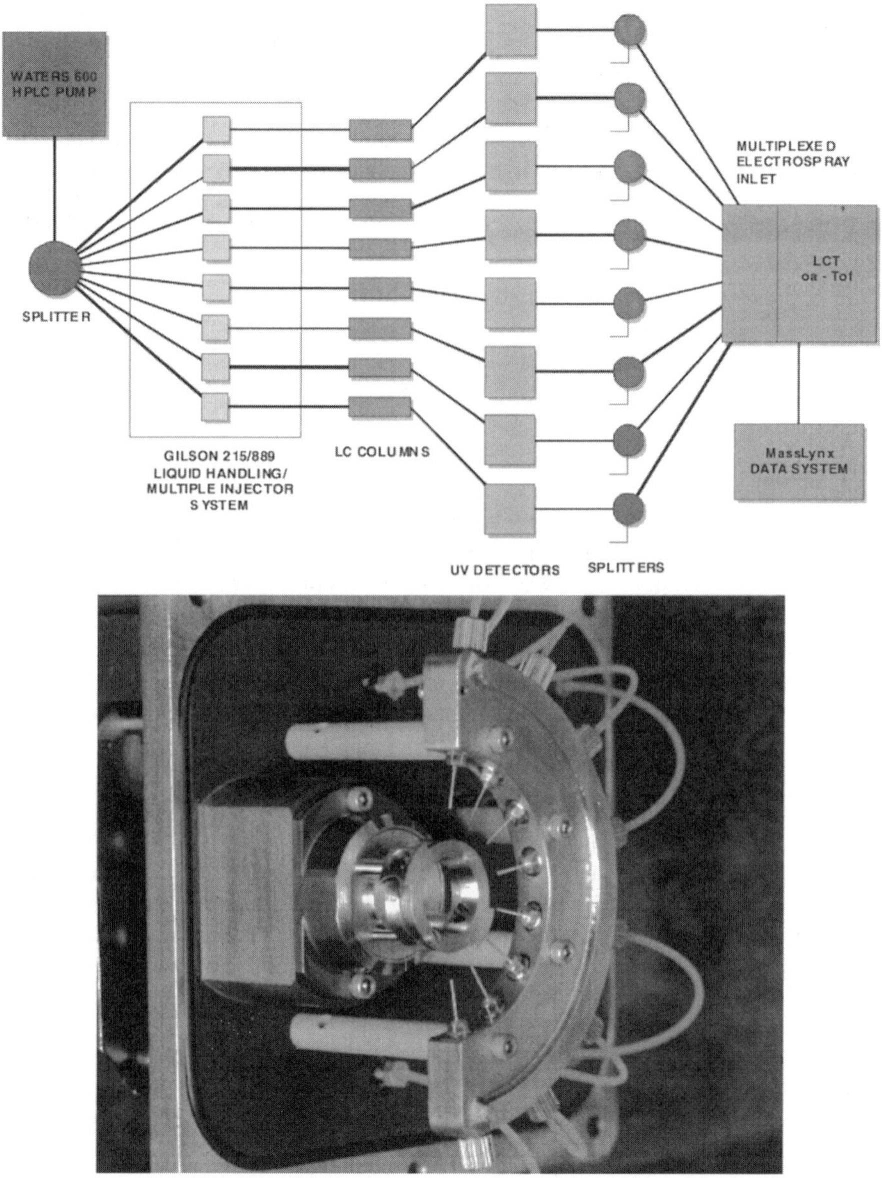

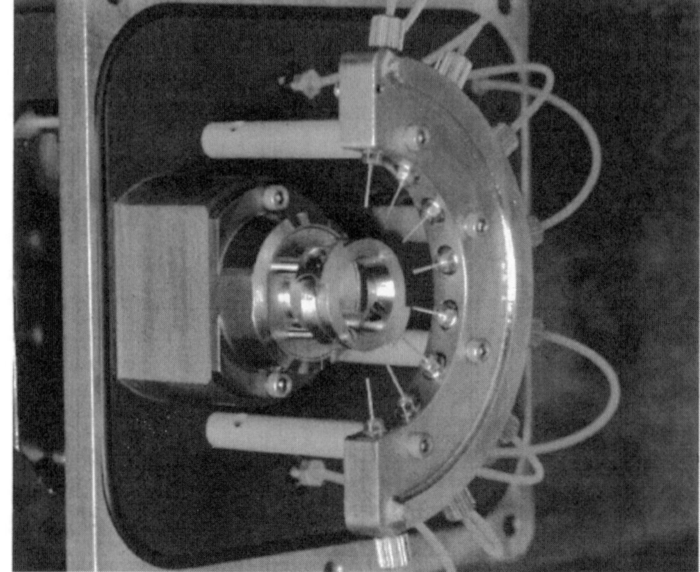

Figure 5 Micromass eight-channel MUX system shown schematically. An eight-way multiplexed electrospray (ESP) interface (photo) is attached directly to the Z-spray source of a Micromass LCT oa-TOF mass spectrometer. Eight nebulization-assisted ESP probe tips are arranged radially about the sampling cone, and the spray from each tip is admitted in rapid succession using a rotating aperture. Heated dry N_2 is used for desolvation, and the cycle time for eight sprays is 1.6 s.

$$R\text{–}N \;+\; O_2 \;\xrightarrow{1050\,°C}\; {}^{*}NO \;+\; \text{combustion products}$$

$$^{*}NO + O_3 \;\longrightarrow\; NO_2^{*} + O_2 \;\longrightarrow\; NO_2 + O_2 + h\nu$$

Figure 6 Chemiluminescent nitrogen detection (CLND).

only be made slowly by visual inspection. Automated triage of samples against QC standards by NMR must be the next significant hurdle to be overcome.

Elemental Analysis

A major issue for library production is accurate quantification, particularly when the analyte lacks an adequate chromophore or is too volatile for ELS detection. As most druglike compounds contain nitrogen, a suitable method may be the use of chemiluminescent nitrogen detection (CLND). This technique, pioneered by Antek Instruments Inc. [29], relies on light emission as shown in Fig. 6. Due to equimolar detector response, which extends over five orders of magnitude, and sensitivity down to 0.1 ng of N, nitrogen can be quantitated without regard to the target compound. The technique is particularly powerful when used in tandem with another detector such as a mass spectrometer. The main drawbacks are that present systems lack speed and stability and the technique may not be compatible with the current common use of MeCN-based chromatography.

2.8 Future Developments

The scope and usefulness of high-throughput synthesis is greatly enhanced if high-throughput purification can also be put in place, but the latter is often the rate-limiting step. However, it is unlikely that column cycle times of either the preparative or the analytic HPLC systems can be further reduced without loss of performance. A possible way forward is to reduce the number of fractions collected and examined by mass triggered collection or some other predictive method. It is unlikely that ''universal'' solvent systems will be found to deal with the increasing range of diversity present across library runs. Currently, nonproductive operator time is wasted recovering and rerunning failed separations. Smarter software that continually monitors and optimizes solvent system performance would be a welcome advance.

A more promising area for advance is the renewal of interest in so-called supercritical fluid chromatography (SFC). This was originally a focus of some interest during the 1980s, but progress was limited due to waning vendor interest and competition from the increasing range of column chemistries available to

HPLC. Now SFC, in which a near supercritical fluid (usually CO_2) containing a small amount of an organic solvent (often methanol) is used as the eluant, usefully promises an order-of-magnitude reduction in run times and equilibration cycles. As the greater part of the mobile phase (about 75%) vents to the atmosphere during sample collection, fraction collection volumes are much smaller than for HPLC and are water free. There are thus further savings to be made during post-chromatographic processing. Provided efficient handling of druglike molecules can be effectively demonstrated, the potential for increased productivity using this technique, coupled with the reduced cost of purchase and disposal of solvents, will challenge the strongly held position of HPLC. At present, most chromatography vendors are working towards commercialized systems. Further payoff from a change to SFC-based purification would be a reduction in the amount of equipment and time devoted to fraction evaporation and, with a nonnitrogen containing solvent, a better prospect for the implementation of N-based quantitation.

3 AUTOMATION SUITES AND WORKSTATIONS

3.1 Personal Scale Equipment and Workstations

In recent years medicinal chemistry has undergone a period of rapid and turbulent technological and methodological change. While some chemists may not have welcomed a diversion from familiar tools and stable practices, most pharmaceutical companies attest to enhanced productivity, reduced cycle times, and the advantageous cost of automation over staff increases, so reversion to the older order is unlikely [4]. The role of the great majority of medicinal chemists is to produce a relatively small number of highly specified compounds as part of a lead optimization campaign. For this group of chemists, a modest expansion of output to speed SAR accumulation and enhance serendipity is seen as a worthwhile goal. In contrast, technologies and methodologies that target ultra-high throughput as a first priority are seen as inappropriate as they can overload advanced biological assay systems, consume premium synthesized reagents, and slow the main objective of an optimized compound for development. Therefore, for automation to be a successful contributor to lead optimization, enhancements to chemist productivity have to come to a greater extent from its chemical performance and reliable unattended use than its throughput.

It is in the history of chemistry to find a substantial interval between the introduction of new techniques and their general adoption. Usually, it is the success of a relatively small number of pioneers who develop the usefulness of constrained early examples of use into general applicability that eventually persuades others. Therefore, for the present, a balance must be struck between having automation equipment readily available to optimization chemists without impinging on their ability to do manual work. For as long as the rewards of automated chemistry are perceived as marginal relative to manual methods, there will be reluctance to modify substantially the infrastructure of a laboratory already optimized for manual chemistry. There is therefore a strong pressure on manufacturers to produce equipment for the different stages of synthesis and purification that will each fit in an unmodified normal hood. The automated ''workstation'' therefore devolves to a series of stand-alone pieces (synthesis, workup, chromatography, evaporation, QC), often from different manufacturers, each in its own hood. While this has the virtue of allowing an individual step to be selected for automation, it does little to help establish an easy-to-use, reliable, unattended overall process for making compounds. Despite this, increasing numbers of successful demonstrations of effectiveness from those who have implemented automation, perhaps coupled with manufacturers' cost reductions to reflect the potentially larger market, will prove to be a watershed in establishing a new paradigm for lead optimization.

3.2 Industrial Scale Equipment and Automation Suites

For lead generation, the concentration is very much on numbers, diversity, and productivity. Greatest effectiveness requires the seamless concatenation of matched instruments to address all stages in the process coupled with open and compatible software architecture to facilitate scheduling and data exchange. Because of the higher space and manpower requirements and the increased complexity of materials and data transfer, enhancing throughput by the deployment of large numbers of identical machines is not a preferable option. Equipment aimed simultaneously at high performance, high throughput, and substantial sample size along with a capability for automated sample transfer and scheduling is desirable. Because of their inherent size and complexity these machines will probably be manufactured in small numbers for specialist use only and will attract a premium price. Because of their criticality to productivity, manufacturers able to deliver reliable fast service and support may be favoured over smaller players. Potentially, this could become a ''technology provider'' market where responsibility is devolved to the vendor to provide, by sale, lease, or licence, access to and support for the latest technology.

As the space requirements for industrial scale runs of equipment and associated circulation and storage areas is very large, the normal chemistry laboratory is no longer suitable accommodation. As a result, among others for example, Ontagen, ArQule, SmithKline Beecham, and Merck have sought to build special facilities, ''automation suites'' in which space, service access, ease of maintenance, and continuous operation are placed ahead of normal human considerations. In building these, the opportunity has often been taken to secure an adjacency with other ''industrial'' scale technologies, such as high-throughput screening and automated sample handling and storage, although reagent management facilities are as yet less well developed than the others. For success, these facilities have had to adopt other aspects of "industrial" infrastructure. Thus version control, logging, and planning for component failure and other aspects of reliability engineering that determine the level of equipment and spares redundancy required to insure against showstopper failures are becoming part of the high-throughput chemist's language. However, there is still much to be learned in this area from established industries such as aviation and electronics, where issues such as error control, maintainability, and rapid support are essential to the business.

It will probably not be the function of these "compound factories" within the pharmaceutical industry to compete in the synthesis of the large random collections currently available through compound resellers. In cost terms, the surplus of an item made for another purpose will probably always be too cheap to undercut. The compound factory will make the diversity that cannot be "cherry picked" from these cheap sources—high quality and intellectual property rich novel diversity and optimization material, with a back-up service.

Clearly, for many smaller companies the cost of setting up and maintaining an in-house industrial "compound factory," although essential, may be too great. This unfulfilled need will drive a diversification in the custom synthesis market to provide a comprehensive outsource to companies seeking the strategic and cost advantages of high-throughput compounds. In fact, the number of strategic alliances involving combinatorial chemistry companies has grown rapidly in recent times and now reaches several hundred [30]. By 2002, it is estimated that US$2.6 billion will be spent on combinatorial chemistry and that 45% of that research will be outsourced, This will include the top 10 Pharma companies who will allocate 20% of their R&D budgets [31].

3.3 Human Factors

For maximum efficiency, speed, and cost reduction, routine tasks are best carried out in an automated centralized production environment. That chemistry is late in embracing this concept can be in part attributed to resistance based on the long persistence of a belief among chemists that every synthetic target should be an individually bespoke and synthesized item. However, just as the products of the combinatorial "revolution" have been ameliorated into an adjunct to, not a replacement for, medicinal chemistry and employed only when beneficial, the productivity gains from judicious and appropriate use of automation will eventually endear it to chemists as an important tool in medicinal chemistry.

In reality, the act of synthesis is only one aspect of medicinal chemistry, and it is subordinate to the design of compounds and syntheses, neither of which will be automated in the foreseeable future. Therefore for most medicinal chemists automation will not substantially change their job but simply remove much of the drudgery of routine repetitive and time-consuming processes such as chromatography, and expand the time available for the nonroutine skills they hold (more) important! It is the relative small number of chemists who have had to manage and run an industrial scale high-throughput chemistry facility who have undergone an almost complete reappraisal of their skills and established practices. Apart from the acquisition of substantial skills in engineering, computing, and fault diagnosis, and ensuring continuity of operation through unsociable hours, these staff also have to achieve levels of teamwork, forward planning, and ruthlessness in dealing with exceptions not normally given to medicinal chemists. This has been marked by recent moves to employ staff with relevant nonchemical skills, but the retained presence of chemists has been found to be important and fundamental for the operation.

4 INFORMATICS AND PROCESS MANAGERS

4.1 The Need for Process Management

It is rare today to find an article that has not been made in multiples. For commonplace items, commercial competitivity and success has been based largely on the uniform quality and cost reductions achieved in the mass production of standardized goods. Although automation has played a part in achieving this higher productivity, a more important factor is the use of a cradle-to-grave protocol that defines the total manufacturing process. Although early mass manufacturing scenarios allowed for no variation in the product [32], improvements in information technology and control systems now allow for variable protocols to allow the production of variants on the same assembly line, for example, the various derivatives of the base model of car. Automated chemistry, where the number of different base models (templates) and the number of variations are each huge, presents a challenge in ''simultaneous'' engineering, and manageable protocols are critical to success. A detailed script for the synthesis and/or purification of a large number of compounds must be written, and only then can it be assured that the necessary resources will be available in the right place at the right time.

This task lies beyond a paper-based system, and nothing would be achieved by automation if scientific duties were simply replaced by clerical ones. Electronic management systems have therefore evolved in a variety of forms. Usually these have attempted to encompass the main areas of library design, enumeration, and logistic control of process. How extensive and extendable these systems are is important. Recent times have seen great activity in this area in an attempt to gain better control of the greatly enhanced productivity suddenly available to the chemist. To be robust and efficient, current protocols for high-throughput chemistry have had to differ significantly in principle and concept from those of iterative manual chemistry. The main difference, particularly in early software, was the emphasis on preordained actions because of the limited ability of the system to deal with exceptions. Modern packages show more dynamic flexibility, and the aim is to increase further their reactivity and resilience when faced with unexpected change. However, this must be achieved without undue increases in complexity, as it is important for promulgation of high-throughput techniques through the chemical community that the process manager software remains user friendly.

4.2 Enumerators

Enumerators provide the basic functionality of high-throughput chemistry management by performing the iteration of all members of the target library. These engines were born out of the necessity to help early combinatorial chemists to plan, manage, and write up a large number of compounds. Efficient data management was essential to the high-throughput chemistry process. The accounting needed lay well beyond the chemist's traditional methods of planning, record keeping, and registration. Initially, an enumerator would provide an exhaustive explosion to give the virtual library from which a working library was selected. Implicit in this process was the creation of database cells containing individual compounds and with which other generated or collected data could be associated. Clearly, how far this device could be moved on to link to chemical inventories, export data in new formats for input elsewhere or, say, provide labeling and submission facilities, was driven by need and creativity.

Most of the early process managers treated compounds in a "template and variables" manner, firstly, because this corresponds to the chemist's intuitive Markush representation of theoretical library structure, and secondly, because coding is simplified. More recently, several systems, e.g., Afferent[®] [33] and RADICAL [34] (Fig. 7), have used a reaction-based approach. This aligns more closely to the reality of how the chemistry will be done and the location and nature of the products that might be obtained, thus providing a better correspondence with the synthesis robotics script. Other advantages are being able to relate and retrieve information not just on a structural basis but also on a procedural basis, being able to deal with unexpected and varied reaction outcomes, and being able to cope with multiple products purified from the same reaction. This also helps to reduce human error as, for example, the real reagents will be written into the synthesis rather than defining off-line "clipped" side chains to a core structure that may not all be derived by the same reagent transformation. Reaction-based generation of a virtual library is also better at a technical level. For example, false products will not be generated for a Diels–Alder library—for a dienophile and two unsymmetrically and differently trisubstituted dienes, the reaction-based system will correctly generate the four products (including regioisomers), whereas the generic approach will exhaustively enumerate 144 options, including impossible combinations. Libraries that cannot be easily described by

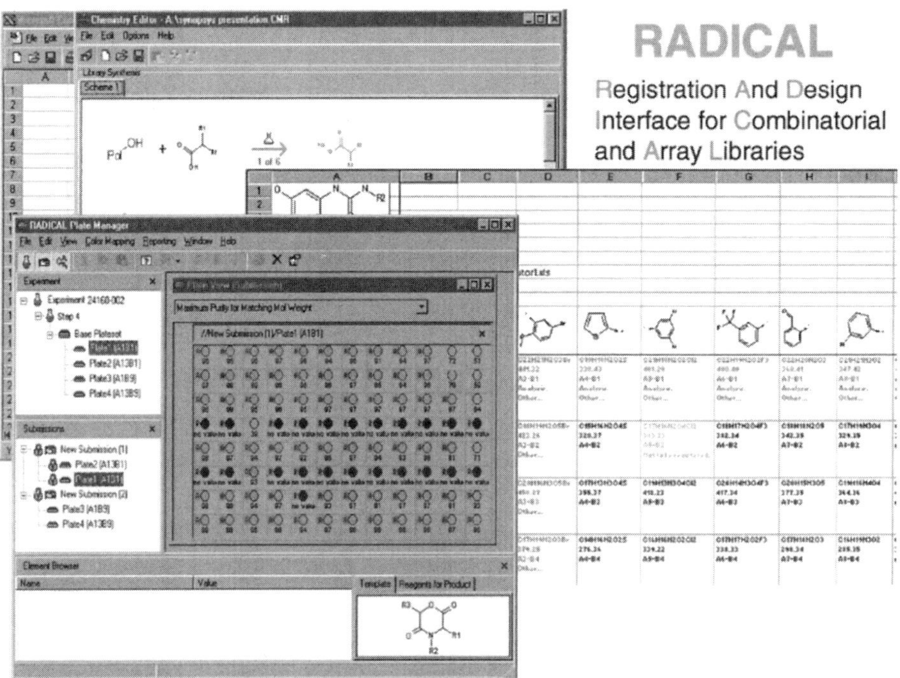

Figure 7 Example of RADICAL reaction-based software. Sample screen shots of the chemistry editor, library enumerator, and plate manager modules from RADICAL (registration and design interface for combinatorial and array libraries), an in-house SmithKline Beecham software work flow management package.

a single template, e.g., a library of Cys disulphide-linked cyclic peptides, which will depend on where and if Cys residues will appear [32], can also be handled.

4.3 Logistic Managers

Logistics managers attempt to move process management beyond the facility of a data repository and source as provided by an enumerator. They provide the methodology for dealing dynamically with nonpredicted circumstances and are usually targeted at stages of the process where human intervention would be error prone because of man's poor multiplexing skills. A recent example is WINNOW (see Fig. 8), which operates within RADICAL to reduce the burden of sample handling during postsynthetic purification by allowing man to select only key fractions on the fly without having to deal with the complex logistics of repeated

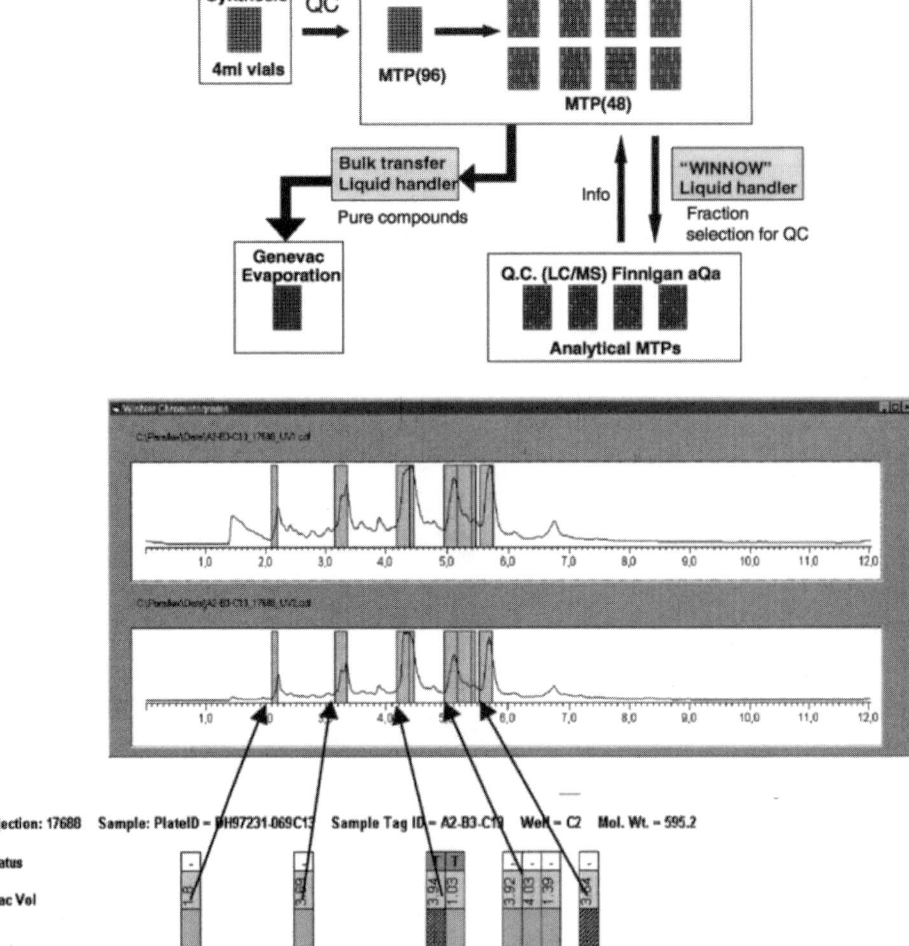

Figure 8 WINNOW fraction prediction software. WINNOW, an excel-based utility developed at SmithKline Beecham, reduces the number of fractions selected for analysis based on absorbance, volume, and LC-MS data of the crude sample. For example, from a library of 4320 compounds giving rise to 26,191 fractions, WINNOW allowed the selection of 7093 for further analysis by LC-MS to give 4012 submittable compounds (93%), of which > 90% had > 95% purity. The exercise also required 72 h of machine time, 350 L of solvent, and 1 man.

plate mapping and sample reformatting. LC-MS data for the crude samples are used to guide fraction collection. In addition, fraction collection criteria (slope, threshold, and wavelength) can be adjusted to allow "intelligent fraction collection," which reduces the number of fractions collected. Because of the high discriminating function of the human eye and brain and the inherent sample management benefits, routines such as WINNOW are competitive and more reliable than complex automated solutions such as mass triggering.

4.4 Scheduling

Enumerators and logistics managers deal with data in a device independent manner. The control of the physical devices has largely been achieved by individual programming using each device's native software. Accordingly, the transfer of data between devices has been manual and often involved reformatting. Recent software, such as ACE (Automated Chemistry Environment) [35] (Fig. 9) can be layered above the vendor software and hide the complexity from the operator. In effect, it enhances the functionality of the process manager suite by providing central scheduled control of all the items of third-party equipment that might be invoked in the high-throughput process and automates data transfer between them. To cope with the variety of hardware that might form part of the environment and the frequency of upgrades, the core of the software is device independent, and individual pieces of equipment are addressed via "agents." As with process managers, to promulgate automation successfully there is a need for simplicity at the interface. In ACE this is achieved by "zoning" so that a user can limit the view to their immediate environment; dependencies on the extended environment are not dropped, just unseen unless there is conflict.

4.5 Shared Information

Informatics software, combining the functions of enumeration, logistics management, and scheduling, provides an excellent platform for storing, sharing, and using information. Ultimately it might not just provide an alternative to the notebook for chemical record keeping but potentially be one of the most effective productivity tools for medicinal chemistry. Thus, while many of the truisms implied in Fig. 10 have always been known in the microcosm of a small discovery chemistry program, only recently have there been tools available to implement them on a universal scale. Currently, discovery chemistry is carried out in three distinct modes. High-throughput chemistry has been used principally for lead generation; low-throughput parallel synthesis of arrays and manual iterative chemistry, when only one compound is required or parallel synthesis is prevented in some way, have been used in mixture for lead optimization. Although lead generation material can be admitted to screening from external sources, the cycli-

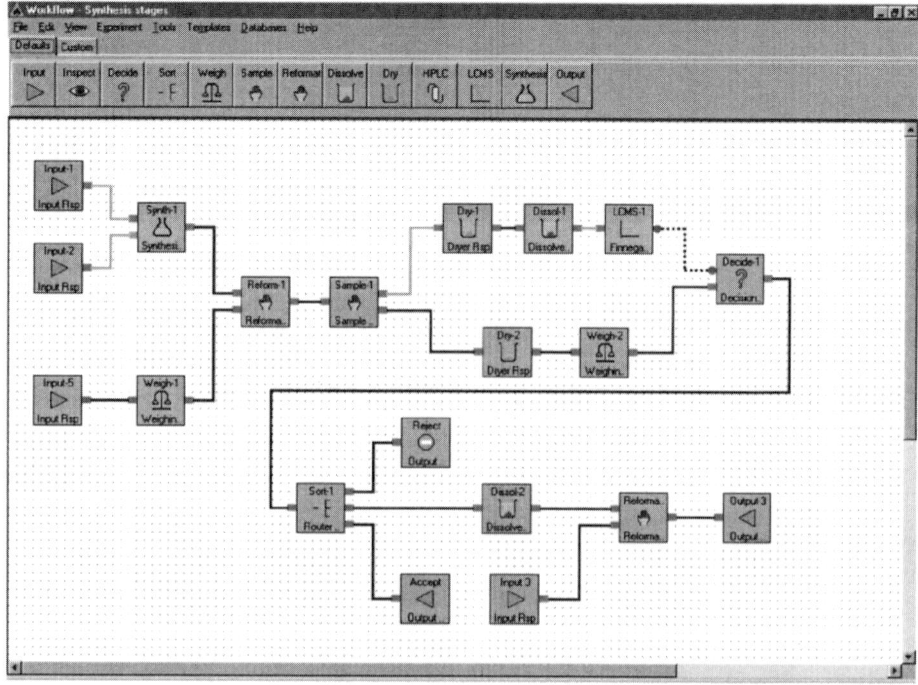

Figure 9 Automated chemistry environment (ACE) software is a product of the The Technology Partnership (UK) currently under trial at SmithKline Beecham Pharmaceuticals. The package allows iconization of automation tasks and effects transparent control and data transfer between a variety of third-party equipment. (From http://www.ttpgroup.co.uk/ttp/software.)

cal structure of the internal process holds many synergies. For example, by sharing information, the asset of having a validated synthetic procedure for preparing even a small late-stage optimization library, with good bioavailability and developability, can be maximized by enabling high-throughput expansion (perhaps of the predicted less active members to avoid future selectivity problems) to provide lead discovery stock. The synergy becomes even greater if software and hardware-compatible devices such as the Myriad MCS and PS or the Biotage Parallex and Flex are in use. Of course, the reverse is also true, that an optimization chemist might find a prevalidated method within the shared procedure bank.

The more types of data and the more universal the input to the process manager from chemists, the greater are the potential benefits of database processing and mining. In addition, powerful linkages can be made to other aspects

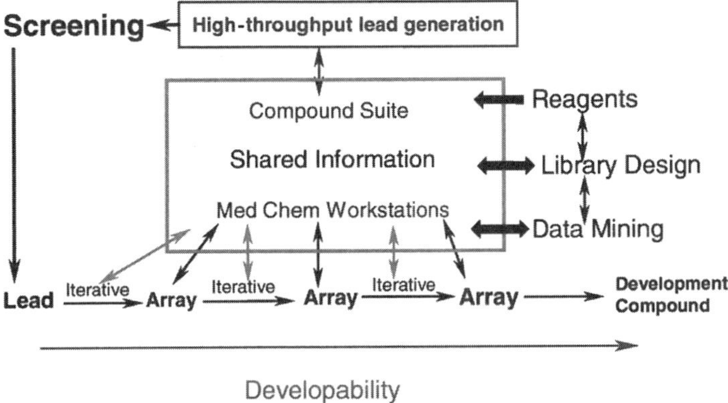

Figure 10 Main features of a shared information system for lead generation and lead optimization.

of the medicinal chemistry process by bilateral porting to other software suites such as library design and screening data mining. For example, a chemist contemplating a synthesis can then query from the same point potential synthesis scripts, failure rates of individual reagents in different procedures, and any screening data for similar compounds. More useful will be the development of advanced systems in which internal agents preemptively identify any relevant patterned behavior. Clearly, in order to get information from a shared information system it is first necessary to spend time inputting data and a query. At present the payoff for the chemist performing iterative chemistry is marginal, but undoubtedly use will increase with the richness of the data source, leading ultimately to the complete transfer of chemical record keeping to a searchable electronic format.

The processing of shared information can also provide useful system management tools to assist in error avoidance (such as highlighting a recent and persistent failure of a particular reagent) and in optimizing queues to achieve efficient processing. An important link is that to the reagent inventory. Without being restrictive, a library design system that minimizes the number of reagents used in routine exploration of diversity not only has advantages for economy and simplicity in reagent management but also assists in data mining by concentrating data. However, the more important function is to maintain an inventory and ensure a seamless supply of reagents including proprietary diversity.

4.6 Learn As You Go

Shortening the times for the optimization of a reaction or the optimization of biological activity in a compound series has impact in the key areas of productivity and cycle time. Although the problem can be handled using statistical methods, particularly by the use of response surfaces, a readily implemented and functional software package does not appear to have appeared. Perhaps the problem is the paradox that if the library members are to be very diverse, each will need to be prepared singly for optimization! The additional reaction monitoring equipment required is also a deterrent.

With the advent of efficient process systems it has been possible to reliably manage the practical complexities of the mixing and multiplexing of a continuum of libraries to implement learn-as-you-go systems such as those described in Fig. 11. In these schemes the synthesis of each library is extended temporally to allow data to be obtained for early members to be used to improve later members. Thus the feedback of a stream of QC data can be used to optimize reaction conditions (or even avert disaster), particularly if the variables in early library slices are well chosen to sample reaction parameters, and in favorable cases prevalidation of the reaction may be minimal or unnecessary. In a similar way, screening data from early library slices containing well-chosen probes of property space can be used as a pathfinder for the design of later slices. Overall, efficiencies accrue from the earlier steering and a reduction in the number of compounds required to achieve the goal. This dynamic retargeting approximates more to the information rich rational searches of traditional iterative medicinal chemistry than single

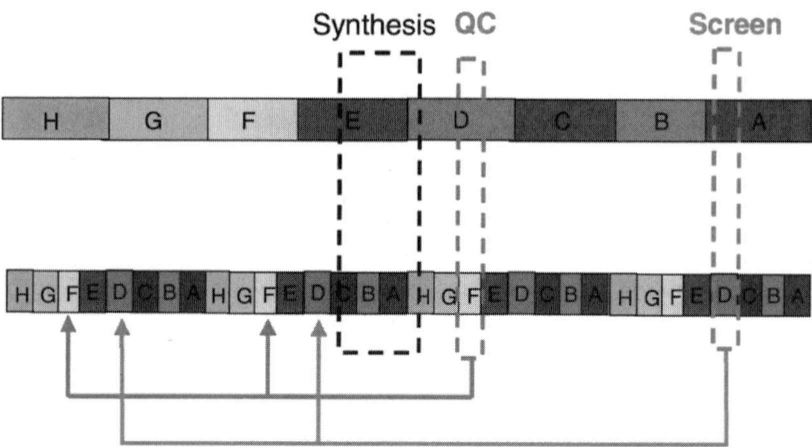

Figure 11 Learn as you go. The horizontal axis is time. By splicing small sections from different libraries from the task list, feedback of the results from early members in analysis can be used to optimize the conditions for preparation of later members of the same library (or even make changes to the planned library). Screening results may be similarly used to enhance the SAR relevance of the library.

session parallel synthesis of entire libraries. In fact, contemporary software and high performance automation may be approaching the point where the "slice" could be an "array of one."

5 NEW CHEMISTRY FOR AUTOMATION

5.1 Background

The development of automation suites and the manufacture of large libraries has largely been the domain of pharmaceutical companies or privately funded combinatorial companies. Adventures in these areas largely have proved too costly (and too repetitive) for academic chemists who instead have tended to concentrate on novel chemical methodology for combinatorial and high-throughput techniques. As a result they have brought forward a series of new synthetic tools and procedures important not only to high-throughput technologies but also with significant impact in defining modern manual chemistry techniques. Against this background, there is nothing to be gained by seeking exclusively a complex mechanical solution to the performance demand of high-throughput chemistry if an alternative chemical strategy or a different choice of reagents will more readily ease the problem.

5.2 Polymer-Supported Substrates

One of the earliest easements to high-throughput chemistry was the use of a solid-phase support for the reaction substrate. By retaining the molecule of interest on a plastic bead, large excesses of reagents could be used to drive reactions towards completion, and product recovery was achieved by a simple filter wash. Although this enabled simultaneous synthesis by "mix and split" strategies not usually viable in solution, such very high reaction and vessel efficiency could be achieved that the process was hardly worth automating. In fact the major inefficiency lay in the long and necessarily manual validations required to ensure that a screenable mixture was obtained. The criticality of manual validation was less, but still substantial, if "mix and split" was carried out using directed sorting ("tea-bag") protocols, such as the IRORI Accutag™ technique, as rescue of impure products was possible. However, the total postsynthetic loss of vessel efficiency with each product requiring individual work up and (often) purification was an issue for manual handling and demanded automation. Library synthesis in full array format (i.e., with one compound per reaction vessel) lacked not only vessel efficiency but also reaction efficiency and benefited even more from automation. Although in this last format the use of polymer-supported substrates is not a requirement, the prospect of carrying out a long linear synthesis without a polymer-supported substrate is probably beyond the capabilities of simple ("wash and filter") robots.

Despite its usefulness in enabling useful high-throughput methodology, the use of polymer supported chemistry suffers severe drawbacks in efficiency and scope. Time must be taken to configure a new solid-phase reaction or to transfer a known solution-phase reaction to the solid-phase, and the process is not always straightforward or successful. In addition, any chemistry that attacks the bead, cannot access it, or is otherwise incompatible with the substrate is unsuitable, and as a result the armory of solid phase reactions available to the chemist is only a small fraction of those available in solution. Thus polystyrene is a much-favored support for the synthesis of potential drug candidates, as its good swelling properties allow high loading and its hydrophobic characteristics suit the synthesis of bioavailable compounds. However, reactions that attack phenyl rings or benzylic protons, involve substantially insoluble components, or require very high or very low temperatures, are unlikely to have a successful outcome. To some extent these deficiencies would be ameliorated by the identification of alternative, more robust high-loading media, and this is already an area of intensive effort but slow progress. Acrylic and acrylamide resins or possibly even silica-based materials may provide a way forward. However, a seminal finding in this area is likely only to partially redirect the current efforts to advance automated solution-phase synthesis, as the majority of medicinal chemists do not work on large libraries for lead generation but on small, highly specified and novel libraries for lead optimization. For them, the overheads of attaching the substrate to the resin, which may involve identifying a new, optimally traceless, linker and developing and validating a solid-phase reaction will be rarely time-optimal.

5.3 Polymer-Supported Reagents

For obtaining a clean and easy isolation of a reaction product, the use of polymer-supported reagents rather than a polymer-supported substrate has many advantages. The need to find a satisfactory linker and to bulk load the substrate onto the resin is avoided, and no cleavage is required to liberate the reaction product. In addition, although the resin matrix may suffer the same reactivity liabilities as substrate support resins, they are usually less severe because of the shorter postreaction exposure, and any damage to the resin does not necessarily impair later steps in the procedure. However, the biggest advantage of all is that once the problems of generating a new polymer-supported reagent have been overcome it accrues as a permanent asset to be used in the construction of many libraries, rather than a transient tool used to assist the preparation of a single library. The only significant disadvantage is that because the material selectively removed as solid phase is not the target material and removal may be incomplete, impurity accumulation in a long linear synthesis may be greater than with a solid-supported substrate approach. The solid-supported materials can be used as reactants per se or as scavengers for unwanted reaction components in solution. An elegant demonstration of both modes of use can be seen in the synthesis of maritidine analogues by Ley et al. [36] (Fig. 12).

Development of new supported reagents has been driven hard not only by the needs of high-throughput lead generation but also by their acceptability as useful tools for low-throughput parallel synthesis and for manual chemistry. As a result, the range of polymer-supported reagents available has increased, though it is still not very extensive. Polymer supported catalysts are even less frequently encountered but are more useful, as they do not suffer the loading restrictions of polymer reagents (Fig. 13) [37]. The microencapsulation techniques described by Kobayashi and Nagayama [38] for the preparation and stabilization of polymer-supported catalysts—physical envelopment of the catalyst within the poly-

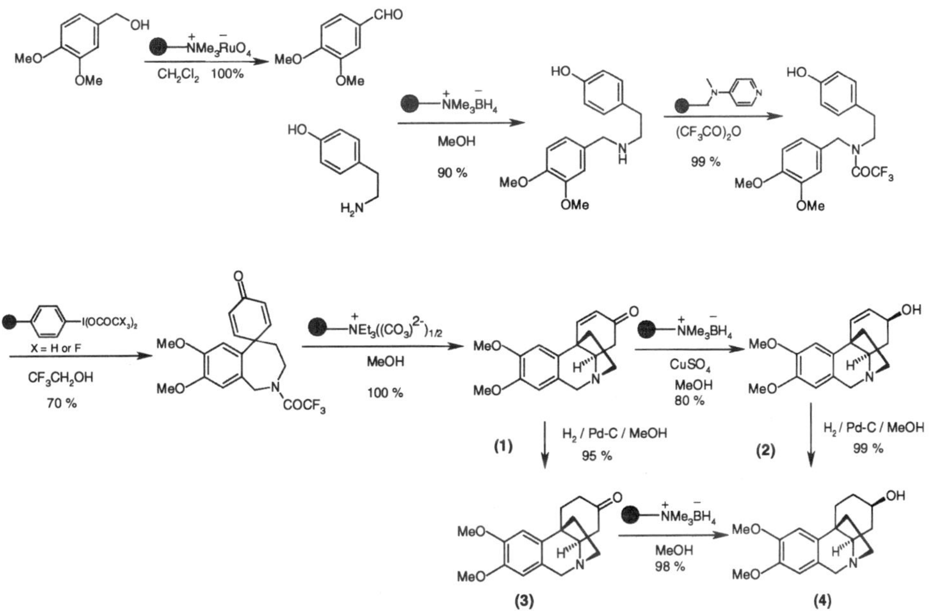

(1) (±)-oxomaritidine
(2) (±)-epimaritidine
(3) (±)-dihyrooxomaritidine
(4) (±)-dihydroepimaritidine

Figure 12 Synthesis of maritidine alkaloids using polymer supported reagents. (1) (±)-oxomaritidine, (2) (±)-epimaritidine, (3) (±)-dihyrooxomaritidine, and (4) (±)-dihydroepimaritidine. (From Ley, S. V., Schucht, O., Thomas, A. W., Murray, P. J. Synthesis of the alkaloids (±)-oxomaritidine and (±)-epimaritidine using an orchestrated multi-step sequence of polymer supported reagents. J. Chem. Soc., Perkin Trans. 1, 1999, 1251–1252.)

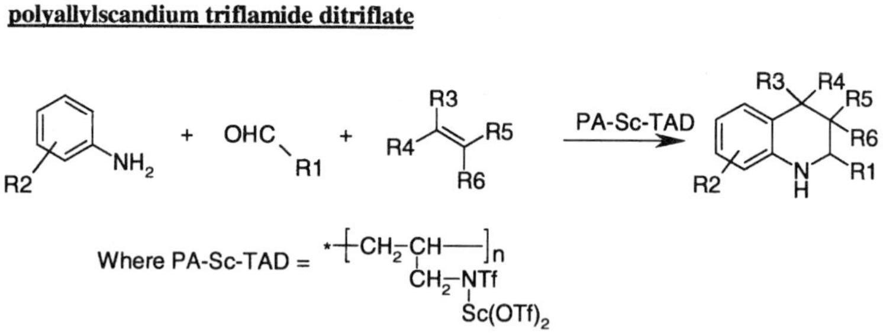

<u>polyallylscandium triflamide ditriflate</u>

Figure 13 Multicomponent coupling reaction to give tetrahydroquinolines catalyzed by polyallylscandium triflamide ditriflate. (From Kobayashi, S., Nagayama, S. A new methodology for combinatorial synthesis. Preparation of diverse quinoline derivatives using a novel polymer-supported scandium catalyst. J. Am. Chem. Soc., 1996, 118, 8977–8978.)

mer matrix—may be more broadly useful. Similar entrapment of reagents, if this can be achieved with useful capacity, may be a way of extending the range of available reagents.

5.4 Chemistry to Fit the Purpose

The heroic syntheses of classical chemistry have usually long, elegant, enantio- and regio specific syntheses of a complex natural product, carried out manually often by teams in relay. The purpose was usually to confirm structure and provide an academic platform for the development of reagents with exquisite functionality or selectivity in the production of a single specific product. In contrast, the aim of high-throughput chemistry is usually the most economical controled creation of the maximum amount of diversity possible, and in optimization chemistry it is usually disadvantageous to synthesize a complex molecule where a simple one will do. Thus, although automated chemistry cannot yet emulate the classical masterpieces, for most practical purposes the automated parallel synthesizer has the potential massively to outstrip manual productivity and does so in many cases. What is required to increase its usefulness is an expansion of chemistry that is geared to the efficient production of diversity, and this requires significant change in the emphasis of organic chemistry research. Fortunately the pendulum of academic respectability (and funding) is aligning more and more with the needs of modern high-throughput chemistry, even to the extent of setting up special "combinatorial" centers.

An obvious target for academic research is to broaden the range of solid-phase aids to synthesis along the lines already indicated. This will provide benefits to manual and automated chemistry whether for lead generation or lead optimization. In addition, work to enrich the armory of economical diversity creating reactions that have been variously described by terms such as "multigeneration," "multicomponent," "cascade," "domino," etc. [39] will contribute importantly to lead discovery libraries. New strategies that develop the power of high-throughput and combinatorial techniques will also be sought. An interesting recent example for developing leads (which may owe some of its parentage to the SAR by NMR techniques described Abbott Laboratories [40]) is Ellman's approach to target-guided ligand assembly Fig. 14 [41].

From the commercial world a growing contribution to the high-throughput systemization of chemistry is expected. It is unlikely that this can be developed to the extent seen for biochemistry or molecular biology, where there appears to be a kit for everything, because synthetic organic chemistry is less predictable

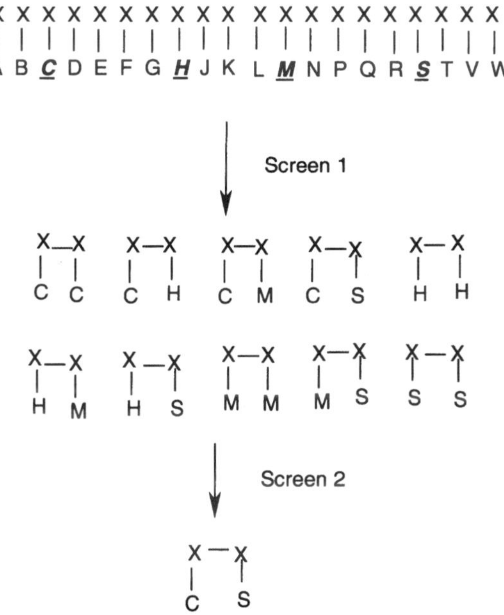

Figure 14 Combinatorial target guided ligand assembly. The aim is to combine structural elements that have relatively low affinity for the target into ligands of high affinity. A set of potential binding elements is prepared with a common chemical linkage X. These are screened (Screen 1) to find which bind the target (not necessarily at the same binding sites). All possible combinations of binding elements are then prepared and screened (Screen 2) to identify the highest affinity ligand. (From Maly, D. J., Choong, I. C., Ellman, J. A. Combinatorial target-guided ligand assembly. Identification of potent subtype-selective c-Src inhibitors. Proc. Natl. Acad. Sci. USA, 2000, 97, 2419–2424.)

in its needs. However, there is substantial scope to improve on the apparently random availability of reagents and packaging currently seen, and this becomes a necessity if automated handling of reagents is to be instituted.

5.5 Chemistry and Procedure Databases

It is likely that automation will create its own cache of specialized chemistry. Already there are numerous examples to be found where the preferred automated procedure for specific synthesis is quite different from the earlier reported manual protocol. Clearly, as more automated chemistry is performed, this knowledge base will grow. However, unlike manual chemistry, where the experimental method is usually retained as a condensed paper record suitable to guide a person "skilled in the art," a synthesizer run creates a storable electronic record of every detail of the process. The opportunity therefore exists to create procedure libraries that can be rerun automatically without further development or interpretation if further supplies are required, or with minor modification to provide analogues. As the synthesizer market develops and standardizes, trade in validated procedures, perhaps coupled with proprietary building blocks, will probably develop.

6 LONGER TERM OUTLOOK

6.1 The Need for Greater Intelligence, Speed, and Economy in Chemical Synthesis

The success of pharmaceutical companies has been based largely on their ability to discover new chemical entities. Unless there is an unexpected shift to a predominately biology-based strategy such as protein therapeutics or gene therapy, proprietary chemistry will probably persist as the main arena for pharmaceutical competition for many years. However, new drugs are being displaced by "fast followers" of the same or different chemical class at an ever-increasing rate (Fig. 2). Therefore the essential target for drug discovery has been to improve the thoroughness and speed with which the "best" chemical lead for a biological target can be identified and developed into the "best" drug in a cost-limited environment.

The derivation of most drugs has involved an element of (often overpuffed) rational design and a nonrational component of either deliberate or (seldom admitted) accidental serendipity. Historically, iterative medicinal chemistry approximated to serial cycles of compound synthesis and bioassay as this permitted near maximal incorporation of the immediate informational payback and accumulated SAR into the design of the next compound. However, the cycling rate was slow, and the main reason for the focus on rationality was actually to minimize the number of compounds and therefore the time required to identify a development candidate. As optimization started often from a generally known lead, differentiation of competitor products was often marginal. High-throughput chemistry has changed this paradigm as, based on a single act of modeling and probably far less accumulated SAR knowledge, a library of compounds is designed and committed for parallel synthesis. Assay results are then received en bloc and processed in one session to create or update SAR. Despite the apparent reduction in rigor, the good news is that the increased speed and the inherent serendipity of this process has led to advantageously reduced optimization times. However, the increasing frequency that the starting point for optimization is a nonintuitive lead from high-throughput screening should be a cause for concern, as there are many clues that there are severe diversity shortfalls in corporate collections. Thus there is failure to find a satisfactory lead for a high proportion of targets, the occurrence of competitor leads from a different structural class and the discovery of recently synthesized nonintuitive leads in cross-screens containing earlier screened targets. Therefore the emergence of competitor leads with very different structures for the same biological target, rather than improving the prospects for product differentiation, implies a potentially costly late-stage failure for all but one of the competitors. Even then, there is no guarantee that the particular selection of compounds available in the armory of the successful combatant actually contains the best lead, and future displacement from a market position is still a possibility.

So, what are the implications for future automation? Better optimization of the ''best'' lead (i.e., at least the right structural class) would not require a significant change in the current aims for the development of high-performance, low-throughput automation for optimization. It is likely that optimization chemists would aim to improve rational design, as the profligacy of serendipity would be too demanding of specially synthesized intermediates (rarely available in bulk without diverted effort) and relatively slow secondary and tertiary assays could end up being grid-locked. However, the probability of the ''wrong'' starting lead (i.e., the wrong structural class) being optimized to deliver the ''best'' development compound is very low and is unlikely to be enhanced by further developments in the field of chemical automation directed at lead optimization. The greater need is for automated technology that will enhance successful lead generation.

In reality the search space to find a drug is impossibly large to exemplify; a truly ''universal'' library would be impossible to make, store, or screen, as the number of possible compounds would always exceed the number of atoms available to the chemist. Traditionally, medicinal chemistry has steered itself around the numerical impossibility of exhaustive random synthesis as a method to find drugs by the application of intelligence. In optimization, despite the enormous potential search space, a lead can usually be translated into a potential development compound by the synthesis of about 10^4 compounds. However, chemists are not usually very successful in creating leads de novo and hence rely heavily on high-throughput screening of incomplete collections to identify nonoptimal lead candidates. Currently there appear to be two approaches to improving the likelihood of identifying a better lead. The first is passively to accumulate compounds or mixtures that become available either for purchase (suitably filtered!) or from internal optimization chemistry output. The second, more active, approach is to create lead discovery libraries according to some plan (e.g., synthesis of prejudiced structures or strategic diversity "hole" filling). However, with both of these approaches, until the materials have been subjected to screening for some considerable time, it is impossible to discern whether they are valuable lead generators or screening dross. Simply increasing the numbers made could make screening less efficient. What would be better for lead discovery, if it can be achieved, is an approach in which the ultra-high-throughput synthesis of diverse compounds is dynamically steered by a close-coupled high-throughput assay. Thus the results can be used on the fly to direct progress intelligently through property space towards hot spots of activity. These can then be selectively exploded, perhaps using a different synthesis technology, thus expediting novel chemistry while reducing dross. A suitable "indicative" assay could be as simple as a "developability" assay such as cell penetrability, as a practical and more accurate alternative to the current use of theoretical developability criteria in the design of unbiased libraries. Alternatively, it could be an activity assay based on a representative of a target class whose ligands tend to show cross-reactivity (GPCRs, kinases) or even a particularly recalcitrant (but high-value) target that has failed to find a lead in screening. In the remaining sections, some potential techniques for fast "prospecting" are examined.

6.2 Important Factors in Faster Prospecting in Drug Discovery

The basic schema for the drug discovery process is shown in Fig. 15; clearly, faster cycling will speed the conversion of reagents to products. The problems raised would be how the time courses for the physical processes can be compressed; how the increased demand on external feeds (such as reagents for chemistry and screening) can be alleviated; and how an "intelligent" process can be maintained. For a large number of other types of technology, including computing, data storage, car engine and collision management systems, entertainment media, etc., the solution for analogous needs ("faster, cheaper, better") has been to miniaturize, integrate, and automate. Material consumption is thereby reduced, direct or close coupling enhances material and data transfer, and the uniformity of execution of a process that may lie outside the range of human skills or competence is assured. It seems likely that the issue for chemistry is not so much whether it will undergo a similar change—Moore's law [42] seems universal and

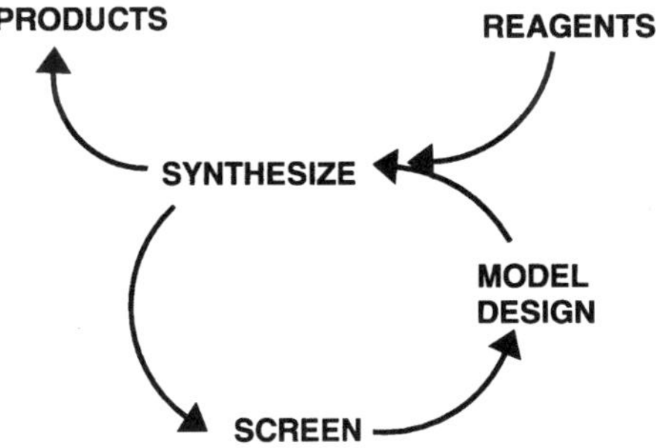

Figure 15 Schema for the drug discovery process.

inexorable—but how the changes will be achieved, because often a shift in the technology or operating principle has been required. For example, in electronics the profligacy with which thousands of active devices can be replicated on a small area of an integrated chip allows the complex multitasking of active devices invoked in integrated macrocircuitry, to contain component count, size, and cost.

In chemistry, simply making the device smaller would be expected to speed up homogeneous reactions used in chemical synthesis or screening. From the approximate diffusive mixing times for various aqueous volumes (Fig. 16), it can be seen that in very small volumes the rate of molecular mass transfer is dramatically increased. However, the time required for the transfer of electrons to effect a reaction is shorter still, so the time required for a reaction to complete or achieve equilibrium is determined mainly by the mass transfer time. There appear to be few disadvantages from miniaturizing chemistry to increase speed and reduce

Volume	Mixing Time (sec)
1 L	10^7
1 mL	10^5
1 µL	10^3
1 nL	10
1 pL	10^{-1}
1 fL	10^{-3}

Figure 16 Diffusive mixing times for various volumes.

the reagent used per reaction as the amount of compound now required for high-throughput screening (where the benefits of miniaturization have already been extensively exploited!) is exquisitely small. In fact, miniaturized chemistry has been with us for some time in the form of pins and crowns and mix 'n split beads, where each bead can be considered to be a minireactor. However, the potential for speed gains have not been realized with these techniques due to the rate-limiting diffusion time of solvent into the solid phase during reaction and the extensive validation and pre- and postsynthesis effort required.

A further attraction of the miniaturization is the very high density of vessels that can be achieved, thus providing a prospect for time compression gained from reduced robotic travel. Thus $2,500 \times 1$ nL or $250,000 \times 1$ pL volumes can be arranged per cm^2. Based on diffusion times, the maximum rates at which chemical syntheses or analyses could be performed would be 250 analysis results per second and cm^2 for the 1 nL wells, or 2,500,000 results per second and cm^2 for 1 pL wells. However, beyond a certain point the gains to be made by using such massively parallel systems can be illusory. High-throughput screeners struggle to dispense and manipulate aqueous aliquots reliably to 3456-well plates, and Sarnoff/Orchid's attempt to carry out chemistry in a miniaturized well-based synthesiser (Fig. 17) [43]. also experienced difficulties with the complexities of liquid transfer required. The poor behavior in both these systems was due in no small part to the increased dominance of surface effects over bulk effects in miniaturized systems and illustrates the importance of working with, rather than against, the characteristic properties of a chosen scale.

As interfacing micro devices to a macro world for the transfer of minute amounts of material is so notoriously difficult and inefficient, it is likely that effort will be made to integrate as many aspects of the drug discovery process as possible within the same device. An early example of this can be seen in the Orchid Biocomputer where, on completion of all of the subprocesses required for reagent management and chemical synthesis, the product is delivered directly into a screening plate. However, in a different form, the integration of synthesis and screening might serve other goals than simply enhancing the speed and ease of material transfer, a key one being that of improving the rate at which assay results can be used to steer and refine chemical design. At one extreme this aspiration can be seen, for example, in Screentec's pragmatic implementation of a screening module as part of the commonplace analytical equipment found within the synthetic chemistry laboratory (Fig. 18), to help speed the chemist's response to biological data. At the other extreme are the attempts of, for example, Huc and Lehn [44] to use evolutionary methodology in which the biological target acts as a template for its own ligand.

The final requirement for an effective miniaturized integrated device is for an automated process. Clearly, programs that control the functional hardware and script the actions to perform a synthesis or an assay will be required. However,

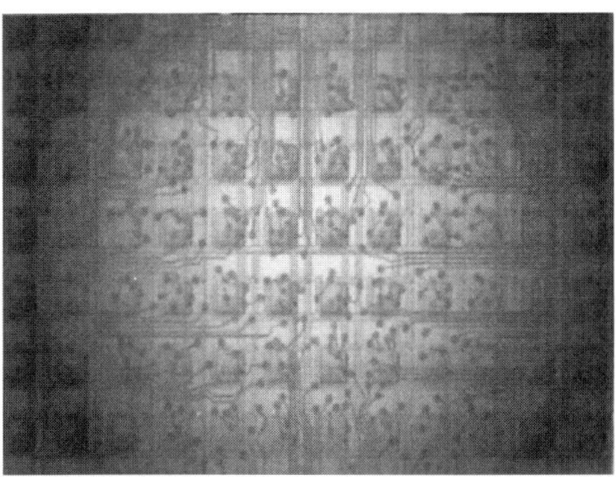 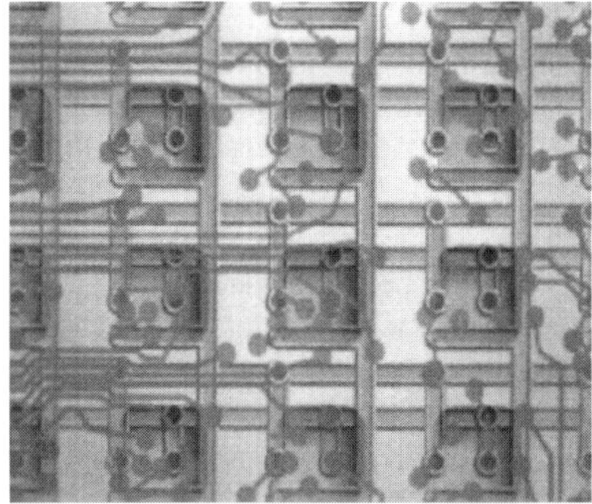

Figure 17 Sarnoff/Orchid synthesizer. Orchid microfluidics chip: multilayered devices consisting of arrayed networks of liquid reagent flow paths in channels or conduits. These chips allow the processing of sequential and/or parallel reactions. The reagents conveyed in the conduits and delivered to the location of the reactions can range in volume from nanoliters to milliliters with a typical reactor volume being from 100 to 800 nanoliters. A variety of materials are used to create the chips, including glass, silicon, and polymers, and the structures are typically flat and layered to create the desired three-dimensional structures with the required network of fluidic channels in the upper reusable portions and an array of reactors in a consumable lower portion.—Proprietary valving technology relies on a capillary break, which halts the reagent flow at a defined expansion point in a fluidic channel. Electrodes in the channel create simple pneumatic or hydraulic pressure and elec-trohydrodynamic pumping. Once the flow is halted, pressure or electric current can reini-tiate the flow. (From http://www.orchidbio.com.)

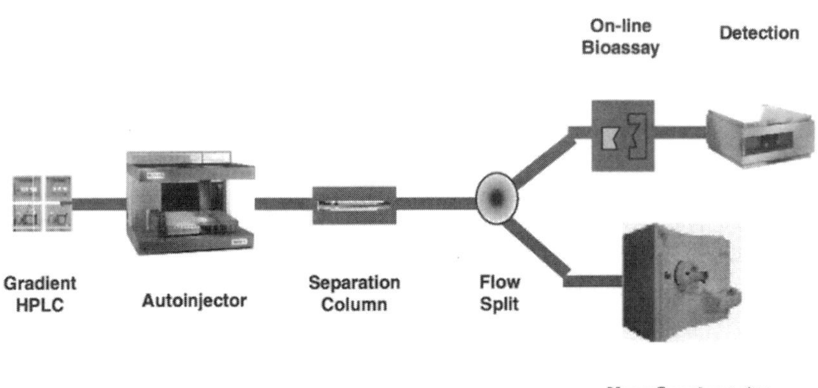

Figure 18 Biochemical assay integrated as postcolumn reaction detection system (Screentec B.V.).

in a fully automated miniaturized system that integrates synthesis with screening, some form of intelligent control will also be needed. Currently, a knowledge-based system in which cumulative SAR, built up as assay data is collected, is used fully to design the next structure, as might be used for low-throughput iterative chemistry, is an unlikely possibility. This is not only because of the current inability to code such a system to run without human intelligent input and management, but also because as an open-ended design tool it would give rise to an impractical demand for unavailable chemistry and reagents, thus preventing any systemization. However, based on the known success of array chemistry, the way forward might be a stochastic "intelligent" system, operating within a finite range of processes and reagents ("library synthesis"). What miniaturization and integration can provide, because of its extremely sparing use of materials, is the potential to access a huge set of reagents without invoking impossible replenishment logistics. If the reagents were categorized in terms of their many relevant properties, they could be selectively and "intelligently" drawn for the serial synthesis of members of a sparse array, leading economically to a "best" member. Perhaps the working algorithm could be something as simple as the unsophisticated, reflexive, but fast automated algorithm shown in Fig. 19. After all, once the published fine words and post hoc rationalization have been stripped away, this has probably been the working paradigm of many successful structure-based optimization programmes. Failure to find a lead or a "best" lead was probably due to too few iterations being performed.

As an illustration, the approximately 1,800 available diamines could undergo reactions with the 27,000 acylating agents, halides, and carbonyl compounds in ACD to give rise to about 10^{13} products. It is also known that activity can be determined by as little as the presence or absence of a methyl group, but there are no reliable rules that allow prediction of this phenomenon, so it is desirable to make all of the possible compounds. Using a macrosynthesizer, screenable amounts (say 1 ng) of all 10^{13} products could be made using on average <10 g of each reagent. Technically this is possible—dispensing small volumes of up to 27,000 solutions on demand is well within the scope of known technology—but the time required to complete the task is too great. However, with an intelligent "make and test" algorithm to steer fast serial synthesis to the "best" goal by the selective use of reagents, the number of compounds required might veer closer to the 10^3 to 10^4 usually made in a medicinal chemistry optimization campaign. Overall, this might provide a better overall combination of intelligence and numbers.

It is probable that improved chemical computation and modelling will yield better methods to direct synthesis towards an ideal ligand. However, if the test for usefulness is the ability to deliver a few paradigm-shifting nonobvious leads in the same context as some reassuringly intuitive candidates, it seems unlikely that the whole process can be conducted in silicon. For example, how could this

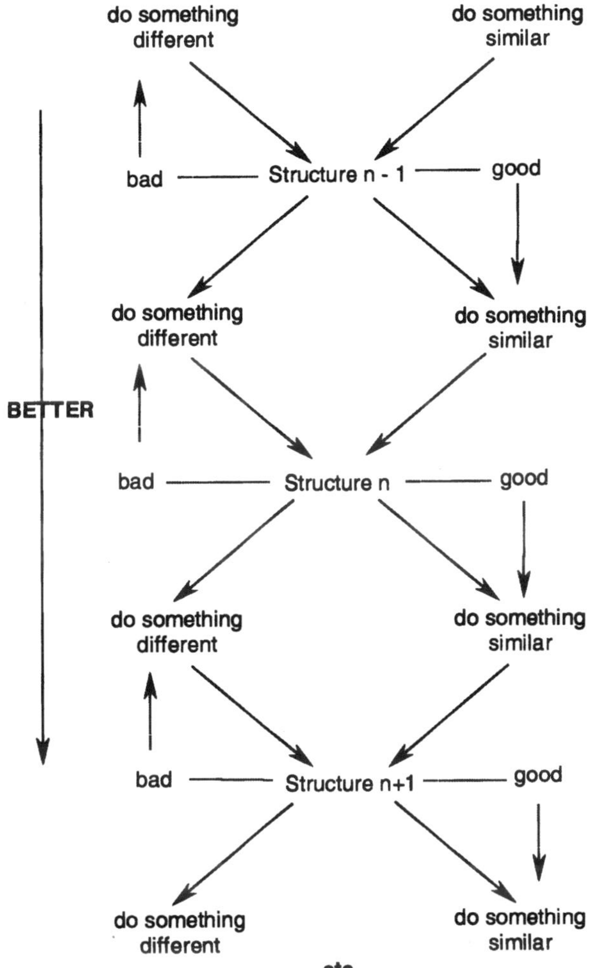

Figure 19 Rudimentary algorithm for dynamic lead discovery. Reagents to give variables on a library template are categorized in various ways (size, lipophilicity, length, etc.). Structure n tests a change in one of these factors. The assay outcome determines what the next operand and process will be. A ''good'' assay result (i.e., higher activity than shown by compound $n - 1$) will determine (dotted lines) that the factor tested via structure n is emphasized further in structure $n + 1$. A ''bad'' assay result for structure n will cause a return to structure $n - 1$ and a different factor to be investigated. Where there is no clear result, a die is thrown.

methodology be verified other than by synthesis of the compounds? In addition, it is more likely that this sort of advance will come from a better understanding of topomeric fields [45] than of neighborhood behaviour [46], as the latter is heavily dependent on similarity comparisons. For the moment, however, the neighborhood behavior implicit in Lipinsky's rules [47] and their derivatives provide a useful way to reduce the diversity space ''in play'' for drug discovery.

6.3   Biological Techniques

Of course it can (and will) be argued that molecular biologists have already blazed a trail in dealing with search algorithms involving extremely large numbers, exquisite selectivity, and even product amplification without invoking a computer, and as usual, it is chemists who lag behind. Phage display [48] and Selex (Systematic Evolution of Ligands by Exponential enrichment) [49] are examples of these adaptive molecular evolution techniques.

Inserting a DNA strand that encodes a peptide into a phage genome causes the phage to display that peptide on the terminus of a selected coat protein. Thus a bacteriophage can be made to display a library of up to 10^{10} different peptides or proteins at one per phage. Typically the library is screened against an immobilized target and unbound or weakly bound phage are washed off. Bound phage are then released, eluted, and then replicated for the next round of screening, and this process is repeated three or four times. Clonally pure phage isolates are then selected for DNA sequencing to give, typically, a number of DNA sequences that encode for structurally closely related peptides. The free peptides can then be synthesized using recombinant or chemical techniques to act as a lead for small organic molecule synthesis.

In a similar paradigm, a variation of parallel SELEX technology uses an extremely large library (10^{15}) of RNA molecules attached to a set of small organic reactants ''A'' as a source of catalysts to facilitate reactions with a set of small organic reactants ''B''. The library is then screened and the minuscule amounts of any A–B ligand bound is recovered, and the RNA tag is amplified using indirect PCR techniques to provide a catalyst that will selectively catalyze the formation of the small molecule ligand in an ''A'' + ''B'' mixture (Fig. 20).

Although highly attractive and exquisite in their operation, these methods do not directly yield molecules that would normally be classed as developable. Notably, although both techniques have been known for over 10 years, there is still no evidence of their successful application in the discovery of developable drugs.

6.4   Randomization Techniques

Natural product screening was seen as a method for exploiting the enormous amount of diversity provided by nature. However, the diversity had to be accepted

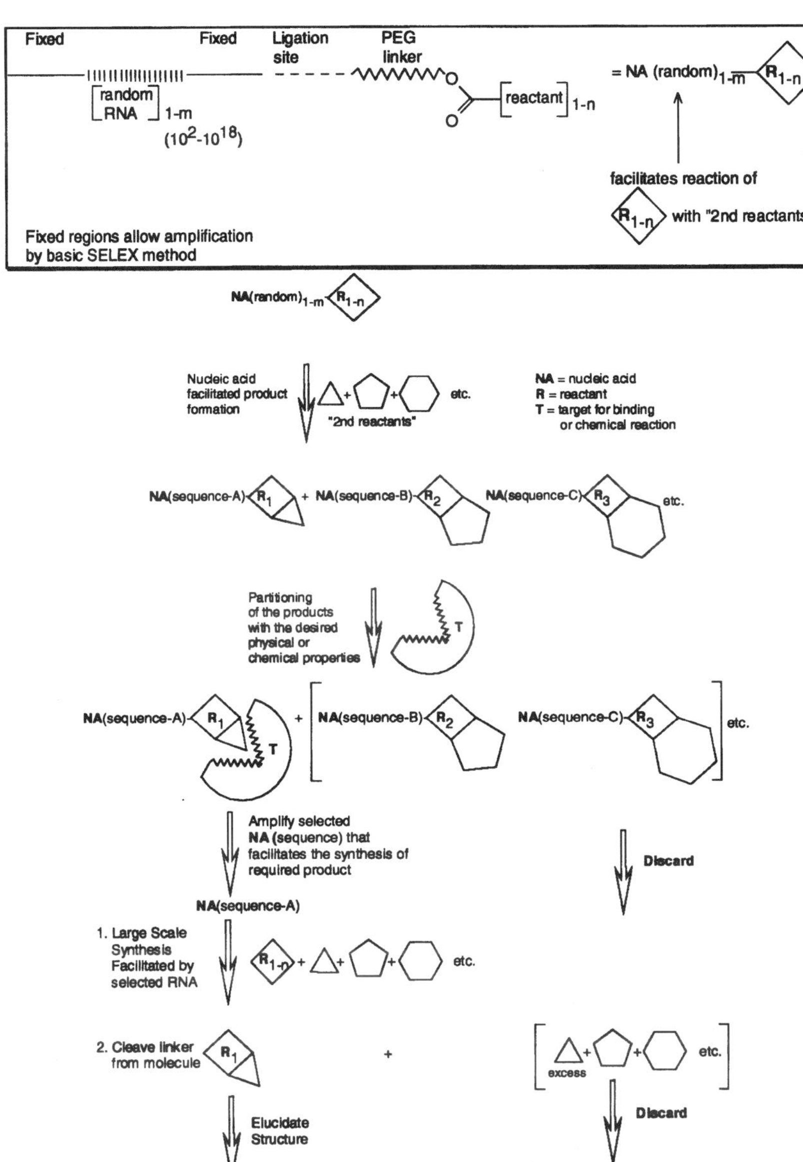

Figure 20 A SELEX-based evolutionary cascade system involving imprinting of RNA at the product formation stage. The system selects randomized RNA fragments on the basis of their ability to ''imprint'' and effectively catalyze the reaction between 1st and 2nd reactants. Following partitioning on the basis of target protein ligation, PCR amplification and sequencing of the ''successful'' RNA fragment provides a route to a selective catalyst to be used on a macro scale to reprepare and identify the ''successful'' ligand.

in the way nature chose to deliver it—very few needles in an enormous haystack along with many potent but undevelopable items to beguile and mislead the unwary. However, it may be possible to create random diversity that is richer in druglike leads. Bend Research [50] has shown how cold plasma can be used to create in an almost random library from a simple prejudiced set of reagents. For example, using a multiple-pass plasma reflux reactor (Fig. 21a), reactants are volatilized into a plasma at low pressure (0.1 to 10 torr). Radicals form, recombine, and quench to give larger products that are less volatile and do not reach the plasma zone. Fractionation of the mixture is required. Figure 21b shows just a few of the outcomes of plasma-mediated reactions between tetrahydroamino-naphthalene (THNA) and resorcinol (RS). Figure 21c shows the κ-opioid receptor binding of the fractionated mixture actually obtained. The main limitations of the method are the very slow rate at which small amounts of material can be processed and the fractionation required before assay. However, as there is much activity currently towards the development of ''nanocolumns'' capable of separating small amounts of mixtures with very high resolution, these difficulties may be only temporary.

6.5 Dynamic Libraries

The concept of using the biological system itself as a template for its own ligand has been explored by means of chemical evolution using dynamic libraries. In the equilibrium shown in Fig. 22, the template T will promote the formation of the product that binds best. This type of behavior has been described for a variety of systems including imine-based inhibitors of carbonic anhydrase [44], transesterifications in a series of ester-linked macrocyclic oligocholates (Fig. 23) with a variety of recognition and reporter elements [51] and reversible assembly of cinchona alkaloid and xanthene building blocks [52]. The problem with this approach is the very limited amount of suitable chemistry to found dynamic libraries in which the potential products are biologically relevant and the solvent system is biocompatible and to which a target protein can be directly exposed. Almost invariably the products from these essentially thermodynamic equilibria will be too unstable and reactive to be druglike.

A move towards systems in which mixture components are remodeled remotely from the biological target and therefore may provide for a more biocompatible system for screening has been described by Eliseev's group [53,54]. Here, arginine receptors are reversibly equilibrated under UV light and then presented for selection by immobilized arginine (see Fig. 24). This system still allows the production of amplified amounts of effective noncovalent binders from pools of compounds that can exist in dynamic equilibrium. More importantly, the mass action effect is achieved fairly rapidly, showing that any coercion by the template on the equilibrating mixture in achieving selective amplification of a ligand is

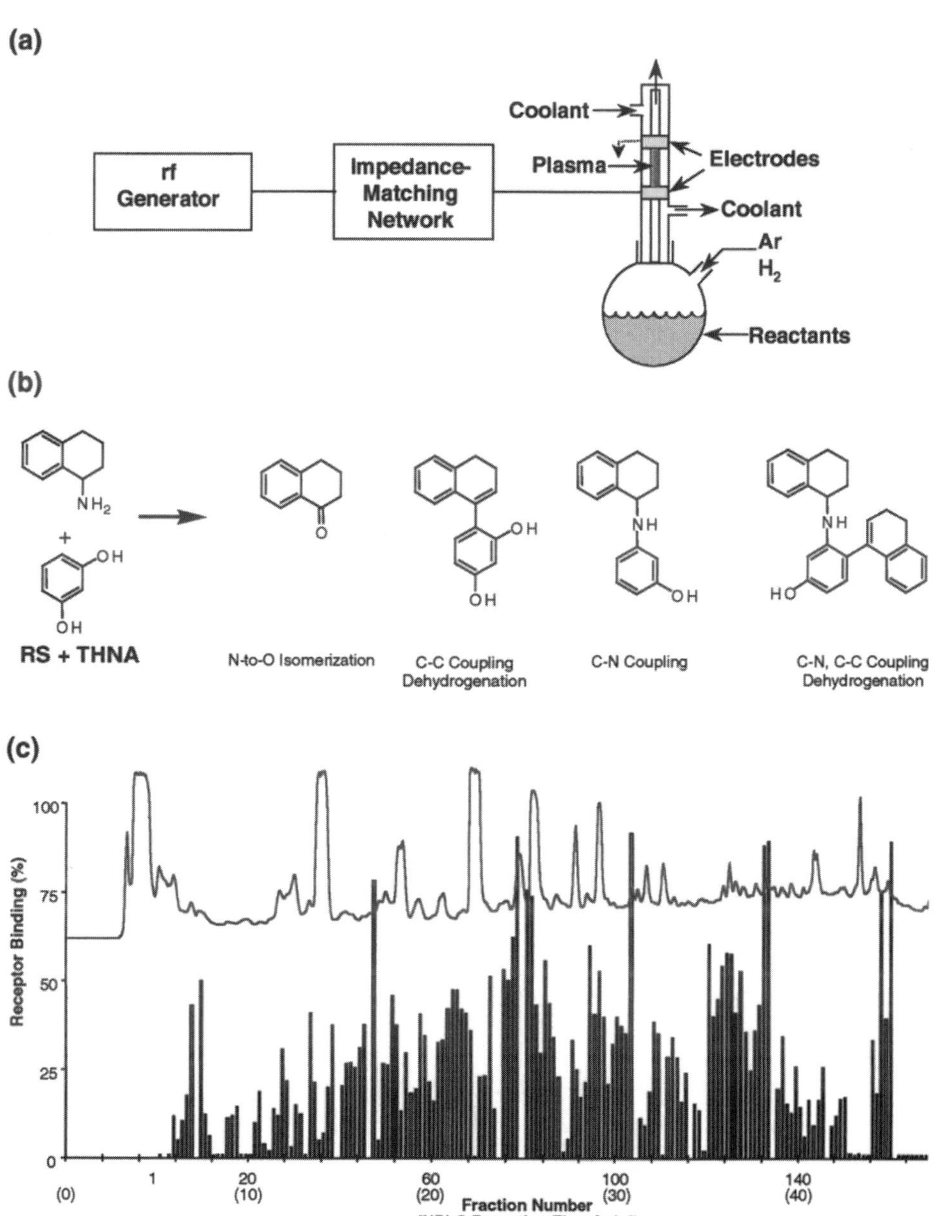

Figure 21 Semirandom synthesis (SeraSyn) from Bend Research Inc., Bend, Oregon. K-Opioid receptor binding (30 nM average concentration).

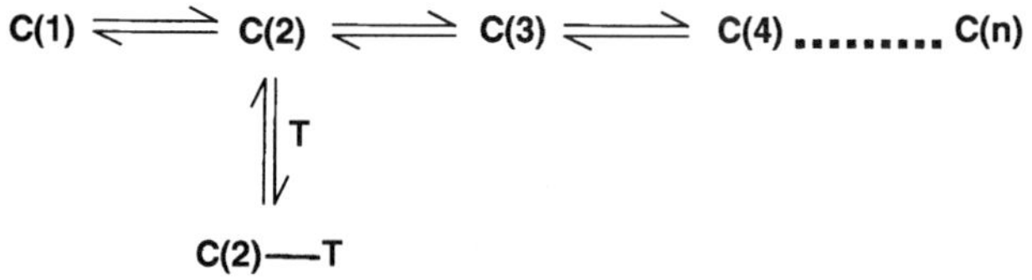

C(1) ⇌ C(2) ⇌ C(3) ⇌ C(4) .... C(n)

C(2)—T

Figure 22 Thermodynamic equilibrium of multiple species C in the presence of template T. (From Brady, P. A., Sanders, J. K. M. Thermodynamically controlled cyclisation and interconversion of oligocholates: metal ion templated 'living' macrolactonisation. J. Chem. Soc. Perkin Trans. I, 1997, 3237–3253.)

not a requirement in achieving a workable time frame. However, the range of chemistry available in biocompatible solvents is still esoteric and restrictive.

6.6 Microchannels

In reality, the attraction of the recycling system of Fig. 24 is not that it nicely demonstrates le Chatelier's principal theory in practice but that the result of screening feeds back to influence the chemistry. Could this be the basis of a rapid-fire drug prospecting device that may even offer hope of intelligent action via automated genetic algorithms [54]? The answer is maybe—but reliance on fully reflexive closed equilibria without evidence for significant active coercion of the substrate equilibrium may not be the best way to achieve this goal. The opposite approach, where screening results are used actively and intelligently to recycle SAR via open systems and stable compounds until the target result is obtained (i.e., medicinal chemistry) may give better results if a rapid-fire prospecting system can be found.

Strong contenders are microchannel systems, which capitalize on the speed and well-behaved nature of solution chemistry in microchannel systems. Within glass microchannels with diameters in the 50–300 μm range, due to electro-

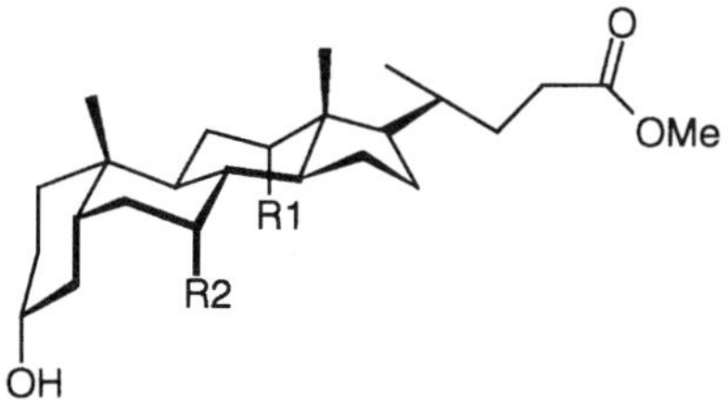

Figure 23 Ester-linked macrocyclic oligocholates. (From Ref. 52.)

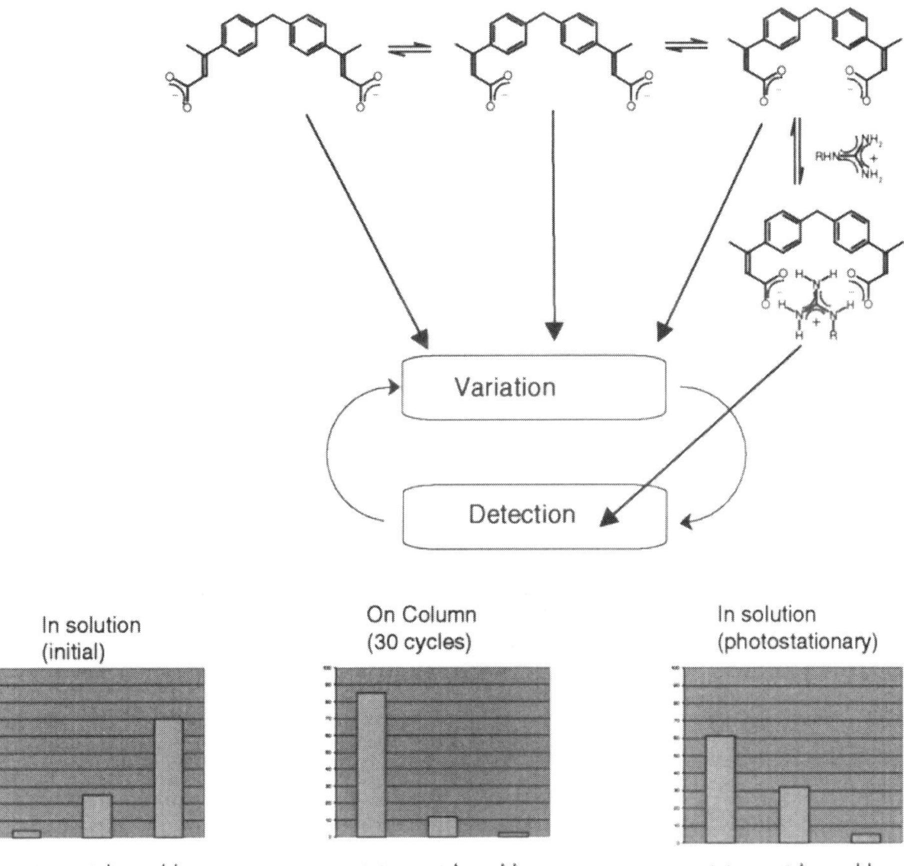

Figure 24 Use of molecular recognition to drive chemical evolution. A recirculating pool contains a mixture of the cis and trans isomers of ligands that can be brought to a dynamic equilibrium by UV light. The selection site contains an immobilized target moiety, in this case arginine on silica, which preferentially depletes the pool of strong binders, in this example the cis, cis variant. In the variation vessel UV light catalyzed reequilibration leading gradually through successive cycles to a preferential amplification of the strong binder in its complexed form and a decrease in its solution concentration. (From Eliseev, A. V., Nelen, M. I., Use of molecular recognition to drive chemical evolution. 1. Controlling the composition of an equilibrating mixture of arginine receptors. J. A. Chem. Soc., 1997, 119, 1147.)

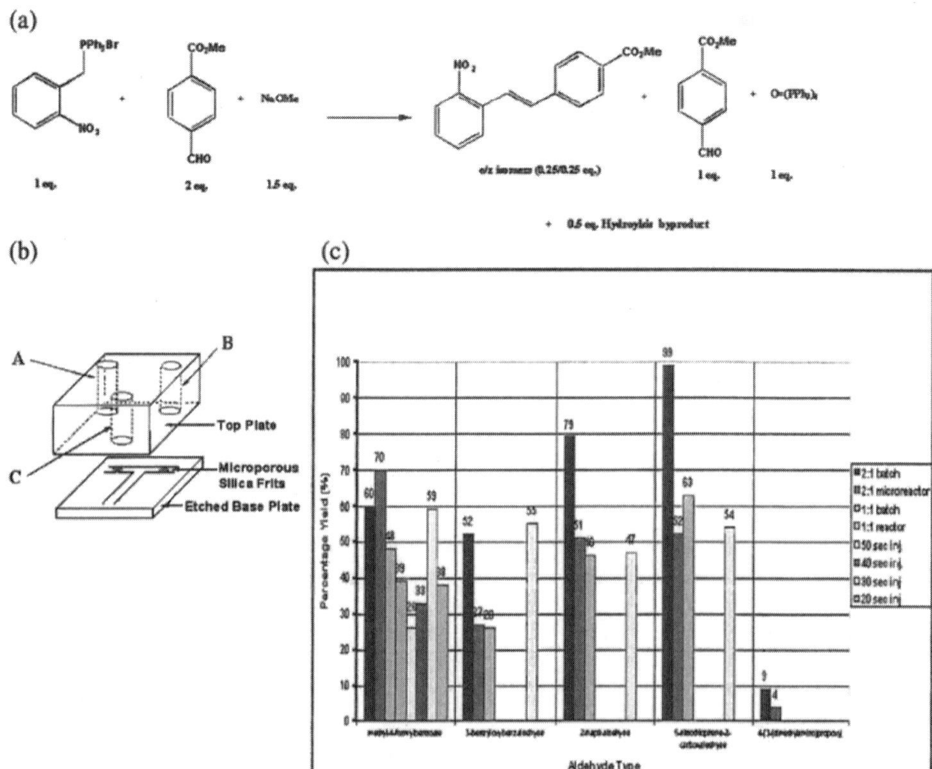

Figure 25 Microchannel chemistry. (a) Synthetic application of a microreactor to a Wittig synthesis. The microreactor (b) was fabricated in borosilicate glass (channel geometries 300 microns wide and 100 microns deep) using photolithographic patterning and modified wet etch techniques. Reagents are introduced in continuous flow to ports A and B, and the effects of reaction stoichiometry variation are described. The resulting optimized method was used with a variety of aldehydes to demonstrate its general applicability (c). (From Skelton, V., Greenway, G. M., Haswell, S. J., Morgan, D. O., Styring, P., Warrington, B. H., Wong, S. Y. F. Preliminary results: the development of a continuous flow injection microreactor for organic synthesis and combinatorial applications using Wittig chemistry. http://www.mdpi.org/ecsoc-3.htm. The Third International Electronic Conference on Synthetic Organic Chemistry, B0006, September 1–30, 1999.)

osmotic or electrohydrodynamic effects, solutions of reactants and products can be caused to flow in a controlled manner by applying suitable potentials, thus eliminating the need for many (or possibly all) mechanical components. The rigidly laminar flow of fluids in the low-Reynolds-number environment of a microchannel also allows the selection and diversion of channel streams and substreams, thus enabling a control of mixing and separation events not available in (turbulent) macro systems. In addition, because of the high ratio of the wall area to the fluid volume and its polarity (particularly after special treatment), chromatographic effects can occur, and the components of mixtures of solutes flowing in the microchannel tend to undergo spontaneous geometric separation.

Using a T-shaped device (Fig. 25) [55,56], short pulses of reagents in flowing solvents have been be brought together at a microchannel junction element of similar dimensions to the diffusion distance. As a result, all molecules can react almost simultaneously. In practice, the times to achieve complete reaction are much shorter (seconds) than those found for the same reaction on the macro scale (minutes, hours). The acceleration appears to be due to the virtual elimination of the need for mass transfer, although the reduction of any reverse reaction due to a spontaneous separation of products may also contribute. Overall, the effect of regularly and sequentially releasing pulses of reagents from a combinatorial set is serially to produce ''plugs'' of a library of products that undergo separation (and purification) as they continue to flow down the main channel. The predicted identity of any product is therefore determined by its time of release. Clearly, products could be collected, but if flow is continued to a suitably configured screening device, the basis for a fast synthesis and screening device, consuming minute amounts of reagent, becomes apparent. Only those compounds showing evidence of activity would be resynthesized for confirmation of the predicted structure and biological activity. These data could be used to direct macro solution synthesizers more economically to the bioactive regions of chemical property space.

Clearly, with the ability to create complex circuitry combining microfluidic and electronic components, a sophisticated level of control of reagent selection and synthesis management is possible. This then provides the basis for intelligent control, perhaps using software of the type that has made its appearance in recent computer games invoking ''intelligent'' life forms. Alternatively, the simple ''electronic chemist'' described in Fig. 19 might carry out the task!

I'll do something.
If the result was good I'll do something similar.
If the result was bad, I'll return to the last good thing I did
and do something different.

REFERENCES

1. Drews, J. Genomic sciences and the medicine of tomorrow. Nature Biotechnology, 1996, 14, 1516–1518.
2. Harris, C., et al. Analyst Report. Life Sciences, Bringing Industrial Scale Science to the Clinical and Research Laboratory. Warburg Dillon Read, New York, 24 Jan. 2000.

3. Phillips, D., Cahill, C., Stanley, C., Delesandro, M. Drivers in outsourcing deals in combinatorial chemistry. Scrip Magazine, December 1999, 6–8.

4. Studt, T. Raising the bar on combinatorial discovery. Drug Discovery and Development, 2000, Jan/Feb, 24–9.

5. Martin, Y. C. Challenges and prospects for computational aids to molecular diversity. Perspect. Drug Res. Discovery 1997, 7/8, 159–172.

6. Lancet, D., Sadovsky, E., Seidemann, E. Probability model for molecular recognition in biological receptor repertoires: significance to the olfactory system. Proc. Natl. Acad. Sci. USA, 1993, 90, 3715–3719.

7. Seligmann, B., Lebl, M., Lam, K. S. Solid-phase peptide synthesis, lead generation, and optimisation. In: Combinatorial Chemistry and Molecular Diversity in Drug Discovery (E. M. Gordon, J. F. Kerwin, eds.). 1998, Wiley-Liss, New York, pp. 39–109.

8. Martin, E. J., Blaney, J. M., Siaini, M. A., Spellmeyer, D. C., Wong, A. K., Moos, W. H. Measuring diversity: experimental design of combinatorial libraries for drug discovery. J. Med. Chem. 1995, 38, 1431–1436.

9. Bures, M. G., Brown, R., Martin, Y. C. Analysing larger databases to increase the diversity of the Abbott corporate collection. Abstracts of papers of the American Chemical Society, 1995, pp. 210, 60-Cinf.

10. Pharma 2005, An Industrial Revolution in R&D. Pricewaterhouse Coopers, 1998.

11. Hird, N. W. Automated synthesis: new tools for the organic chemist. Drug Discovery Today, 1999, 4, 265–274.

12. http://www.mtmyriad.com.

13. Brooking, P., Doran, A., Grimsey, P., Hird, N. W., Maclachlan, W. S., Vimal, M. Split/split: a multiple synthesiser approach to efficient automated parallel synthesis. Tet. Lett. 1999, 40, 1405–1408.

14. http://www.personalchemistry.com.

15. http://hamiltoncompany.com.

16. http://www.zinsser-analytic.com.

17. http://www.biotage.com.

18. Kibbey, E. An automated system for the purification of combinatorial libraries by preparative LC/MS. Laboratory Robotics and Automation 1997, 9, 309–321. Zeng, L., Kassel, D. B. Anal. Chem. Developments of a fully automated parallel HPLC/Mass spectrometry system for the characterisation and preparative purification of combinatorial libraries. 1998, 70, 4380–4388. Kiplinger, J. P., Cole, R. O., Robinson, S., Roskamp, E. J., Ware, R. S., O'Connell, H. J., Brailsford, A., Batt, J. Structure-controlled automated purification of parallel synthesis products in drug discovery. Rapid Commun. Mass Spectrom., 1998, 12, 658–664.

19. http://www.jones-chrom.co.uk.

20. http://www.anachem.co.uk.

21. http://www.micromass.co.uk.

22. http://www.thermoquest.com.

23. Bateman, R. H., Jarvis, S., Giles, K., Organ, A., de Biasi, V., Haskins, N. J. Multiple LC/MS: parallel and simultaneous analyses of liquid streams by LC/TOF mass spec-

trometry using a novel eight-way interface. Proc. 47th ASMS Conf. Mass Spectrom. Allied Topics, Dallas, TX, June 13–17, 1999, WPE 121.

24. http://abs-lcms.com.

25. http://www.pecorporation.com.

26. http://www.bruker.com.

27. http://www.varianinc.com.

28. Keifer, P. A., Smallcombe, S. H., Williams, E. H., Salomon, K. E., Mendez, G., Belletire, J. L., Moore, C. D. Direct Injection NMR (DI-NMR). A flow NMR technique for the analysis of combinatorial chemistry libraries I. J. Comb. Chem, 2000, 2, 151–171.

29. http://www.antekhou.com.

30. Pharma 2005. Silicon rally: the race to e-R&D. Pricewaterhouse Coopers, 1999.

31. Warr, W. Strategic management of drug discovery. Scrip Reports, 1998.

32. Henry Ford (1863–1947) on the choice of color for the Model T Ford: "Any color— so long as it's black."

33. http://www.afferent.com. Afferent Software has now been acquired by MDL Information Systems Inc., http://www.mdli.com.

34. RADICAL. Registration and Design Interface for Combinatorial and Array Libraries. SB internal workflow management software package.

35. http://www.ttpgroup.co.uk/ttp/software.

36. Ley, S. V., Schucht, O., Thomas, A. W., Murray, P. J. Synthesis of the alkaloids (±)-oxomaritidine and (±)-epimaritidine using an orchestrated multi-step sequence of polymer supported reagents. J. Chem. Soc., Perkin Trans. 1, 1999, 1251–1252.

37. Kobayashi, S., Nagayama, S. A new methodology for combinatorial synthesis. Preparation of diverse quinoline derivatives using a novel polymer-supported scandium catalyst. J. Am. Chem. Soc., 1996, 118, 8977–8978.

38. Kobayashi, S., Nagayama, S. A microencapsulated Lewis acid. A new type of polymer-supported Lewis acid catalyst of wide utility in organic synthesis. J. Am. Chem. Soc., 1998, 120, 2985–2986. Kobayashi, S., Endo, M., Nagayama, S. Catalytic asymmetric dihydroxylation of olefins using a recoverable and reusable polymer-supported osmium catalyst. J. Am. Chem. Soc., 1999, 121, 11229–11230.

39. For an excellent review of multicomponent, multigeneration, cascade, and domino strategies see Obrecht, D., Masquelin, T. Synthetic strategies in combinatorial and parallel synthesis. In: Solid-Supported Combinatorial and Parallel Synthesis of Small-Molecular-Weight Compound Libraries (D. Obrecht, J. M. Villalgordo, eds.). Tetrahedron Organic Chemistry Series Vol. 17, 1998, Elsevier Science, Oxford, UK, pp. 105–126.

40. Shuker, S. B., Hajduk, P. J., Meadows, R. P., Fesik, S. W. Discovering high affinity ligands for proteins: SAR by NMR. Science 1996, 274, 1531–1534.

41. Maly, D. J., Choong, I. C., Ellman, J. A. Combinatorial target-guided ligand assembly. Identification of potent subtype-selective c-Src inhibitors. Proc. Natl. Acad. Sci. USA, 2000, 97, 2419–2424.

42. The observation (from Gordon Moore, cofounder of Intel) that the logic density of silicon integrated circuits has closely followed the curve (bits per square inch) = $2^{(t-1962)}$ where t is time in years. Thus the amount of information storable on a given amount of silicon, and for a given price, has roughly doubled every year since

the technology was invented. Unfortunately in many circumstances this effect is negated by Gate's law (Bill Gates, Microsoft), which states that the speed of software will halve every 18 months.

43. http://www.orchidbio.com.

44. Huc, I., Lehn, J.-M. Virtual combinatorial libraries: dynamic generation of molecular and supramolecular diversity by self-assembly. Proc. Natl. Acad. Sci. USA, 1997, 94, 2106–2110.

45. Cramer, R. D., Clark, R. D., Patterson, D. E., Ferguson, A. M. Bioisosterism as a molecular diversity descriptor: steric fields of single topomeric conformers. J. Med. Chem., 1996, 39, 3060–3069.

46. Brown, R. D., Martin, Y. C. Use of structure activity data to compare structure-based clustering methods and descriptors for use in compound selection. J. Chem. Inf. Comput. Sci. 1996, 36, 572–584.

47. Lipinsky, C. A., Lombardo, F., Doming, B. W., Feeney, P. J. Experimental and computational approaches to estimate solubility and permeability in drug discovery and development settings. Advanced Drug Delivery Reviews, 1997, 23. 3–25. Fecik, R. A., Frank, K. E., Gentry, E. J., Menon, S. R., Mitscher, L. A., Telikepalli, H. The search for orally active medications through combinatorial chemistry. Med. Res. Rev., 1998, 18, 149–85. Walters, W. P., Stahl, M. T., Murcko, M. A. Virtual screening—an overview. Drug Disc. Today, 1998, 3, 160–78. Martin, E. J., Critchlow, R. E. Beyond mere diversity: tailoring combinatorial libraries for drug discovery. J. Comb. Chem., 1999, 1, 32–45.

48. Cwirla, S. E., Peters, E. A., Barrett, R. W., Dower, W. J. Peptides on phage: a vast library of peptides for identifying ligands. Proc. Natl. Acad. Sci. USA, 1990, 87, 6378–6382. Kay, B. K. Biologically displayed random peptides as reagents in mapping protein–protein interactions. Persp. Drug Disc. Des., 1994, 2, 251–326. Scott, J. K., Smith, G. P. Searching for peptide ligands with an epitope library. Science, 1990, 249, 386–390.

49. Ellington, A. D., Szostak, J. W. In vitro selection of RNA molecules that bind specific ligands. Nature, 1990, 346, 818–822. Tuerk, C., Gold, L. Systematic evolution of ligands by eponetial enrichment: RNA ligands to bacteriophage T4 DNA polymerase. Science, 1990, 249, 505–510. Gold, L. Oligonucleotides as research diagnostic and therapeutic agents. J. Biol. Chem., 1995, 270, 13581–13584. Gold, L., Alper, J. Keeping pace with genomics through combinatorial chemistry. Nature Biotechnol. 1997, 15, 297.

50. http://www.bendres.com/.

51. Brady, P. A., Sanders, J. K. M. Thermodynamically controlled cyclisation and interconversion of oligocholates: metal ion templated 'living' macrolactonisation. J. Chem. Soc. Perkin Trans. I, 1997, 3237–3253.

52. Rowan, S. J., Hamilton, D. G., Brady, P. A., Sanders, J. K. M. Automated recognition, sorting and covalent self-assembly by predisposed building blocks in a mixture. J. Am. Chem. Soc., 1997, 119, 2578–2579. Rowan, S. J., Brady, P. A., Sanders, J. K. M. Building thermodynamic combinatorial libraries of quinine macrocycles. Chem Commun 1997, 1407–1408.

53. Eliseev, A. V., Nelen, M. I. Use of molecular recognition to drive chemical evolu-

tion. I. Controlling the composition of an equilibrating mixture of arginine receptors. J. A. Chem. Soc., 1997, 119, 1147.

54. Eliseev, A. V., Nelen, M. I. Use of molecular recognition to drive chemical evolution: mechanisms of an automated genetic algorithm implementation. Chem. Eur. J. 1998, 4, 825–834.

55. Skelton, V., Greenway, G. M., Haswell, S. J., Morgan, D. O., Styring, P., Warrington, B. H., Wong, S. Y. F. Preliminary results: The development of a continuous flow injection microreactor for organic synthesis and combinatorial applications using Wittig chemistry. http://www.mdpi.org/ecsoc-3.htm, The 3rd International Electronic Conference on Synthetic Organic Chemistry, B0006, September 1–30, 1999.

56. Greenway, G. M., Haswell, S. J., Morgan, D. O., Skelton, V., Styring, P. The use of a novel microreactor for high throughput continuous flow organic synthesis. Sensors and Actuators B: Chemical, 2000, 63, 153–158.

7

Laboratory Information Management Systems for Laboratory Automation of the Chemical Industries

Christine Paszko
Accelerated Technology Laboratories, Inc., West End, North Carolina

1 INTRODUCTION

Laboratory information management systems (LIMS) are software systems that allow laboratories to automate, manage, and control laboratory procedures and organize data and information. LIMS utilize relational database technology to manage laboratory information such as tests, methods, results, personnel, and instruments, which allows users to access that information for analysis, reporting, tracking, and other purposes. Commercially available LIMS solutions became popular in the late 1970s primarily with the large pharmaceutical firms. These early systems were cost prohibitive to many other industries that could also greatly benefit from their use. However, with the decreasing cost of computer technology, increasing need for productivity, rapid access to information, and increasing growth of legislation, many other industries are implementing LIMS. Today, these systems are in widespread use in a number of laboratories where people study organic and inorganic chemistry, the environment, agriculture, biotechnology, utilities, research and development, government and many others areas. LIMS offer laboratories many advantages including increased productivity, regulatory compliance, and rapid access to information.

Most LIMS are set up in a client/server configuration as shown in Fig. 1. In this configuration, the database tables reside on the server and the graphical user interface resides on the client machines. The advantage of this configuration

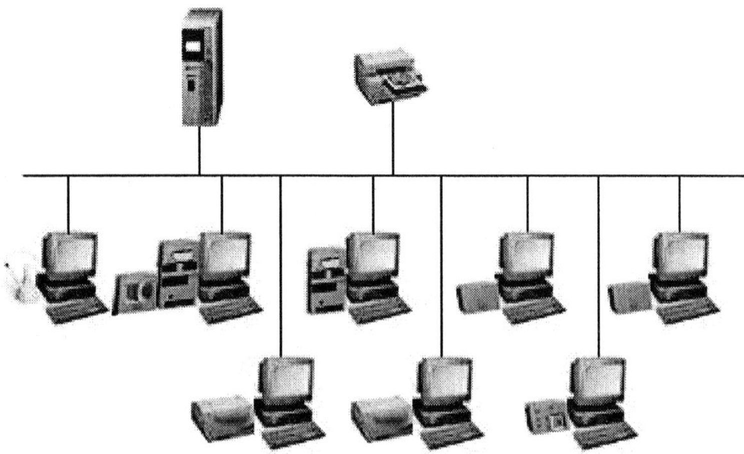

Figure 1 Typical client/server configuration.

is that the majority of the data processing occurs on the server, allowing the laboratory to utilize "thinner" client machines.

LIMS allow laboratories to manage data from tracking samples and raw materials through analysis, importing data directly from analytical instrumentation, generating worklists, performing QA/QC, automatically scheduling runs, e-mailing reports, entering formulations, performing simple calculations on raw data, flagging of results that are high/low, and creating numerous reports including certificates of analysis, quality control, turnaround times, production, and many more functions. Quite often these systems are integrated with enterprise systems such as SAP and ERP (enterprise resource planning) to allow firms to have a complete understanding of their processes and how they interact. Figure 2 depicts the information flow through a typical laboratory.

In today's chemical laboratory a LIMS is no longer a luxury but a necessity. With the introduction of combinatorial chemistry and high-throughput screening, computers, robotics, and LIMS are integral for successful data management in a chemical laboratory. In the past, LIMS were rigid programs that ran on mainframes that were only available to a handful of large pharmaceutical or chemical firms that could justify the high costs. Today, things have changed significantly; desktop personal computers (PCs) that cost under a $1000 U.S. have significantly more computing power than the early VAXs that cost thousands more. LIMS is no longer inflexible software that is run on mainframes; rather it provides extreme flexibility to conform to laboratory operations with the ability to import directly

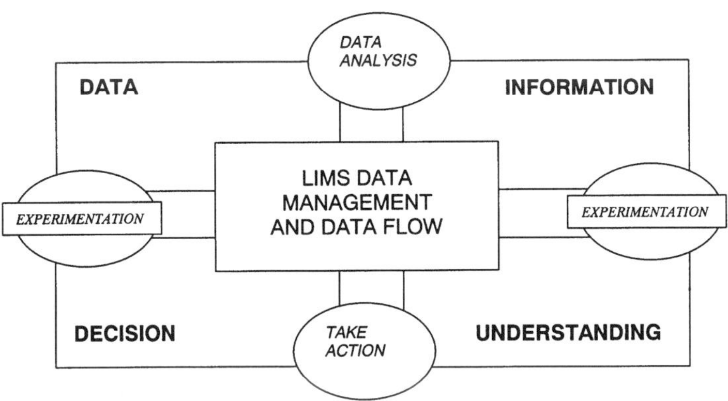

Figure 2 LIMS information migration.

vast quantities of instrument data, schedule additional analyses based on the initial run, and disseminate information to laboratory managers via e-mail. As hardware advances have progressed, enhancements have also been made in the software arena. With the introduction of the GUI (graphical user interface), users now have an intuitive interface with pull-down menus, radio buttons, hot lookups, and the ability to share information via the Internet. This chapter will attempt to provide the reader with an overview of LIMS, instrument integration, and the current advances in chemical data management.

With the introduction of genomics, sequencing, and combinatorial chemistry, and the high-throughput synthesis and screening of chemical substances to identify agents with useful properties, LIMS systems play a crucial role in information management. Chemistry laboratories share many of the same requirements of other laboratories, and some of these requirements are described below to provide to reader with a solid overview of LIMS use and functionality.

System security that assures only authorized users are able to log into the system and once in the system will only have access to those areas of the LIMS functions that they have permissions for; this can be at a department-by-department basis or even down to the test level. A full audit trail is also critical to ensuring that data changes are recorded and the original results are maintained with the who, why, what, and when information of each change to the data recorded. Integrated security prevents unauthorized access to laboratory information; electronic signatures (such as biometric ID) and validation provide authentication and compliance.

Sample tracking and management provides source information on the sample, processing information (standards, references, nonroutine samples), and a full chain of custody of that sample (who handled the sample and where it was stored during its laboratory life-cycle—from receipt through disposal).

Complete data management allows end-users to access all the laboratory results, methods, standard operating procedures, instruments, costs, projects, and other items stored in the database.

Cost and time tracking features are also integral to a LIMS providing managers with a means of better understanding resource allocation with actual data, often complemented by production and backlog reports, by test, instrument, and analyst.

Personnel and instrument management functions allow end-users to track and centralize calibration records, maintenance records of instruments, as well as training certificates of end-users.

Rapid information delivery is perhaps the most important feature providing decision makers with information on which to base their decisions. LIMS provide real-time data access and graphical capabilities that allow users to examine trends and anomalies in the data that may not be obvious. LIMS also allow users to apply calculations and export the data to other packages for specialized analysis or complex statistical analysis.

Many LIMS features are common to a wide range of laboratories. Requirements of a typical LIMS system in a chemical testing laboratory might include the following:

Sample login
Sample tracking/barcode support/quoting
Scheduling/time tracking
Chain of custody
Instrument integration
Result entry/audit trail
QA/QC/specification checking
Result reporting
Web integration/links to enterprise software
Chemical and reagent inventory
Personnel training record tracking/instrument maintenance
Archiving/data warehousing

2 DATABASE CHOICES

There are typically three different types of LIMS. The first is a proprietary system. This is a system in which all the source code is known only to the vendor or the development group that created that particular language. An advantage of a proprietary language is speed. A major disadvantage is that any modification that the users may require must involve the original vendor or programmers. If the original programmers are no longer available and the source code is locked, updates will not be possible and users will be left with no support or migration path.

The second type of database is an ISAM (indexed sequential access method) database, which includes Microsoft Access, Paradox, and FoxPro. These are all relational databases. In ISAM databases, requests are made to the server via the personal computers (client machines), in the information request that is sent to the server, the entire table that holds the piece of information requested is sent back to the client machine to be searched on the client's hard drive. As the tables get larger, the performance will deteriorate, requiring the users to archive data out of the active database. The advantage of ISAM databases is that a full-time LIMS administrator is not required to manage and maintain the LIMS. Another advantage, in the case of Microsoft Access, is that many chemists are familiar with this application and are able easily to modify reports, create custom queries, run automated tasks, and add functionality to commercially available packages that have the core LIMS functionality.

The third type is a SQL (structured query language), which is ideal for managing large volumes of data. Examples of SQL databases include Microsoft SQL Server, Oracle, Informix, Sybase, and Ingress. These database engines are ideal for large databases or for users that would like to keep several years of data in the active database. These database engines were designed to manage large amounts of data like that generated in high-throughput screening operations as encountered in combinatorial chemistry or sequencing. For many years, the sample preparation or the analytical instrumentation that analyzed the samples and generated the raw data was the limiting factor or bottleneck for researching new compounds or identifying new sequence data. Today, too often the limiting factor is the organization, management, and analysis of the data generated, especially, with the introduction of robotics, auto samplers, multiple plate readers, and high-throughput instrumentation, many systems being designed for continuous operation. Many analytical instruments are equipped with raw data analysis software that will provide specific information on the run, results of internal controls, instrument calibration, statistical analysis of the data, and often charting capabilities. A few packages have evolved to include some LIMS functions. Many of these software programs are based on a proprietary language, but there is a strong push to standardize, and most instruments export to a common output file that can be imported to other applications including LIMS. It is important to select a database application that is ODBC (open database connectivity) compliant so that it can be linked to other databases for purposes of data sharing.

Although many systems utilize the SQL database engine, they may utilize Visual Basic, Microsoft Access, ASP (Active Server Pages), Pearl, or other programs for the front end or user interface of the application, often permitting users the ability to add bolt-on functionality to the database.

3 TO BUILD OR TO BUY?

This is the first decision that the laboratory has to make. It often sounds appealing to create a customized laboratory information management system for your laboratory. However, most designers underestimate the time required to create a LIMS and if they are not experienced in database design they can make some unrecoverable errors in data structure. In addition, the alpha and beta site to test the new LIMS is the laboratory; this can be quite disruptive if it is not properly implemented. It can also be very costly taking this approach, since the technology is changing rapidly and a LIMS needs to keep up with these changes, which means that the internal development group must be dedicated to maintaining, supporting, and upgrading the LIMS. Ultimately, the laboratory must consider a number of variables in making its decision; the results of a laboratory need assessment; internal resources and expertise, a project timeline, and return on investment calculations will all be included in the final decision.

An off-the-shelf solution may not provide 100% out-of-the-box functionality required by the laboratory; it may only provide 70 or 85%, since each laboratory may have some special requirements or interfaces that are needed. Based on the laboratory's functional assessment, if the majority of the required functionality is contained in the commercially available product, the laboratory may want to utilize the internal IT resources to incorporate and integrate bolt-on functionality to the LIMS. With the LIMS vendor providing the migration path to product upgrades, end-user support, service packs, training, and any customization that the laboratory does not have the resources for, this scenario often offers the best solution. Users have a commercially available product with a team of database professionals keeping the product current with upgrades, service packs and enhancements, and user group meetings where users can share ideas, while maintaining the ability to incorporate custom features to meet their laboratories specific needs. Major considerations in selection of a LIMS include:

> Database engine (ISAM, SQL, proprietary) choices
> System features to match user requirements
> Current hardware and software available in the laboratory
> Ease of use, on-line help, technical support, and availability of upgrades
> Internal software and laboratory expertise
> Flexibility to accommodate a changing laboratory environment
> Cost/ROI (return on investment) considerations

3.1 The LIMS Selection Process

The process of purchasing an off-the shelf LIMS solution versus developing an in-house data management system can be consolidated into six phases as outlined below and as shown in Figure 3.

1. *Project Definition.* During this phase, the LIMS team creates a brief statement of what is to be accomplished via the LIMS acquisition and why it is needed.

2. *Functional Requirements.* During this step, all inputs to and outputs from the system are described in detail. In this phase, any instrumentation or enterprise systems that are to be linked are also identified and described.

3. *Functional Design.* The output of this phase of planning will include a detailed description of the system; it may be represented in a detailed flow chart.

4. *Implementation Design.* In this phase, hardware and software components are selected. Implementation, training, operation, and maintenance procedures are developed in this phase if the system will be created internally; if not, the vendor will offer an implementation package.

5. *System Implementation.* During this phase, the system components (hardware and software) are built (if the internal development route is selected), acquired (if purchased), and tested. The system is assembled, integrated, installed, and placed into operation for final acceptance testing.

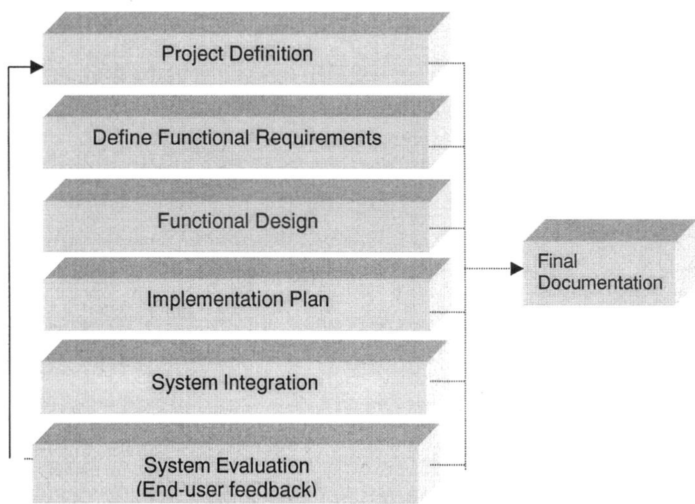

Figure 3 Traditional development of a software system.

6. *System Evaluation*. In this final phase, the system is compared with its requirements to determine how well the requirements are met, to determine if there are any possibilities for growth and improvement, and to remember the lessons learned during the course of this project for application to future projects.

4 TYPICAL LIMS FUNCTIONALITY

Table 1 outlines some typical LIMS functionality that is shared across industries. The core of any LIMS is the sample login functionality or sample tracking. If this is not streamlined to permit rapid entry of samples and the tests to be run, it can become a bottleneck. The utilization of automatic ID, such as bar codes, allows sample login to be an automated process in that the sample number can be directly scanned from a sample. This number corresponds to that sample and that sample only and includes information about that sample such as its origin, tests to be run, controls, limits, contact information (in terms of whom the results are to be reported to), any prep data, and quantity. The USEPA has spearheaded GALP (good automated laboratory practices) for LIMS, to establish a uniform set of procedures to assure that all LIMS data are reliable and credible. GALP encompasses good laboratory practices (GLP) and computer security. There are many additional industry standards that LIMS must conform to; some are industry specific.

Most laboratories look to LIMS for the following benefits:

A reduction in paperwork and manual entry
Increase in accuracy and data quality
Better management of sample or patient demographics and test requests
Ability to perform statistical analysis and to generate reports
Reduced turnaround times and autoreporting
On-line SOPs to enforce standardization
Automated test scheduling and test ordering
Reduction in misplaced samples
Assistance in interpreting test results and automatically scheduling repeat
    tests based on set result criteria
Increased productivity

5 SAMPLE TRACKING

Sample tracking functionality in a LIMS allows analysts to know the status of a sample throughout the sample lifecycle, from sample login to analysis to sample disposal. Once samples are logged into the LIMS and a unique identifier has been

Table 1 Typical LIMS Functionality

Typical LIMS Functions	Description
Sample tracking	This function allows laboratories to track their samples through different departments in the laboratory with a computer-generated unique sample identification number and provides a complete chain of custody.
Data entry	Data entry function allows analysts to enter results into the LIMS and to assign QC run batches. Reporting to clients via fax, e-mail, or hard copy.
Sample scheduling	Automatically logs in samples, receives them into the laboratory, prints labels, and assigns the tests for projects on a routine basis.
QA/QC	This function allows users to perform calculations, generate control charts, and view trend analysis graphs. Control charts can encompass blanks, duplicates, spikes, surrogates, standards, etc.
Electronic data transfer	A feature that allows automatic transfer of data from analytical instrumentation into the LIMS. This feature increases productivity and greatly decreases the potential for transcription errors.
Chemical and reagent inventory	Functionality that tracks the purchase and usage of supplies in the laboratory and manages lot and order numbers, shelf life, costs, etc. Assisting in supply management.
Personnel and equipment management	A feature that allows users to track employee training records for ISO purposes and also track instrument calibration, repairs, costs, trends, etc.
Maintenance	A function that allows the database administrator to manage the database, keeping track of client lists, employees, tests, methods, parameters, permissions, priorities, etc.

assigned (often linked to an existing sample number), the tests to be performed are assigned, as well as the method, any QC batch information and any additional information (i.e., lot number, priority, etc.) that is important to the analysis of that sample. Typically the sample moves through the different departments of the laboratory, such as login, preparation department, analysis, reporting, and perhaps disposal. High-throughput laboratories typically make extensive use of bar coded samples to speed and automate login. The LIMS software allows users to know the exact location of the sample as it passes through the laboratory.

6   UTILIZATION OF BAR CODES AND AUTOMATIC IDENTIFICATION TECHNOLOGIES

Laboratories that have streamlined their operations have typically implemented a number of different technologies to meet their automation, identification, and security challenges. This section will briefly describe the utilization of bar coding and the use of automatic identification devices. Today's high-throughput laboratories rely on bar coded samples, sample plates (such as the 96-well variety), and chemicals for inventory and disposal purposes, to name a few. The major benefits of bar code entry include speed (approximately 20 times faster than keyboard entry), increased productivity (higher throughput), and increased accuracy. Bar codes are available in many different fonts; the font that best suits your laboratory will vary by application. In its simplest form, the bar code system consists of a scanner, a decoder, and a computer that has a bar code font installed and a printer. The scanner is used to read the bar code, and the main types of scanners include hand-held wands, CCD (charged couple devices), laser scanners, and fixed focus optics (FFO). Again the application will typically determine the best type of scanner. Other scanners that are becoming popular include optical character recognition (OCR), intelligent character recognition (ICR), and biometrics identification. Some laboratories scan in sample login forms to automate further the login process and keep the original form on file. For computer-generated forms, the scanning is quite accurate, but for handwritten forms it is best that a person review the information that has been scanned for accuracy.

Biometric ID technology provides the ability to digitize an individual's physiological trait and use it as a means of identification. This technology is used in high-security areas to protect restricted information or access to data. Biometric verification techniques include fingerprint recognition, retinal scanning, voice pattern recognition, and blood vessel pattern recognition in wrists or hands. The most popular is fingerprint identification, but voice and face recognition packages are gaining in popularity. Possibly the most accurate and expensive is the retinal scan. In the FDA's 21 CFR Part 11, the FDA has approved the use of electronic signatures. Although not required by the FDA, the biometric ID devices offer a higher level of data security and electronic signatures than is required by law.

7   DATA ENTRY

This function of the LIMS allows users to input test results manually or to view test results that have been automatically uploaded from an instrument into the LIMS. Just as bar codes are utilized within the sample tracking functionality, they can also be utilized in data entry, in that a fixed scanner can record the sample number as it passes the scanner and insert the appropriate number of result holders into the database for the results to be entered upon analysis. This also provides additional checks on the sample by date and time stamping when the sample passed through each department. This provides management with valuable information on laboratory turnaround time and resource utilization. Within this module users are able to view results, examine the QC results for those samples, and insert any comments that they may have and electronically tie that comment to the result. Instrument integration is a tremendous time-saver and it is one of the reasons that laboratories purchase LIMS. Many analytical instruments have tremendous data output files that would take some time to enter manually into a LIMS. Some laboratories export this data to an Excel spreadsheet for review. However, once the volume of data becomes too large and analysts can no longer manage the data, it should be migrated to a relational database such as a LIMS. Most instrument output files are in the form of text files (ASCII) or .csv files that can easily be imported to a LIMS. Some systems allow users to upload the data as an automated task with a user-definable time. The advantages of instrument integration include enhanced data quality (reduction in transcription errors) and increased productivity. Good candidates for instrument integration are instruments that can be prolific in their outputs such as ICPs, AAs, GC/MS and tandem MS.

8 MANAGING THE DATA STREAM: AUTOMATED INSTRUMENT DATA ENTRY

Electronic data transfer from analytical instrumentation to the LIMS basically replaces the keyboard entry of test and QC result data. The data transfer entails the LIMS reading a file (often placed in a directory that the LIMS scans). The instrument file is typically parsed or massaged to be in a format that ensures that the data is moved to the correct table and field in the LIMS. The data download is user configurable and can be configured in several ways, to import a single data file at a time or to import an entire data directory or combinations. Once the data is placed into the database, the LIMS can apply the same validity and limit checking that the data would pass through if it were manually entered. Once the data is imported, an analyst can also choose to review and validate the data and QC results manually. The sample status of the imported results can also be manually or automatically updated. If results are out of range (possibly via an out-of-calibration instrument), the LIMS can flag these samples and provide analysts the ability to delete the run, before bad data is entered into the LIMS. When importing multiple determinations for a test, the LIMS should be flexible to allow analysts to delete individual determinations or the entire run. Data files are available in electronic format: a parsing file should be written to convert that data into a format that is recognized by the LIMS to make full use of the LIMS and avoid transcription errors. If an instrument is on a local area network, the data can be transferred directly to the LIMS over the network. For instruments that are not networked, electronic data transfer is still the best method, only files will need to be transferred to the LIMS via a sneaker net.

9 ROUTINE SAMPLE SCHEDULING FOR LIMS

Many laboratories utilize the LIMS software to schedule automatically experiments and runs. This can be a tremendous time saver. Not only does the system login the samples and place ''place holders'' for the appropriate amount of results in the database but it can also be configured to e-mail users when it is time to pull or collect samples for analysis, as in stability studies, so that no samples are missed.

There are two types of scheduled or repetitive samples: first, those involved with a study of some sort, such as a stability study, which have a finite number of samples coming in, and second, those involved with a process plant where there is no foreseeable end to the samples coming into the lab. The first type is best handled by utilizing a scheduling functionality of a LIMS and following a study protocol that generates a specific login list of dates and times and test methods for the entire stability study period. The system can generate a log of all test dates, methods, sample volumes, test methods, etc. This is a useful tool to the analyst performing the analysis.

The automated login functionality of the LIMS can scan for any records in the current period and login the required samples. Once samples are logged in, those records are deleted from the prelogin table. Users can easily generate a worklist of samples to be run, and by which test protocol, with this LIMS functionality. The second type, often called routine sample schedule, involves a perpetual schedule rather than a finite study, so it would be impractical to generate a prelogin list of records to login samples from. Many LIMS systems also provide on-line SOPs (standard operating procedures), such as test methods allowing end-users to obtain all the information about the test method that they need on-line in a read-only format so that the QC manager need only update procedures and protocols in one location.

10 QUALITY CONTROL (QC) AND QUALITY ASSURANCE (QA)

10.1 Quality Control

Quality control and quality assurance functionality is integral to any LIMS. This functionality monitors all controls and allows users to create control charts and to perform a trend analysis of results. Many systems allow users to configure the type of controls that they are using, such as blanks, positive controls, spikes, duplicates, MSD (matrix spike duplicates), standards, CCV (continuous calibration value), ICV (initial calibration value) and many other combinations. Controls typically fall into four types, blanks, standards (controls), duplicates, and spikes.

Quality control can be defined as the operational techniques and activities that are required to maintain and improve product or service quality. When most people hear the term QC in relation to LIMS, they think of the statistical quality control charts that have the values of sampling data over time, with upper and lower warning limits shown on a graph. These graphs provide excellent feedback to users on how the process or product is performing and establish confidence levels. Some users are interested in performing sophisticated statistical determinations on the data. Because most LIMS cannot handle these analyses, users often export their data to a specialized statistical analysis package, such as SAS. It is important to select a LIMS that is based on an open architecture and is OBDC compliant so that it can communicate with other applications. Users should avoid proprietary databases.

10.2 Quality Assurance

The LIMS typically contains a great deal of information on laboratory operation, data quality, and performance. In spite of this, few users effectively mine the information in their LIMS. Many LIMS contain "query builders", basically screens where users can query the information that they are interested in retrieving. For example, a laboratory manager can obtain information on each analyst, and the sample volume that they analyze per day, per month, per year, by test, by client, by the number of audits they signed off, etc. Users can also examine workload by department or by instrument or review QC data and turnaround times for each department (from receipt in the laboratory through final reporting). By measuring overall laboratory performance, laboratory managers can identify areas for improvement and also commend areas that are performing well.

A LIMS can play a significant role in an operation's overall quality. Many of the reports that are generated from the LIMS, such as the analyses reports, statistical reports, SPC (statistical process control) charts, and trend analyses provide significant insight into overall product quality.

11 SPECIFICATION CHECKING

Specification checking is confirmation that a material conforms to properties as defined by the consumer of the material. For example, for raw oil, the laboratory may measure the viscosity. There can be several different specifications for the same product based on the customer's requirements and the uses of that product. LIMS are ideal at automatically notifying an analyst if a product does not meet certain specifications and also at providing a match to customers whose specifications are met, so that the material may be used in another process by another customer. This functionality is widely used in the manufacturing industry.

There are several LIMS quality assurance programs. These include ISO 9000 international standards, government-regulated quality assurance programs, and guidelines such as the EPA's GALP, ASTM (American Society for Testing and Materials), HIPPA, CAP, and CLIA. Other government agencies that participate in regulated QA programs include the FDA, the Public Health Service, the military, the General Services Administration, and state and municipal agencies. Of all the programs offered, the most comprehensive is probably the ISO 9000 certification that requires biannual or annual audits, depending on the customer's request.

12   CHEMICAL AND REAGENT INVENTORY

Many LIMS have expanded their functionality to offer users more features to enhance laboratory productivity. Once such feature is chemical, reagent and even supply inventory. These systems are configured to allow users to set up their tests and record how much of each chemical or reagent is consumed for each test. The LIMS will automatically calculate the quantity on hand by performing reconciliation after each analysis and provide the user with a warning box when they are running low on certain chemicals and reagents. The warning limits and lead time for each chemical are user definable. Keeping an electronic record of each vendor, ordering information (catalog number, quantity, grade, amount, cost, shelf-life, MSDS information (Material Safety Data Sheet), and special shipping and handling instructions, etc.), and even a link to a specific vendor's web site significantly expedites ordering and will save time in reordering. The major advantage of this functionality is the tight integration with the LIMS and automatic warnings.

13   PERSONNEL TRAINING RECORD TRACKING/ INSTRUMENT MAINTENANCE

Many systems also include functionality to manage personnel training certifications and equipment maintenance. Many quality management systems require tracking and maintaining employee training records/certificates to ensure that employees are properly trained on various laboratory procedures, instruments, and methods. This is also required by a number of laboratory certifying agencies (ISO, NELAC, etc.) in an attempt to ensure data quality.

Additional LIMS functionality allows employee-training records to be stored in the LIMS. An advantage of this feature is that if an employee training has expired on a particular method, and if the LIMS has recorded current training as a required field, then the analyst whose training or certification has lapsed will not be allowed to enter information into the LIMS utilizing that method until the training is updated. The LIMS can also be configured to provide message box warnings as the training for specific users is about to expire (for example at 4- and 2-month intervals). These types of reminders can be sent via e-mail to the analyst and the laboratory manager simultaneously and help the laboratory to stay in compliance and maintain its certifications.

Analytical instrument maintenance records can also be stored in the LIMS. This actually makes a lot of sense, since storing all instrument repairs, replacement, or routine maintenance information in the LIMS can be invaluable for troubleshooting. The LIMS can also examine the performance of the instrument on control samples over time so that users are alerted to any potential problems with an instrument, such as drift, before it becomes a major problem. In addition, the LIMS can allow instrument maintenance to be a required field in the software, so that if the instrument is out of calibration users will either need to have the instrument calibrated or use another instrument before they will be able to enter results for a particular test. This provides quality assurance that data entered into the LIMS was not generated by instruments that were out of calibration. This also provides the end user with a comprehensive summary of instrument performance over time and hopefully early warnings to any problems that might arise.

14 ARCHIVING AND DATA WAREHOUSING

Archiving is the process by which old data that has previously been reported and is no longer examined can be removed from the active database and moved to an archived database. The archived data is still readily accessible, but it is no longer in the current database. There are several reasons for archiving:

1. To clean up a LIMS database, for example the laboratory may have retired particular test methods and all associated analyses (previously reported) performed by this method. This keeps the database clean with only current methods and avoids confusion.
2. To enhance system performance, if there is a decrease in system performance due to large amounts of information (stored in tables) transferred over a network with limited bandwidth.
3. Or the laboratory may wish to archive data based on certain time pe-

riods, such as annually or based on government (regulatory agency) timelines for data retention.

Whatever the reason, archiving should only be performed on samples that have passed through each department and when final reporting has been completed. Most LIMS will not allow database administrators to archive active samples. The database administrators should maintain an archive log according to the laboratory's archived data history.

15 ROLE OF THE INTERNET

The Internet has revolutionized the speed at which information is shared and exchanged. By allowing users to publish their data to a site on the Internet, other users can access the data and push their contributions to the site as soon as the data is available. Many LIMS systems can be web-enabled, which allows remote users to visit a site (through a firewall), log into a copy of the LIMS data with a username and password, and view the LIMS data. There are a number of databases on the Internet that provide users helpful information for analyzing data or evaluating experimental design. Table 2 lists just a few useful chemical and biotechnology sites.

16 LIMS VALIDATION

The validation of LIMS is becoming increasingly important for various types of laboratories. Over the past several years, numerous guidelines such as GLP, GALP, and GCP and regulations such as GMPs and Quality System Regulation have been issued. The objective of the validation process is to insure that a system does what it purports to do and will continue to perform in the manner originally intended. Validation not only satisfies regulatory requirements but also provides laboratories with confidence in the system, knowing that is has been tested and the results verified and documented.

There are multiple ways in which to plan and implement a validation process. Validation activities are not a one-time-affair; they should be performed throughout a LIMS lifecycle. Functionality may be modified or new functionality added, and components require revalidation. Validation should begin with the planning of the requirements of the new LIMS and continue through specification, testing, implementation, operation, and retirement of a system. A typical example would include the specifications for a new system, as these are developed; the validation process will later test the system, perhaps with test scripts to determine whether the LIMS will meet the defined specifications.

Table 2 Selected Chemistry and Biotechnology Databases and Websites

Website	Description
http://www.accelrys.com	This site informs chemists of previously performed chemical reactions that have failed. There are over 5,000 reactions in the database, but it is constantly growing with new entries.
http://www.synopsys.co.uk	Methods in organic synthesis (MOS). This site provides chemists with the latest information on recent developments in organic synthesis. It contains abstracts from over 100 key organic chemistry journals. Topics covered include functional group changes, carbon–carbon bond forming reactions, new reagents, and more.
http://www.webreactions.org	This site provides users with direct access to reaction precedents. The search system of web reactions is based on reaction types and bonding change in any reaction.
http://genelogic.com	This site allows researchers to manage, integrate, and analyze their own gene expression and associated clinical data with external data (including proprietary and publicly available databases: GeneBank, SwissProt, LocusLink, and UniGene).
http://www.spotfire.com	This site is used to identify expression patterns. It allows researchers/groups to collaborate on data, computations, and information to locate and validate expression patterns. This site utilizes a common interface that supports SQL Server, Oracle, Sybase, Excel, and DB2.
http://www.lionbioscience.com	This site offers transcript reconstruction software that allows researchers to assign functions to uncharacterized expressed sequence tags (ESTs).

17 GENERAL PRINCIPLES OF SOFTWARE VALIDATION

There are ten general validation principals that are considered to be applicable to LIMS validation.

1. *Timing.* Validation is an ongoing event. It should begin at the start of the project when planning and input begin. Validation occurs throughout the life-cycle of the LIMS.

2. *Management.* Proper validation of a LIMS includes the planning, execution, analysis, and documentation of appropriate validation activities and tasks.

3. *Plans.* Before validation begins, there should be an established design and development plan that describes how the software validation process will be controlled and executed.

4. *Procedures.* Validation procedures should be developed. The validation process should be conducted according to the established procedures.

5. *Requirements.* Before a LIMS validation can begin, there must be a predetermined and documented requirements list. Typically, a request for a proposal will contain the requirements necessary for system validation.

6. *Testing.* Verification includes both static and dynamic techniques. Static techniques include paper/document reviews, while dynamic techniques include physical testing, demonstrating the system's run-time behavior in response to selected inputs and conditions. Dynamic analysis alone may be insufficient to show that the system is fully functional and free of avoidable defects. Static techniques are used to offset limitations of dynamic analysis. Inspections, analyses, walk-throughs, and design reviews may be more effective in finding, correcting, and preventing problems at an earlier stage of the development process.

7. *Partial validation.* A system cannot be partially validated. As LIMS systems are tightly integrated and can be quite complex, a change in one portion may affect another. For this reason, when a change is made to any part of the system, the validation status of the entire system should be addressed.

8. *Amount of effort.* The magnitude of the validation effort should be commensurate with the risk associated with dependence on critical function. The larger the project and number of staff involved, the greater the need for formal communication, extensive written procedures, and management control of the process.

9. *Independence.* It is preferred that validation activities should be conducted using the basic quality assurance concept of ''independence of review.'' For example, the software engineers that wrote the program should not be validating the system.

10. *Real world.* It is fully recognized that LIMS are designed, developed, validated, and regulated in a real-world environment. Environments and risks cover a wide spectrum. Each time a validation principle is used, the implementation may be different.

18 DATA STORAGE

Moore's law states that the density of transistors (therefore processing power) doubles every eighteen months. This is impressive until you compare it with the capacity increases of storage media; hard disk capacity doubles every twelve months on average. The greater data storage capacity has allowed software developers more space for their code. Unfortunately, this is not always beneficial. In the past, programmers were conscientious about writing succinct code, but extra capacity means that queries and code do not have to be recycled and that the programmer can relax and ''fill'' the excess space. This has been quite evident in the explosion of functionality and new features in operating systems alone. In 1981, Microsoft's DOS (Disk Operating System) required less than 160 K of disk space. Eleven years later, in 1992, Microsoft introduced Windows 3.1, which was somewhat familiar to Macintosh users, and required 10 MB for a standard installation. The release of Microsoft Windows 2000 in 1999 required 100 MB for a full installation. It is interesting to note that as hard disks have become smaller and smaller the amount of information that they contain has exploded. Table 3 reviews the history of storage devices; the increased storage has enabled current LIMS technology.

Table 3 Evolution of Storage Devices

Year	Size	Disk space	Description
1956	24″	160 K	IBM introduced the 350 Disk Storage Unit, the first random access hard disk. The size of a dishwasher, it had fifty 2 foot platters that held 5 million bytes of data (approx. 1,500 typewritten pages).
1973	8″	17.5 MB	IBM released the 3340 (the Winchester hard disk). This device had a capacity of 70 MB on four platters.
1980	5.25″	1.25 MB	Seagate introduced the first 5.25 inch hard disk.
1985	5″	650 MB	First CD ROM drive introduced on PCs with 650 MB readonly capacity.
1988	2.5″	10 MB	Release of the Prairie Tek 220, the first 2.5 inch hard disk for portables.
1998	5″	2.6 GB	The debut of the DVD-RAM drive, 5.2 GB rewritable capacity.
1999	1″	340 MB	IBM's new Micro Drive, for portable devices, fits 340 MB.
2001	3.5″	10 GB	Quinta's Optically Assisted Winchester (OAW) drive appears; it is expected to store 20 GB of data per square inch.
2005	3.5″	58 GB	Typical desktop hard disk that holds 280 GB on five platters.

Source: Adapted from PC magazine (May 25, 1999).

19 CONCLUSIONS

A successfully implemented LIMS will increase laboratory productivity, improve data accuracy and quality, and increase the overall effectiveness of the laboratory. A LIMS can organize all the information that is pertinent to the laboratory and allow for rapid data retrieval and reporting. It also allows data to be accessible to others, promoting collaboration among different departments. In addition, many laboratories utilize either a local area network (LAN) or a wide area network (WAN) that allows users to share network printers, instrument files, and laboratory information. Once the LIMS is fully integrated into the laboratory, then users typically have the need to integrate the LIMS to other enterprise applications [customer relationship management (CRM), ERP, accounting, etc.] that will allow managers to mine the databases for information that may not be readily apparent

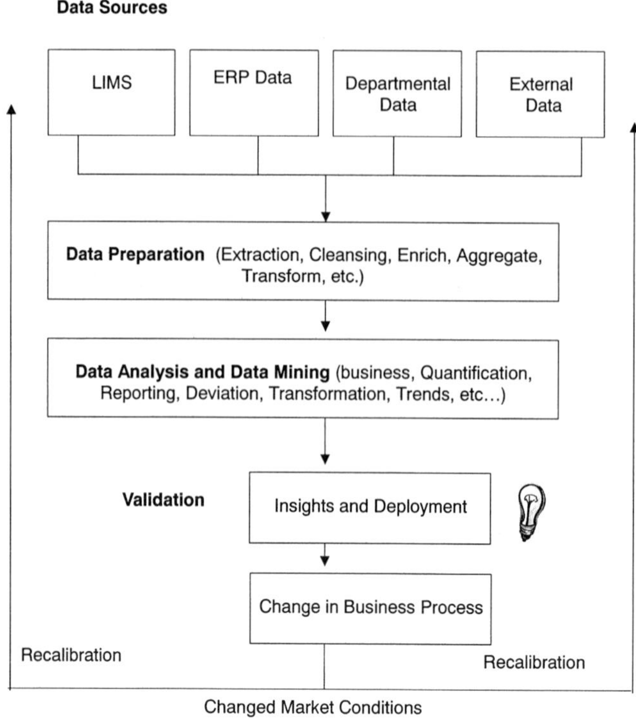

Figure 4 A general review of the data mining process and the role of LIMS.

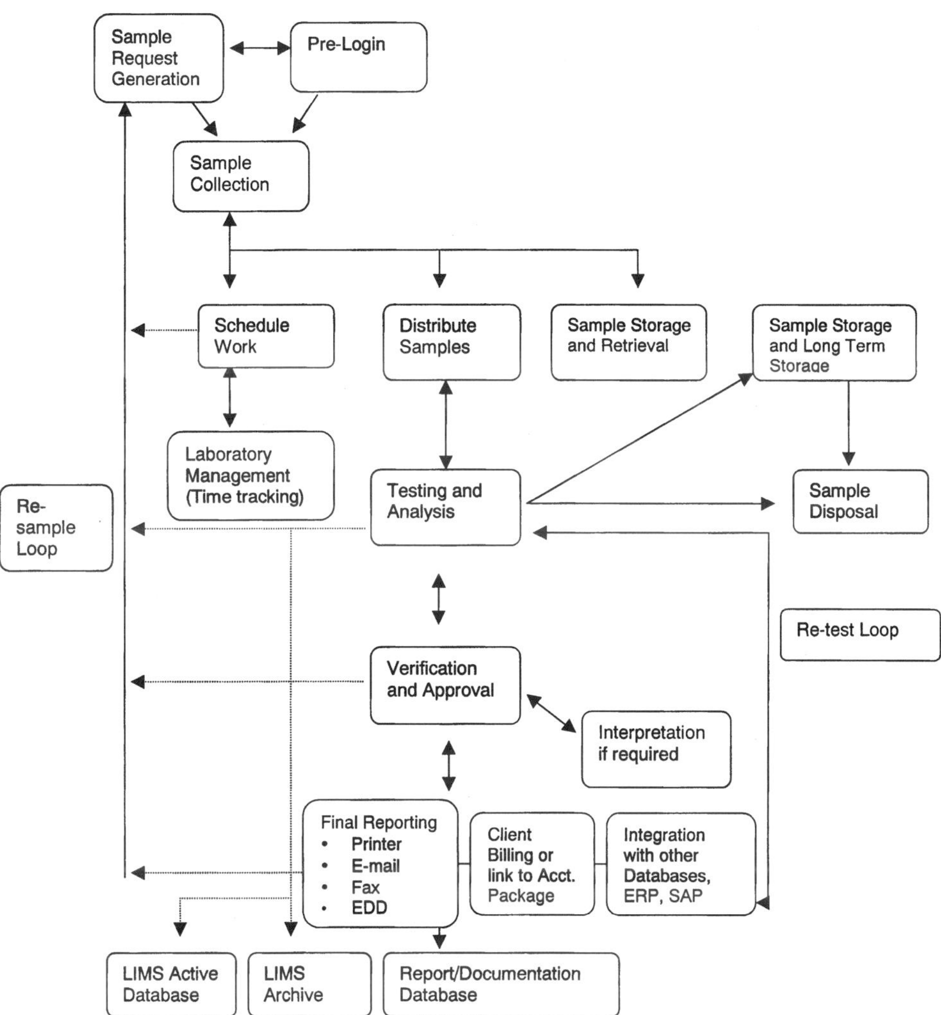

Figure 5 Schematic representing generic LIMS workflow.

from-review of a single database alone. Figure 4 provides a general review of the data mining process, while Fig. 5 depicts a generic LIMS workflow in much greater detail.

A LIMS can be a powerful tool that gives the laboratory a competitive advantage over other laboratories, saving time and money. With the decreasing hardware and software costs, and growing acceptance of the Internet, the time for laboratories to move from paper tracking systems or spreadsheets to a LIMS has never been better. Not all laboratories operate in the same way, so it is extremely important that a LIMS match the laboratory flow and have the flexibility to accommodate future changes in laboratory operations.

ABBREVIATIONS

ASP	active server pages
ASTM	American Society for Testing and Materials
CAP	College of American Pathologists
CCD	charged couple devices
CCV	continuous calibration value
CFR	Code of Federal Regulations
CLIA	Clinical Laboratory Improvement Amendments
CRM	customer relationship management
DOS	disk operating system
ERP	enterprise resource planning
FDA	Federal Drug Agency
FFO	fixed focus optics
GALP	Good Automated Laboratory Practices
GLP	Good Laboratory Practices
GMP	Good Manufacturing Practice
GUI	graphical user interface
HIPPA	Health Insurance Portability and Accountability Act
HL7	Health Level 7
ICR	intelligent character recognition
ICV	initial calibration value
ISAM	indexed sequential access method
ISO	International Organisation for Standards
LAN	local area network
LIMS	laboratory information management systems
MSD	matrix spike duplicates
MSDS	material safety data sheet
NELAC	National Environmental Laboratory Accreditation Conference
OCR	optical character recognition
ODBC	open database connectivity
PC	personal computer
QA	quality assurance
QC	quality control
ROI	return on investment
SOP	standard operating procesures

SPC   statistical process control
SQL   structured query language
USEPA   United States Environmental Protection Agency
VAX   virtual address extension
WAN   wide area network

REFERENCES

Albu, Michael L. ISO 9000 and the Analytical Laboratory. *American Laboratory*, Sept. 1996.

Atzeni, P., and De Antonellis, V. *Relational Database Theory*. Benjamin Cummings, 1993.

Avery, George, McGee, Charles, and Falk, Stan. Implementing a LIMS: a how-to guide. *Analytical Chemistry*, 72:57A–62A, 2000.

Baum, M. How public key infrastructure can support self-authenticating records. Proc. Natl. Conf. Managing Electronic Records, Nov. 2–4, 1998.

Dollar, C. The Fermo report on authentic electronic records; strategies for long-term access. Proc. Natl. Conf. Managing Electronic Records, Nov 2–4, 1998.

Garner, Willa Y., Barge, Maureen S., and Ussary, James P. *Good Laboratory Practice Standards*. Washington DC: American Chemical Society, 1992.

Gillespie, Helen. ISO 9000 in scientific computing. *Scientific Computing and Automation*, Feb. 1994.

Hines, Eric. LIMS: the laboratory's right arm. *Pharmaceutical Formulation and Quality*, May/June 1999, 37–43.

Hinton, Mary D. *Laboratory Information Management Systems—Development and Implementation for a Quality Assurance Laboratory* New York: Marcel Dekker, 1995.

Johnson, Tony, and Hanson, Ron. Training is the key to success. *Scientific Computing and Automation*, April 1999.

Karbo, Michael. The PC technology guide. *WWW.PCTechGuide.com*, 1999.

Liscouski, Joe. *Laboratory and Scientific Computing: A Strategic Approach*. Wiley-Interscience, 1994.

Lysakowski, R. and Doyle, L. Electronic lab notebooks: paving the way of the future in R&D. *Record Management Quarterly*, 32(2):23–30, 1998.

Mahaffey, Richard R. *LIMS Applied Information Technology for the Laboratory*. New York: Van Nostrand Reinhold, 1990.

Miller, Tom. Windows-based LIMS provides flexible information management. Scientific Computing and Automation, July, 27–30.

Nakagawa, Allen S. LIMS implementation and management. Cambridge: Royal Society of Chemistry, 1994.

McArthur, Douglas. Client/server issues for scientific information systems. *Scientific Computing and Automation*, February 1997.

McDowell, R. D. Operational measures to ensure continued validation of computerized systems in regulated or accredited laboratories. *Laboratory Automation and Information Management*, 31, 1995.

Paszko, Christine. Extending LIMS functions over the Internet. Inside Laboratory Management, AOAC International, April 1998, p. 19.

Perrit, H. J. Self-authenticating records: what are they and why are they legally important? Proc. Natl. Conf. on Managing Electronic Records, Nov. 2–4, 1998.

Preston, Curtis. Storage management technology. *Windows NT Systems*, July 1999.

Roberts, Jason. Choosing a LIMS: a crucial decision for the laboratory. *Scientific Computing and Automation*, October 1999, 54–55.

Rothenberg, J. Ensuring the longevity of digital documents. Sci. Amer., 272(1):42–47, 1995.

Sanin, Leo, and Chen, Renzhong. Client/server programming with Access and SQL Server. Ventura Communications Group, 1997.

Segalstad, Siri. LIMS validation requires a systematic approach. *Scientific Computing and Automation*, May 1997, 33–34.

Simovici, D., and Tenney, R. *Relational Database Systems*. Academic Press, 1995.

Tetzlaff, Ronald F. GMP Documentation requirements for automated systems: Part I. *Pharmaceutical Technology*, April 1992, 60–72.

Tetzlaff, Ronald F. GMP Documentation requirements for automated systems: Part II. *Pharmaceutical Technology*, April 1992, 60–72.

Ullman, J. *Principles of Database and Knowledge-Base Systems, Volume 1*. Classical Database Systems. Computer Science Press, 1988.

8
Design of Chemical Processes for Automated Synthesis

Junzo Otera
Okayama University of Science, Okayama, Japan

1 INTRODUCTION

Chemical processes are usually composed of a number of operations. The first reaction transforms a starting material to an intermediate, which after isolation and purification is subjected to the second reaction leading to the next intermediate. Repetition of these operations drives the process to the final goal. The automation of such multistep processes is of great importance from both economical and ecological points of view. Full automation has been achieved in industrial chemical plants, as it is rather easy to realize automation in commercial plants where the routine operations are repeated on a large scale. By contrast, bench work still remains far behind such innovation, despite increasing demands for high efficiency and reliability in laboratory operations. Moreover, the importance of laboratory automation will constantly grow, in order to reduce labor costs and create cleaner environments in the laboratory. Nevertheless, the experimental techniques for synthetic reactions have experienced virtually no improvement for a long time, in contrast to the outstanding progress in analytical instrumentation. The reasons for such delay in innovation are as follows. (1) In laboratories, various types of reactions must be employed, and thus the apparatus must be designed for multipurpose uses. The design is thus much more complicated than that for the specialized use in commercial plants. (2) Small-scale operations encounter technical difficulties in various aspects. For instance, many organic reactions should be run under a rigorously inert atmosphere, but it is rather difficult to seal the apparatus completely from ambient air in small-scale reactions. Also, it is not easy to weigh and charge small amounts of reagents with high precision.

Finally, the control of the reaction temperature is more difficult, in particular, for exothermic reactions, as the scale is smaller. Despite these limitations, some synthetic machines have been developed that overcome the difficulties to some extent [1]. It is noted, however, that no machines are available so far that allow consecutive reactions under a completely inert atmosphere at low temperature.

Two strategies are available for automation of multistep processes. One is so-called "robotics," a miniaturized form of industrial plants, in which the intermediate products are automatically transferred from one reactor to another. In this case, the automation of the process is highly dependent on engineering, whereas the role of chemistry is trivial once the synthetic route has been determined. These systems, however, are expensive to build, and air-sensitive or thermally labile substances cannot be easily transferred in laboratory-scale operations. The other strategy utilizes a one-pot process where multifold reactions are conducted in a single flask. The process design totally relies on chemistry in this case [2]. Although this treatment is cheap, it is virtually impossible to consolidate all the reactions needed for a whole synthetic route. Nevertheless, it is highly useful even if only a part of the synthetic sequence is consolidated.

Numerous one-pot reactions have been put forth so far [3]. I would like to classify the precedents into two categories: (1) the substrate-based strategy and (2) the reagent-based strategy. Strategy (1) is well used for domino, tandem, or cascade reactions. In this treatment, the reaction proceeds elegantly once the precursor is formed. However, much elaboration is frequently needed to prepare the precursor, and thus the whole process is not necessarily compacted if the preparation steps of the precursor are taken into account. Strategy (2) is newer and represented by the consecutive Heck reaction as a typical example. This reaction also requires the preparation of the precursor. More seriously, the reactions employable in both strategies are limited mostly to the addition reaction and Michael/aldol reactions in some cases. The functional group transformations usually cannot be consolidated. Consequently, a lot of difficulties must be overcome to make use of these strategies in practical chemical processes. We were intrigued to devise new concepts for one-pot processes in which ordinary reactions can be conducted under ordinary reaction conditions so that any kind of chemical transformation could be integrated [4].

2 INTEGRATED CHEMICAL PROCESS

2.1 Concept

Before describing our new concept, let us consider the conventional treatment for designing multistep processes. In Scheme 1, a process to arrive at product P starting from A is depicted. The first operation is to design a synthetic route.

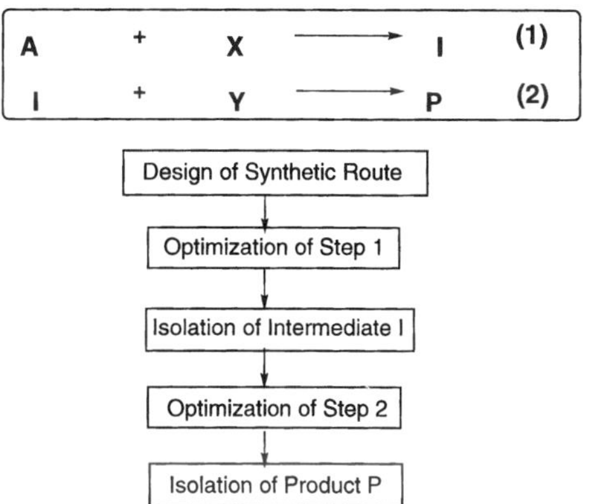

Scheme 1

This may be conveniently done by retrosynthetic analysis. Suppose that a two-step process via intermediate X has been proposed in the present case. Then the first reaction is optimized to obtain the highest yield under the most convenient conditions. The intermediate X is isolated and purified. The second step follows in a similar way to provide the final product P. Most importantly, it should be pointed out that the optimizations of the respective reactions are made regardless of the other reactions. The result of these operations is that there are very few chances of consolidating the two steps, because the reactions are set to run under different conditions in most cases. The integration of the reactions is feasible only when both reaction conditions are compatible by chance.

The above treatment is typical of the reductionism which has met with great success in modern science: the whole is divided into the parts and the parts correctly sum again into the whole. Therefore the reactions giving rise to the best outcomes, in terms of the respective steps, have been sorted out and combined. The new concept, which is described here for the first time, stemmed from the question; Does the combination of the best reactions always lead to the best overall outcome? To answer this question, a holistic approach has been invoked. Scheme 2 illustrates such treatment for the same process as discussed above. The idea is quite simple. The initial treatment is again the design of the synthetic route analogous to the conventional approach. But the most characteristic feature of our approach is that prior to the optimization of each step, the reaction conditions that are to be employed for both reactions are fixed. Under these conditions, each reaction is then optimized. The reactions thus selected may not be the best in terms of optimization of the reaction itself but it necessarily follows that the reactions can be integrated. Although such treatment might have been unintentionally employed on occasions thus far, no conceptualization has been advanced.

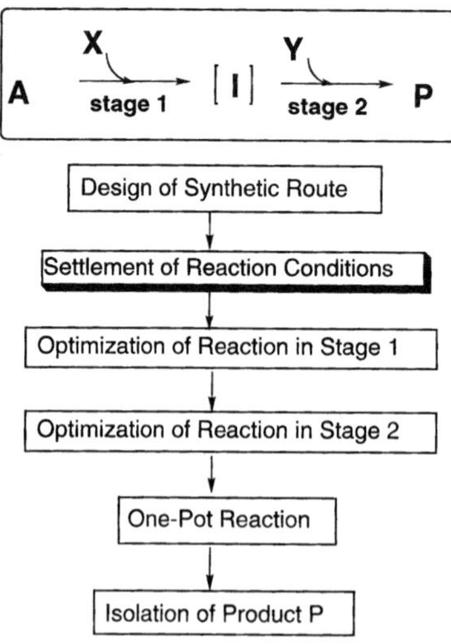

Scheme 2

I designate the new concept as the ''integrated chemical process.'' It will be shown in the following paragraphs that the advantages arising from avoiding the need for isolation and purification of intermediates are reflected in not only the simplification of the chemical process but also the increase of overall yields.

2.2 Double Elimination

2.2.1 Stepwise Process

In 1984, we put forth a double elimination process that provided a convenient access to polyenes and acetylenes (Scheme 3) [5]. This process consists of three reactions: (1) the aldol-type reaction of an α-sulfonyl carbanion with an aldehyde, (2) the attachment of a leaving group on the hydroxyl of the adduct, and (3) the successive elimination of the protected hydroxy and sulfonyl groups. In the latter step, when a proton is available at the allylic position, the vinyl sulfone resulting from the first elimination undergoes isomerization to an allylic sulfone. Then the subsequent 1,4-elimination gives a diene. On the other hand, when no allylic proton is available in the vinyl sulfone, the 1,2-elimination occurs to give an acetylene. In this process, the first and third reactions are conducted under basic conditions, while the second requires acidic conditions. According to our ''integrated chemical process,'' a one-pot process may be feasible if the second step is changed to run under basic conditions. This is indeed the case, as will be shown in the following examples.

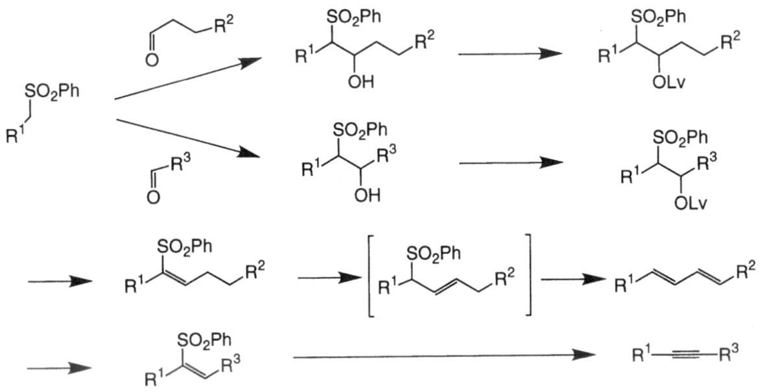

Scheme 3

2.2.2 One-Pot Process for Vitamin A

The original double elimination process for vitamin A is shown in Scheme 4 [6]. Cyclogeranyl sulfone **1** is coupled with aldehyde **2** derived from geranyl acetate. The resulting aldolate **3** is converted to its MOM or THP ether **4**. Treatment of **4** with an excess amount of MeOK or ^{t}BuOK affords vitamin A. This process is of great promise in terms of the following aspects: (1) it is the first strategy based on two readily available C_{10} building blocks and it is much simpler than the existent processes, and (2) the high all-E content ($\sim$ 95%) of the vitamin A that is obtained. Although it is already simple to a considerable extent, the integration of these three steps would lead to an extremely compact process. The one-pot process developed on the basis of an integrated chemical process is illustrated in Scheme 5 [7]. The initial step is the same as that of the stepwise process. Then the resulting anion of the aldolate **3′** is treated with methoxymethyl (MOM) bromide in situ to furnish the MOM ether **4**. Addition of MeOK to this reaction mixture provides vitamin A in 58% overall yield based on **1**. Further elaboration has led us to find that addition of NaI to the reaction mixture improves the yield (Scheme 6). More interestingly, the use of this salt enables the use of the cheaper MOMCl in place of MOMBr. As a result, a significant increase in the overall yield has been attained (78%) compared to the stepwise process (67%). Assuming that the final double elimination in the integrated chemical process took place in the same yield (78%) as in the stepwise process, it is reasonable to conclude that both the first two steps (aldol and MOM ether formation) proceeded in 100% yield in the integrated process. Apparently, the advantage of the one-pot process for avoiding material losses during isolation and purification of the intermediates, which inevitably emerges in the stepwise process, is exemplified. It should also be noted that the all-E content of vitamin A is as high as for the original one.

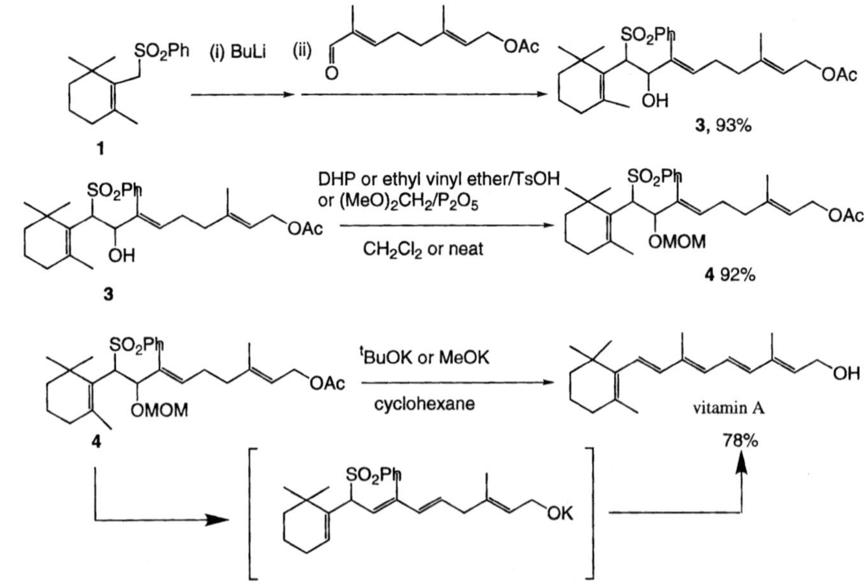
SO2Ph
(i) BuLi
(ii)
OAc
O
1
SO2Ph
OAc
OH
3, 93%
DHP or ethyl vinyl ether/TsOH
or (MeO)2CH2/P2O5
CH2Cl2 or neat
3
OMOM
4 92%
tBuOK or MeOK
cyclohexane
4
OH
vitamin A
78%
OK

Scheme 4

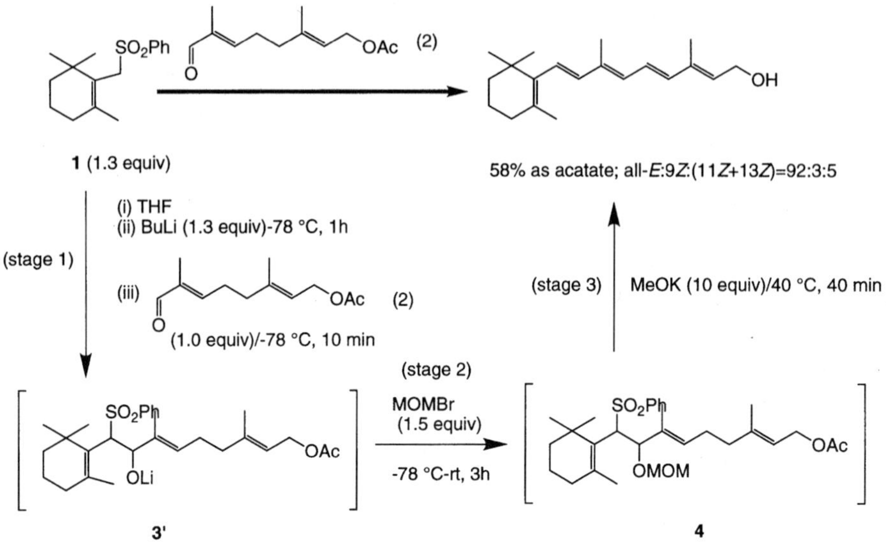
SO2Ph
OAc (2)
OH
1 (1.3 equiv)
58% as acatate; all-E:9Z:(11Z+13Z)=92:3:5
(i) THF
(ii) BuLi (1.3 equiv)-78 °C, 1h
(stage 1)
(iii)
OAc (2)
O
(1.0 equiv)/-78 °C, 10 min
(stage 3)
MeOK (10 equiv)/40 °C, 40 min
(stage 2)
MOMBr
(1.5 equiv)
-78 °C-rt, 3h
OLi
3'
OMOM
4

Scheme 5

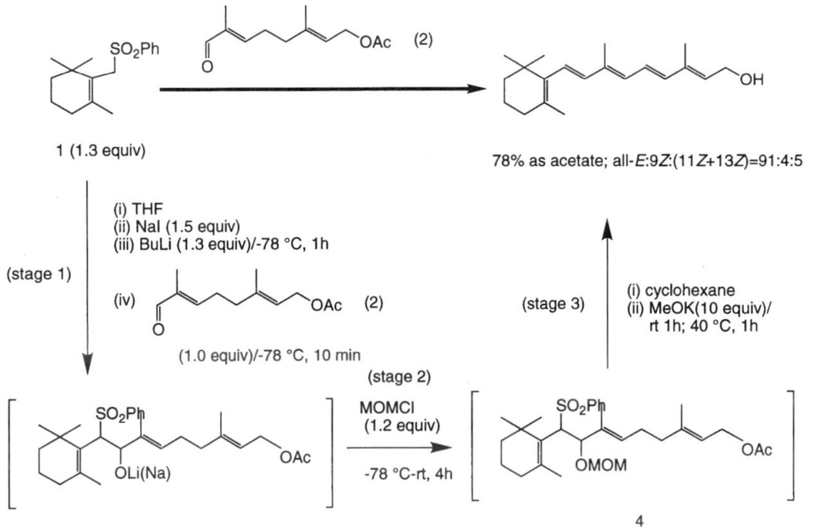

1 (1.3 equiv)

78% as acetate; all-*E*:9*Z*:(11*Z*+13*Z*)=91:4:5

(i) THF
(ii) NaI (1.5 equiv)
(iii) BuLi (1.3 equiv)/-78 °C, 1h

(stage 1)

(iv)   (2)

(1.0 equiv)/-78 °C, 10 min

(stage 2)

MOMCl
(1.2 equiv)

-78 °C-rt, 4h

(stage 3)

(i) cyclohexane
(ii) MeOK(10 equiv)/
rt 1h; 40 °C, 1h

4

Scheme 6

2.2.3 One-Pot Aromatization of Cyclic Enones

The use of cyclic enones in place of the aldehyde in the double elimination process provides a novel methodology to obtain alkyl aromatic compounds (Scheme 7) [8]. Treatment of α-sulfonyl carbanions with cyclic enones results in exclusive 1,2-addition. The resulting tertiary alcohols are efficiently trapped in situ with benzoyl chloride. Addition of t-BuOK to this reaction mixture furnishes the desired alkylation products. A variety of cyclohexanones undergo smooth aromatization (Table 1). α-Tetralones and their higher homologues can also be employed to arrive at polycyclic aromatic hydrocarbons. It should be noted that the primary alkyl groups can be incorporated on the aromatic rings with complete regiocontrol, which is a great advantage over the Friedel–Crafts reaction. The Friedel–

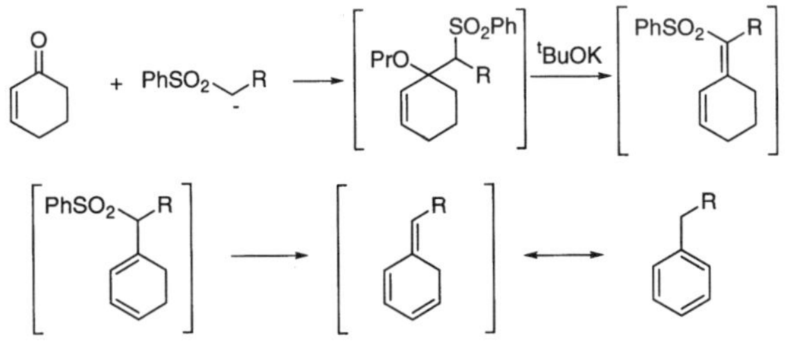

Scheme 7

Table 1 One-Pot Synthesis of Aromatic Compounds from Cyclic Enones

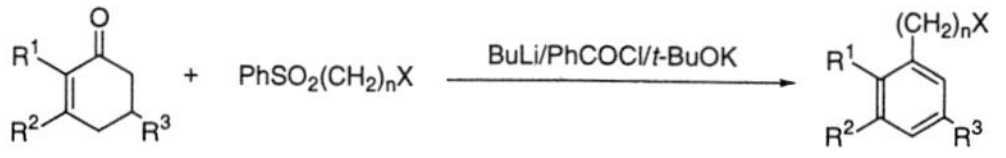

Entry	Enone	Sulfone	Product	Yield (%)
1	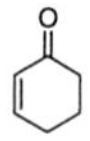	$PhSO_2(CH_2)_5CH_3$	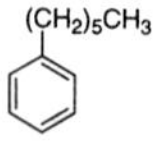	75
2	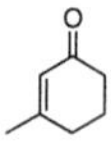	$PhSO_2(CH_2)_5CH_3$	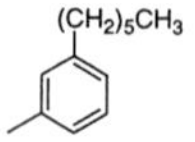	72
3	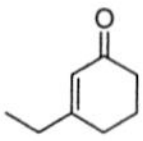	$PhSO_2(CH_2)_5CH_3$	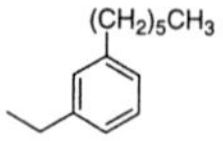	69
4	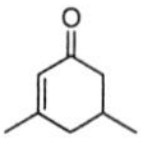	$PhSO_2(CH_2)_5CH_3$	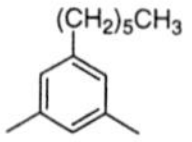	84
5	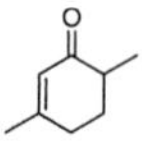	$PhSO_2(CH_2)_5CH_3$	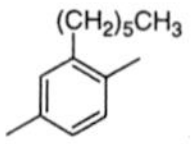	79
6	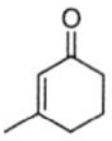	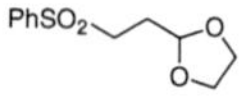	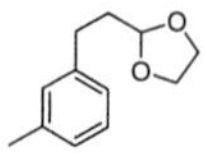	78
7	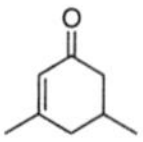	$PhSO_2(CH_2)_6OTBS$	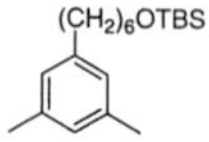	72
8	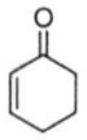	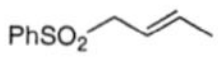	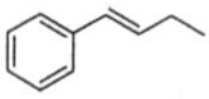	59
9	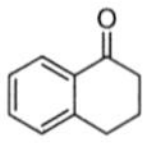	$PhSO_2C_6H_{13}$	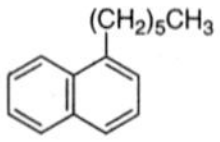	82
10	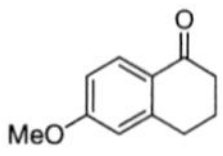	$PhSO_2C_6H_{13}$	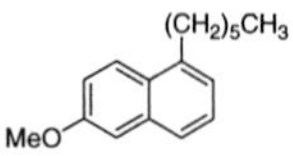	74
11	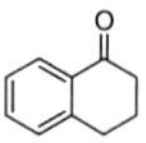	$PhSO_2CH_3$	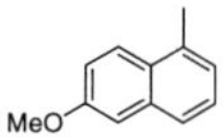	76
12	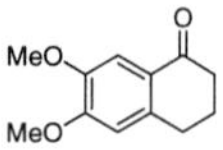	$PhSO_2C_6H_{13}$	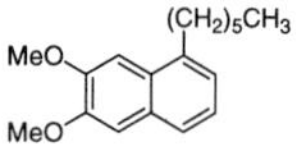	79
13	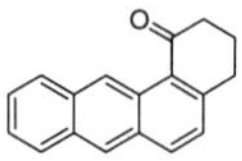	$PhSO_2C_6H_{13}$	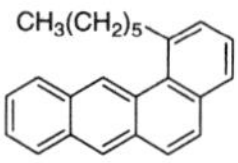	78
14	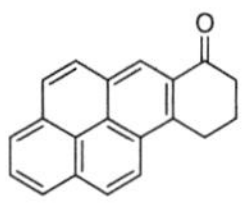	$PhSO_2C_6H_{13}$	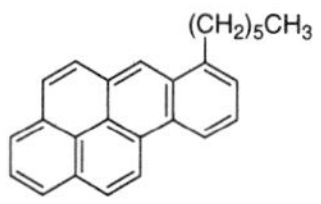	88

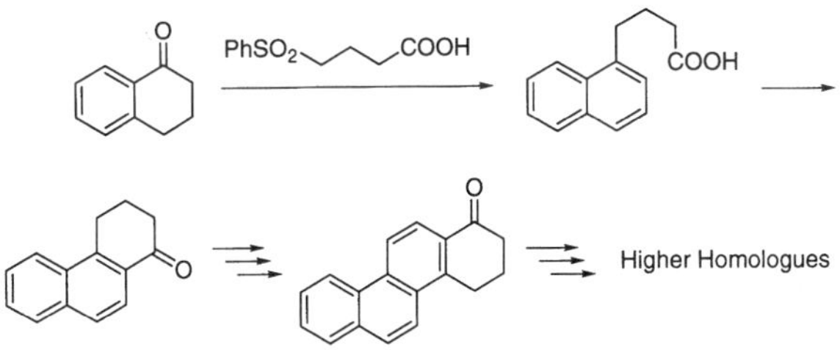

Scheme 8

Crafts reaction suffers poor regioselectivity, and alkyl groups are incorporated in branched forms even if primary alkyl halides are used in the reaction. More seriously, the degree of alkylation is difficult to control: polyalkylation predominates over monoalkylation because the alkylation products are more reactive than their precursors. The present protocol is entirely free from such drawbacks and hence has allowed us to prepare various new polylalkylated aromatic compounds that are not easy to obtain by conventional methods.

Further application has been directed toward the development of a procedure for polycyclic aromatic hydrocarbons, which are the target of current interest as new materials [9]. The use of ω-carboxy sulfones should lead to α-tetralones or their higher homologues, and the iteration of this procedure, in principle, results in annulation of aromatic skeletons (Scheme 8). First, we employed sulfones with a carboxy or its ester group at the γ-position, but the reaction of their anions with ketones failed to give the desired aldol products. However, we have found an alternative convenient procedure with ω-acetal sulfones **5** (Scheme 9). Treatment of **5** with enones accomplishes the one-pot aromatization to give **6**. Then treatment of **6** with OXONE (potassium peroxomonosulfate) in THF/H$_2$O effects deprotection followed by oxidation of the acetal to furnish carboxylic acids **7** directly. Exposure of **7** to trifluoromethanesulfonic acid provides the desired ketones **8** in quantitative yields.

2.2.4 One-Pot Synthesis of Acetylenes

As described already, the double elimination is also applicable to acetylene synthesis. A typical procedure is similar to that employed for polyene synthesis (Scheme 10) [10]. An α-sulfonyl carbanion is treated with aldehyde followed by trapping the resulting aldolate with acetic anhydride or trimethylsilyl (TMS) chloride. Addition of ^{t}BuOK or lithium diisopropylamide (LDA) to this reaction mixture furnishes acetylenes. The successful use of TMSCl implies that β-siloxy sulfone intermediates are formed. It is reasonable to postulate that the same species can be generated by Peterson elimination. In fact, this reaction could be integrated smoothly (Scheme 11). Thus a sequence of silylation, aldol-type cou-

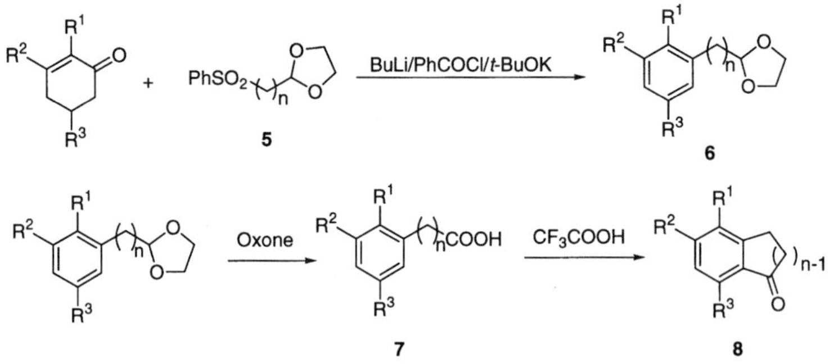

Scheme 9

pling, Peterson elimination, and sulfone elimination has been integrated to provide a variety of enyenes and diynes in reasonable yields (Table 2). It should be pointed out that the integrated process gives higher overall yields than the stepwise process.

A triyne is also synthesized in one pot. Linear polyynes are potentially useful as novel materials [11], and hexatriynes are of particular interest as intermediates for synthesis of enediyne natural products. Nevertheless, the previous methods used to obtain triynes suffered from considerable limitations [12]. The one-pot double elimination protocol, on the other hand, provides a very convenient access (Scheme 12). 3-Phenylpropargyl sulfone (**9**) and 3-phenylpropynal (**10**) is coupled by use of MeOK. The resulting aldolate is trapped as a silyl ether that is then treated with $LiN(SiMe_3)_2$ to give triyne **11** in 78% yield. Furthermore, addition of the trapping agent, TMSCl, from the beginning of the reaction leads to an extremely simple process: successive addition of MeOK and $LiN(SiMe_3)_2$ to a mixture of sulfone, aldehyde, and TMSCl affords **11** in 64% yield.

Aryl acetylenes [13] are another class of substances obtained by the present protocol. Extensive studies have appeared on these compounds, and the strategies that have been applied for their syntheses mostly rested on the Sonogashira-type

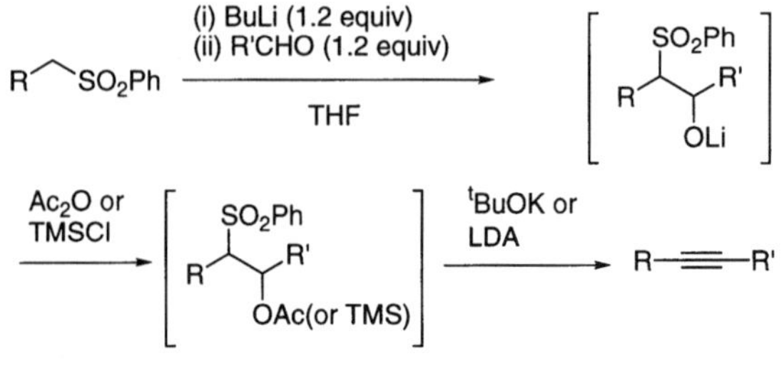

Scheme 10

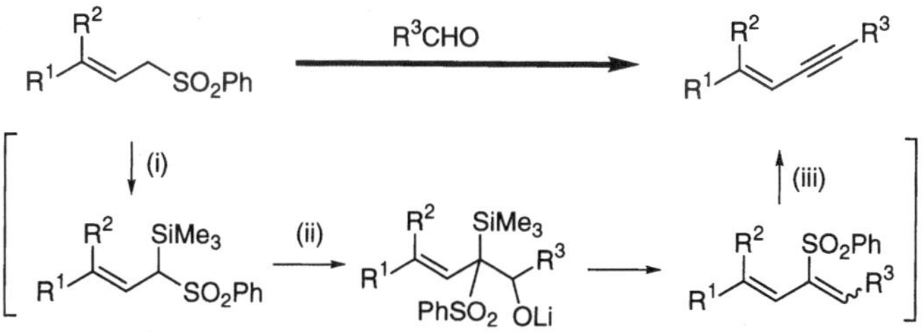

(i) BuLi (1.1 equiv), -78 °C, 0.5 h, then Me$_3$SiCl (1.1 equiv), -78 °C - rt, 1 h.
(ii) BuLi (1.2 equiv), -78 °C, 0.5 h, then R^{3}CHO (1.2 equiv), -78 °C, 1 h and
0 °C, 1 h. (iii) (Me$_3$Si)$_2$NLi.

Scheme 11

Table 2 One-Pot Double Elimination (Peterson Route)

Entry	Sulfone	Aldehyde	Acetylene/yield (%)
1	PhCH$_2$SO$_2$Ph	furyl-CHO	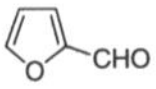 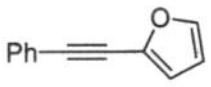 Ph—≡— furyl 66 ($89 \times 45 = 40$)[a]
2  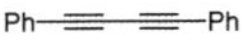		Ph—≡—CHO	Ph—≡—≡—Ph 78
3	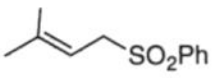 Me$_2$C=CH—CH$_2$SO$_2$Ph	PhCHO	 77
4 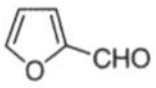 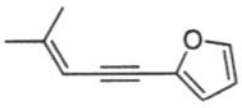		furyl-CHO	75
5		Ph—≡—CHO	  88 ($92 \times 71 = 65$)[a]
6	PhCH=CHCHO	PhCHO	 85 ($85 \times 67 = 57$)[b]
7 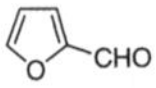 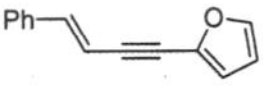		furyl-CHO	71
8		Ph—≡—CHO	  79 ($84 \times 59 = 50$)[b]
9 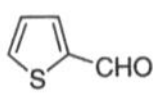 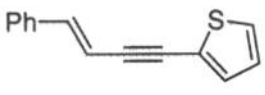		thienyl-CHO	71
10	CH$_3$CH=CHCHO	PhCHO	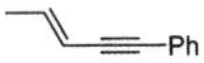 63
11 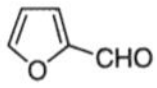 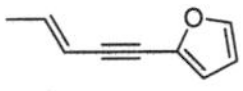		furyl-CHO	57
12		Ph—≡—CHO	 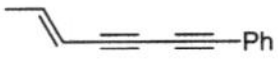 66 ($72 \times 50 = 36$)[b]

[a] Overall yield by the stepwise acetoxy process (% yield of acetate × % yield of acetylene) given in parentheses.

[b] Overall yield by the stepwise Peterson-sulfone elimination process (% yield of vinyl sulfone × % yield of acetylene) given in parentheses.

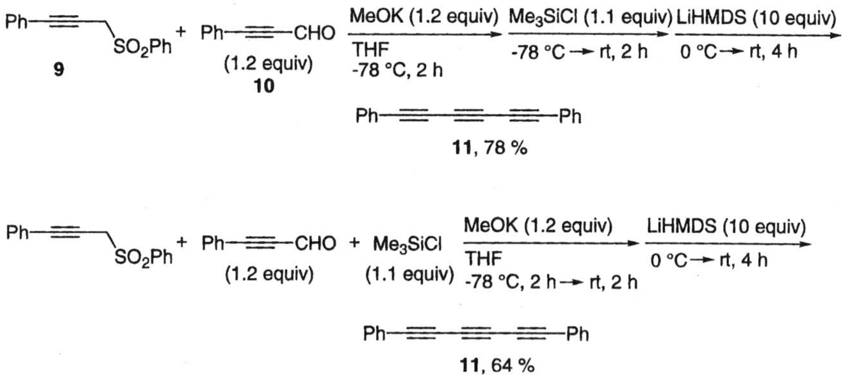

Scheme 12

coupling between aryl halides and terminal acetylenes [14]. Accordingly, aryl acetylenes with ring halogen(s) are of great promise as building blocks for constructing well-defined aryl acetylene scaffolds. Indeed, the present method provides an easy access to various types of functionally substituted aryl acetylenes (Scheme 13). *o*-Substituted and *o,o*-disubstituted diphenyl acetylenes and their higher homologues are obtained by a simple operation.

Iteration of coupling between aryl aldehyde and acetal sulfone modules provides a convenient approach to aryl acetylene oligomers (Scheme 14). Notably, this method is unique in that the chain is elongated concomitant with generation of an acetylenic bond. This is a great synthetic advantage over the Sonogashira reaction because handling of chemically labile terminal acetylene is avoided.

2.3 One-Pot Alkylation–Desulfonylation

Alkylation of allylic sulfones followed by desulfonylation provides a convenient access to olefins [15]. Since the normal carbanion technology is invoked for the alkylation, a one-pot process would be feasible if the desulfonylation could proceed under the conditions compatible with those for the alkylation. Among vari-

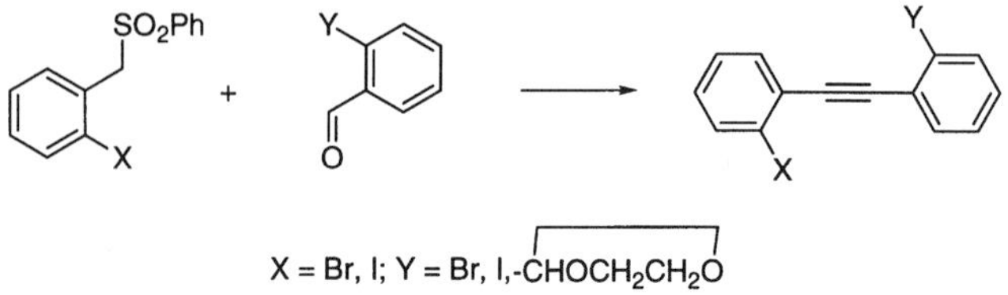

Scheme 13

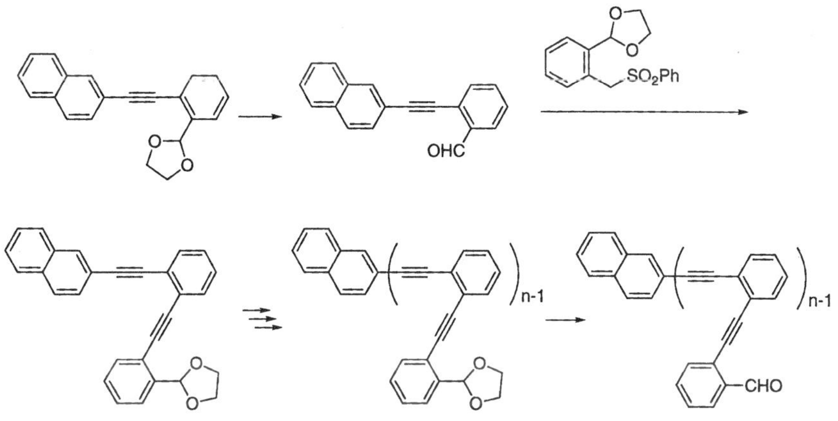

n	yield/%
2	98
3	78
4	92
5	76

n	yield/%
2	84
3	88
4	96
5	96

Scheme 14

ous desulfonylation methods so far advanced, the palladium-catalyzed $LiBHEt_3$ reduction seems the most promising because of its compatibility with basic conditions [16]. Scheme 15 outlines the integration of the alkylation of sulfones followed by reductive desulfonylation to arrive at olefins [17]. Allylic sulfones **12** are treated with BuLi and subsequently alkyl halides **13** to furnish alkylation products **14** (step i). Addition of $Pd(OAc)_2$, phosphine, and $LiBHEt_3$ to this reaction mixture effects the desulfonylation (step ii). The results are summarized in Table 3. As a consequence of screening several phosphines (entries 1–3), 1,3-

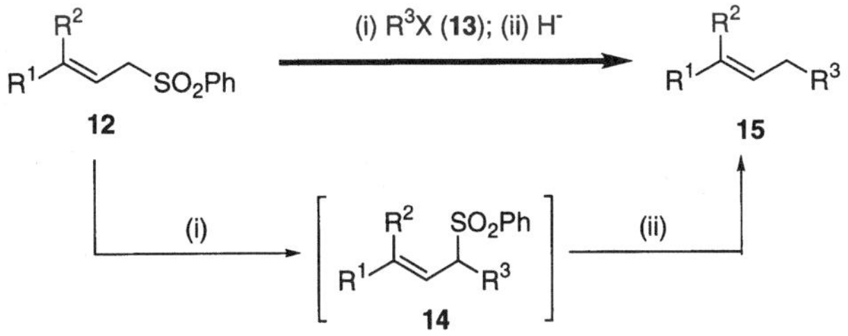

Scheme 15

Table 3 One-Pot Alkylation/Desulfonylation[a]

Entry	12	13	Reaction conditions of step (ii)[b]	15		Yield (%)[c]
1	12a	13a	dppp (0.1 equiv); 0°C, 2 h		15aa	85 (81 × 79 = 64)
2		13a	dppb (0.1 equiv); 0°C, 2 h; r.t., 5 h	15aa		78 (81 × 97 = 79)
3		13a	Bu₃P (0.4 equiv); 0°C, 2 h; r.t. 20 h	15aa		57 (81 × 84 = 84)
4		13b	dppp (0.1 equiv); 0°C, 2 h		15ab	79 (91 × 57 = 52)
5		13c	dppp (0.1 equiv); 0°C, 2 h		15ac	81 (93 × 54 = 50)
6		13d	dppp (0.1 equiv); 0°C, 2 h		15ad	76
7	12b	13a	dppp (0.1 equiv); 0°C, 2.5 h		15ba	66 (74 × 66 = 49)
8		13a	Bu₃P (0.4 equiv); 0°C, 2 h	15ba		83 (74 × 77 = 57)
9		13b	dppp (0.1 equiv); 0°C, 4 h		15bc	66 (89 × 66 = 59)
10		13c	dppp (0.1 equiv); 0°C, 2 h		15bc	46 (70 × 46 = 32)
11	12c	13b	dppp (0.1 equiv); 0°C, 2 h; r.t., 20 h		15cb 16cb	85[d]
12		13c	dppp (0.1 equiv); 0°C, 2 h; r.t., 5 h		15cc 16cc	93[e]

[a] Reaction conditions for (i): **1** (1.05 equiv), BuLi (1.05 equiv), **2** (1.0 equiv), THF, −78°C ~ rt, 2 h.
[b] Pd(OAc)₂ (0.1 equiv.); LiBHEt₃ (2 equiv.).
[c] Isolated yield. The overall yield in stepwise reaction [(% yield in (i)) × (% yield in (ii))] is given in parentheses.
[d] **15cb:16cb** = 18:82.
[e] **15cc:16cc** = 39:61.

bis(diphenylphosphino)propane (dppp) is the ligand of our choice, since it gives rise to higher yields than 1,4-bis(diphenylphosphino)butane (dppb) and Bu_3P, yet a better outcome is obtained with Bu_3P in one exceptional case (entry 8). A variety of allylic sulfones **12** and alkyl halides **13**, whose structures are shown here, are employed, and, in general, reasonable yields of the desired olefins **15** are obtained. The alkyl group is incorporated exclusively at the α-position to furnish a single isomer except **12c**. With this substrate, the hydride attack does not take place regioselectively affording olefin **16** as major products (entries 11 and 12). For comparison, the stepwise approach also has been performed where the alkylated sulfones are isolated and, after purification, subjected to desulfonylation under the same reaction conditions as the integrated process. Yields of the respective steps as well as overall yields are shown in the parentheses in the table. It should be noted that at least three trials have been run for each reaction to assure that the yield and the averaged values given are accurate within the range of $\pm 5\%$ deviation. Remarkably, the integrated process always affords higher yields than the stepwise process, except in two cases (entries 2 and 3).

Another characteristic feature of the carbanion chemistry of allylic sulfones is the ease with which two alkyl groups can be incorporated (Scheme 16). After double alkylation of sulfone **12**, $Pd(OAc)_2$, phosphine, and $LiBHEt_3$ are added to the reaction mixture. A high level of regioselectivity is attained for both alkylations to give α,α-dialkylation products exclusively. Unfortunately, however, the products are a mixture of **17** and **18** that result from a nonregioselective hydride attack. The superiority of the integrated process to the stepwise process in terms of yield is valid in these cases as well.

A sequence of electrophilic alkylation and palladium-catalyzed reductive desulfonylation underlies the above procedures. The π-allylpalladium chemistry enables an alternative means for the direct carbon–carbon bond formation concomitant with desulfonylation of allylic sulfones by use of organometallic nucleophiles. Organozinc compounds serve for this purpose [18], and a protocol for more concise electrophilic alkylation/nucleophilic arylation has been developed (Scheme 17). Thus treatment of the mono alkylation products produced by the

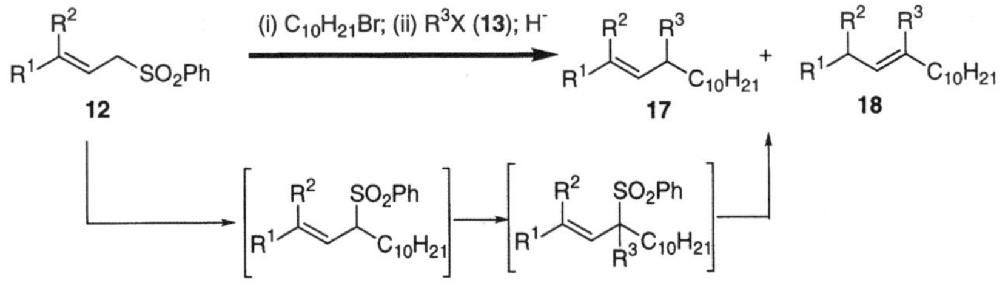

Scheme 16

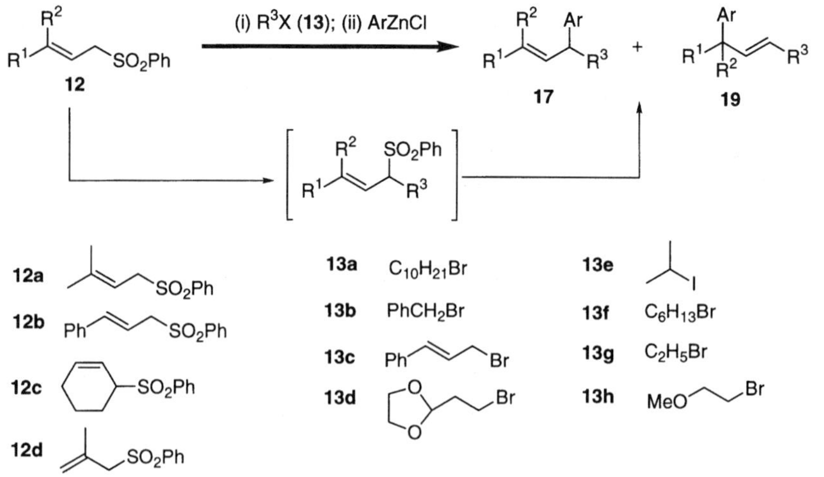

Scheme 17

anionic technology as described above with $ZnCl_2$, aryl magnesium bromide, and $Pd(PPh_3)_4$ affords the desired olefins. The results are given in Table 4. The mode of reaction is dependent on the sulfones. Prenyl, cinnamyl, and methallyl sulfones, **12a**, **12b**, and **12d**, give satisfactory yields, while only modest yields are obtained with cyclohexenyl sulfone **12c**. The regiochemistry of nucleophilic arylation is not straightforward. Both **12b** and **12c** undergo arylation in an exclusive manner but in the opposite sense (entries 4–9). The γ-selectivity with **12c** can be accounted for in terms of steric hindrance, but the analogous explanation cannot be applicable to **12b** which constantly leads to carbon–carbon bond formation at the α-position irrespective of the sulfone substrates. A mixture of α- and γ-regioisomers is obtained from **12a** and **12d**. In addition, E,Z-isomers are formed in γ-arylation products, **19de** and **19df**, from **12d**. In keeping with the previous trends, the integration has produced an increase in the overall yield compared to the stepwise route. In entries 8 and 9, the overall yield is primarily governed by the arylation, due to its poor yield, so that the integration could not give rise to any appreciable advantage over the stepwise process. By contrast, in the other cases where the arylation is relatively clean, circumvention of the intermediate loss is rewarded by improved yields for the integrated process.

3 PARALLEL RECOGNITION PROCESS

While the previous paragraph has dealt with the serial connection of multiple reactions, a parallel treatment is the subject of this paragraph. As shown in

Table 4　One-Pot Electrophilic Alkylation and Desulfonylative Nucleophilic Arylation

Entry	1	2	ArMgX	17	(17:19)	19	Yield (**17** + **19**) (%)[a]
1	12a	13a	PhMgBr	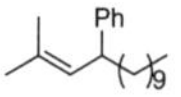 **17aa**	(66:34)	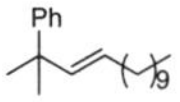 **19aa**	79 (81 × 86 = 70)
2	12a	13f	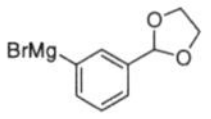	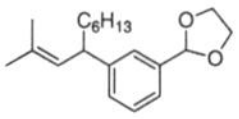 **17af**	(59:41)	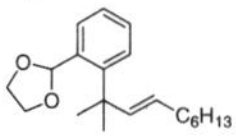 **19af**	71
3	12a	13g	p-MeOC$_6$H$_4$MgBr	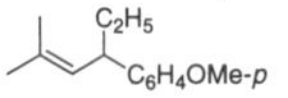 **17ag**	(63:37)	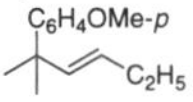 **19ag**	88
4	12b	13a	PhMgBr	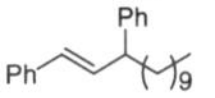 **17ba**	(100:0)		74 (76 × 82 = 62)
5	12b	13e	PhMgBr	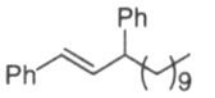 **17be**	(100:0)		80 (83 × 80 = 66)
6	12b	13f	PhMgBr	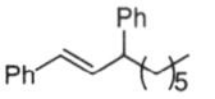 **17bf**	(100:0)		68 (77 × 75 = 58)
7	12b	13h	PhC(MgBr)=CH$_2$	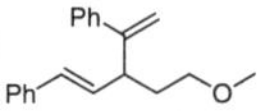 **17bh**	(100:0)		62 (61 × 84 = 51)
8	12c	13e	PhMgBr	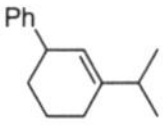 **17ce**	(100:0)		43 (82 × 46 = 38)
9	12c	13f	PhMgBr	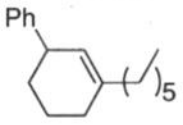 **17cf**	(100:0)		50 (83 × 54 = 45)
10	12d	13e	PhMgBr		(0:100)	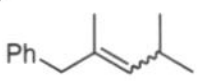 **19de**	72 (84 × 76 = 64)
11	12d	13f	PhMgBr	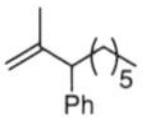 **17df**	(19:81)	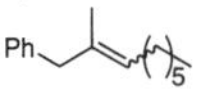 **19df**	88

[a] Isolated yield. The overall yield in stepwise reaction [(% yield of the first alkylation) × (% yield of the second alkylation)] is given in parentheses.

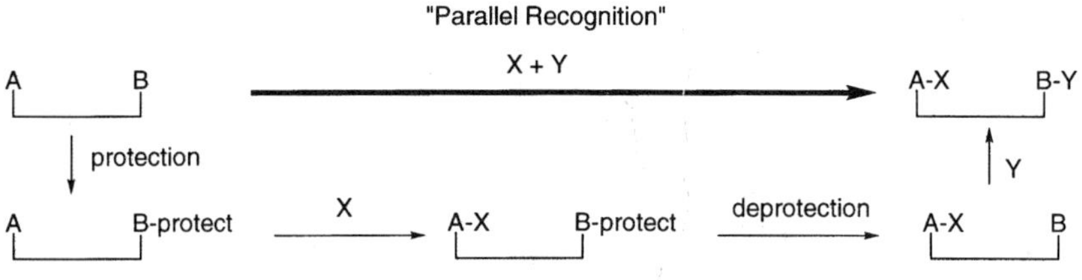

Scheme 18

Scheme 18, we often encounter the necessity of performing different reactions on separate reaction sites, A and B, in a single substrate. The protection–deprotection procedure is usually invoked. B is protected initially and A is converted to A-X. Then, after deprotection, B is converted to B-Y. A new concept, "parallel recognition," stemmed from the idea that if these transformations could be accomplished simultaneously, a highly expeditious and convenient process would be achieved to arrive at the final goal in one pot and one operation. The protection–deprotection steps are no longer necessary, and reaction time is saved as the manifold reactions proceed simultaneously. For this concept to be realized, a new type of chemoselectivity should be explored in which a mixture of substrates A and B reacts with a mixture of reagents X and Y to furnish products A-X and B-Y exclusively (Eq. 1). The requirements would be satisfied if productA-X could be formed predominantly over B-X in the competition reaction of substrates A and B with reagent X (Eq. 2) and B-Y in preference to A-Y in the reaction with reagent Y (Eq. 3). For reaction (1) to occur efficiently the substrates A and B should be similar to each other in chemical reactivity, so that reactions (2) and (3) can proceed under the identical reaction conditions. They should simultaneously undergo different reactions irrespective of the other. The reagents X and Y also need to work in this way.

$$A + B + X + Y \longrightarrow A\text{-}X + B\text{-}Y \tag{1}$$

$$A + X + Y \longrightarrow A\text{-}X \left(+ A\text{-}Y\right) \tag{2}$$

$$B + X + Y \longrightarrow B\text{-}Y \left(+ B\text{-}X\right) \tag{3}$$

In the course of our studies on synthetic applications of organotin Lewis acids, we have found the unique chemoselective catalysis of $(C_6F_5)_2SnBr_2$ (**20**) [19]. When a mixture of ketone **21** and aldehyde **22** is exposed to ketene silyl acetal **23** in the presence of a catalytic amount of **20**, the ketone reacts preferentially over the aldehyde (Table 5). This is completely opposite to the normally accepted selectivity between these two carbonyl compounds. The analogous preference for the ketone is also true in competition with the acetal **26** (Table 6).

Table 5 Competition Reaction Between Ketone and Aldehyde

$$R^1COR^2 \; (21) + R^3CHO \; (22) + CH_2=C(OTBS)OEt \; (23) \xrightarrow{C_6F_5SnBr_2 \; (20)} R^1C(R^2)(OTBS)CH_2COOEt \; (24) + R^3CH(OTBS)CH_2COOEt \; (25)$$

			Yield (%)			Recovery (%)	
Entry	**21**	**22**	**24**	**25**	**24:25**	**21**	**22**
1	**21a**	**22a**	**24aa** 75	**25aa** 7	91:9		
8	**21a**	**22b**	**24aa** 72	**25ab** 8	90:10	16	88
9	**21b**	**22b**	**24ab** 85	**25ab** 5	94:6	0	84
10	**21c**	**22b**	**24ac** 86	**25ab** 4	94:6	0	93
11	**21e**	**22b**	**24ae** 68	**25ab** 9	88:12	21	76
12	**21f**	**22b**	**24af** 72	**25ab** 3	96:4	16	88

Thus we have obtained the reactions that satisfy the condition for Eq. 2. By contrast, the selectivity is normal in the competition reaction of **21** vs. **22** or **26** with enol silyl ethers derived from ketone **28** (Scheme 19). A smooth reaction occurs with aldehydes or acetals while ketones do not react at all. Accordingly, another selectivity which meets the condition for Eq. 3 has been attained. The combination of these two types of competition reactions leads to the desired recognition (Table 7) [20]. A clean parallel recognition takes place upon exposure of a mixture of substrates, **21** and **22**, to a mixture of reagents, **23** and **28b**, in

Table 6 Competition Reaction Between Ketone and Acetal

$$R^1COR^2 \; (21) + R^3R^4C(OMe)_2 \; (26) + 23 \xrightarrow{(C_6F_5)_2SnBr_2} R^1C(R^2)(OTBS)CH_2COOEt \; (24) + R^3C(R^4)(OMe)CH_2COOEt \; (27)$$

			Yield (%)		
Entry	**21**	**26**	**24**	**27**	**24:27**
1	**21a**	**26a**	**24aa** 91	0	100:0
2	**21a**	**26b**	**24aa** 96	0	100:0
3	**21a**	**26c**	**24aa** 93	0	100:0
4	**21a**	**26d**	**24aa** 95	0	100:0
5	**21d**	**26a**	**24ad** 81	0	100:0
6	**21d**	**26d**	**24ad** 88	0	100:0
7	**21f**	**26a**	**24aa** 87	0	100:0
8	**21f**	**26c**	**24af** 79	0	100:0

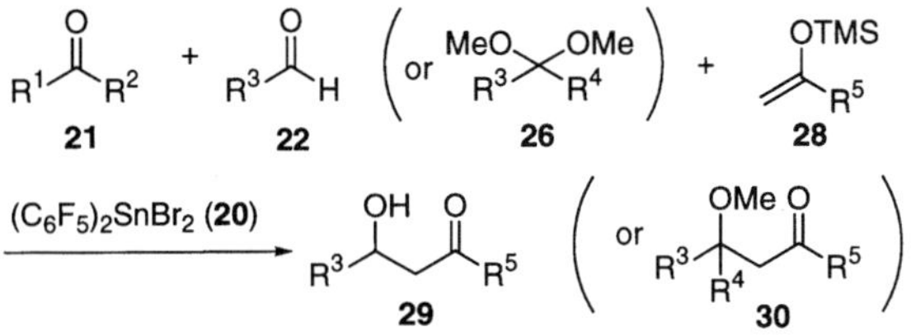

Scheme 19

the presence of a catalytic amount of **20**. Only two products are obtained by the reaction between ketone and ketene silyl acetal and between aldehyde and acetophenone-derived enol silyl ether without formation of crossover products (entries 1 and 2) or together with a trace amount of **25** (entries 3–5). The same recognition also holds between ketone **21** and acetal **26** (Table 8).

Thus, the parallel recognition of doubly functionalized substrates is shown in Scheme 20. Keto aldehydes **31–33** exhibit clean recognition to give sole products: exclusive reactions take place between **23** and ketones and between **28b** and aldehydes, respectively. Analogously, a single product is obtained with keto acetal **34** or **35**.

"Parallel recognition" can be applied to the competition between the Mi-

Table 7 Parallel Recognition Between Ketone and Aldehyde with Ketene Silyl Acetal and Enol Sily Ether

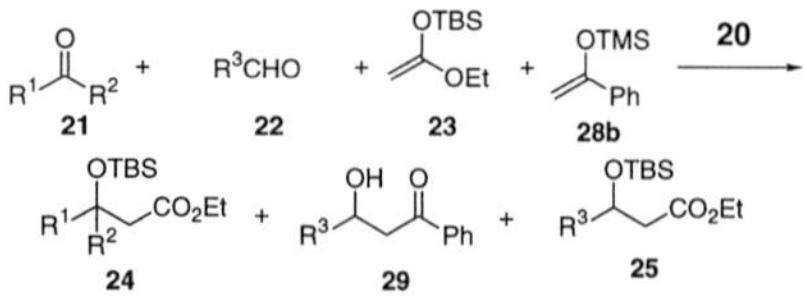

			Yield (%)		
Entry	**21**	**22**	**24**	**29**	**25**
1	**21a**	**22a**	**24aa** 72	**29ba** 61	0
2	**21b**	**22a**	**24ab** 82	**29ba** 70	0
3	**21c**	**22a**	**24ac** 73	**29ba** 74	**25aa** 1
4	**21d**	**22a**	**24ad** 59	**29ba** 54	**25aa** 3
5	**21f**	**22a**	**24af** 73	**29ba** 70	**25aa** 2

21a: $R^1 = Ph$; $R^2 = Me$

  b: $R^1 = 4\text{-}MeOC_6H_4$; $R^2 = Me$

  c: $R^1 = 2,4\text{-}(MeO)_2C_6H_3$; $R^2 = Me$

  d: $R^1 = C_2H_5$; $R^2 = Me$

  e: $R^1 = n\text{-}C_4H_9$; $R^2 = Me$

  f: $R^1, R^2 = (CH_2)_5$

22a: $R^1 = n\text{-}C_5H_{11}$

  b: $R^1 = n\text{-}C_7H_{15}$

26a: $R^1 = Ph$; $R^2 = H$

  b: $R^1 = n\text{-}C_7H_{15}$; $R^2 = H$

  c: $R^1 = n\text{-}C_6H_{13}$; $R^2 = Me$

  d: $R^1, R^2 = Me$

Table 8 Parallel Recognition Between Ketone and Acetal with Ketene Silyl Acetal and Enol Sily Ether

$$R^1R^2C{=}O\ (21) + R^3R^4C(OMe)_2\ (26) + \text{CH}_2{=}C(OTBS)OEt\ (23) + \text{CH}_2{=}C(OTMS)R^5\ (28a:\,^tBu,\ b:\,Ph) \xrightarrow{(C_6F_5)_2SnBr_2} R^1R^2C(OTBS)CH_2CO_2Et\ (24) + R^3R^4C(OMe)CH_2C(O)R^5\ (30)$$

				Yield (%)	
Entry	28	21	26	24	30
1	28a	21a	26a	24a 89	7aa 73
2	28a	21a	26b	24a 85	7ab 24
3	28a	21a	26c	24a 85	7ac 21
4	28a	21b	26a	24b 79	7aa 63
5	28a	21d	26a	24d 75	7aa 77
6	28a	21e	26a	24e 74	7aa 85
7	28a	21e	26b	24e 84	7ab 24
8	28a	21f	26a	24f 74	7aa 80
9	28b	21a	26a	24a 83	7ba 62
10	28b	21a	26b	24a 83	7bb 83
11	28b	21a	26c	24a 84	7bc 83
12	28b	21b	26a	24b 80	7ba 64
13	28b	21b	26b	24b 83	7bb 65
14	28b	21b	26c	24b 76	7bc 75
15	28b	21d	26a	24d 78	7ba 57
16	28b	21d	26b	24d 81	30bb 78
17	28b	21d	26c	24d 86	30bc 82
18	28b	21e	26a	24e 81	30ba 60
19	28b	21e	26b	24e 82	30bb 80
20	28b	21e	26c	24e 83	30bc 83
21	28b	21f	26a	24f 82	30ba 64
22	28b	21f	26b	24f 83	30bb 83
23	28b	21f	26c	24f 86	30bc 78

chael and aldol reactions (Scheme 21). Under the catalysis of **20**, ketene silyl acetal **23** suffers Michael addition with **36** while enol silyl ether **28a** reacts with acetal **26a**: no crossover reactions are observed. The same recognition holds in case of intramolecular versions (Scheme 22). Substrates **37** and **38** that have both an α,β-unsaturated ketone moiety and an acetal function exhibit the explicit recognition of **23** and **28a** to afford sole products, **39** and **40**, respectively.

A further example of ''parallel recognition'' is provided by use of the reagent mixture, ketene silyl acetal and allylsilane (Scheme 23). In parallel with the exclusive reaction of **23** with acetophenone, allyltrimethylsilane furnishes a quantitative yield of allylation product upon reaction with acetal **26a**.

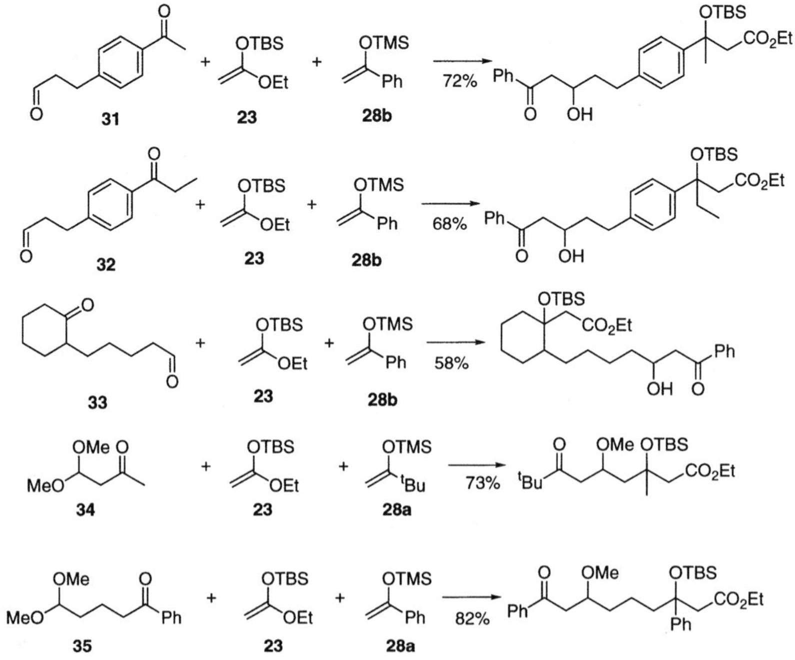

Scheme 20

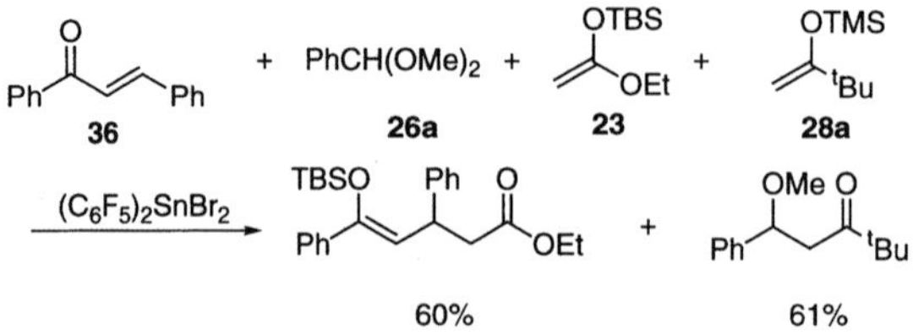

Scheme 21

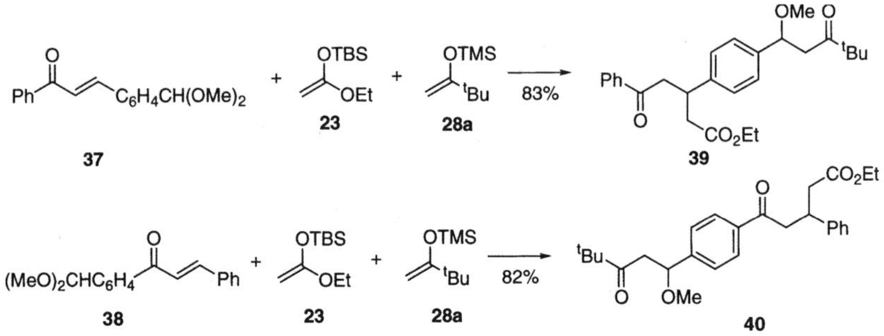

Scheme 22

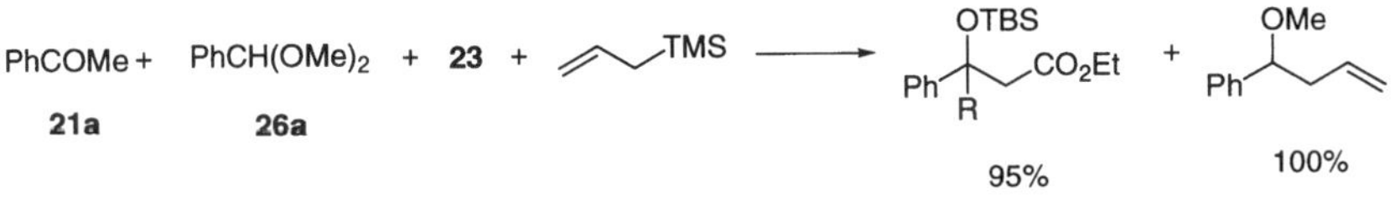

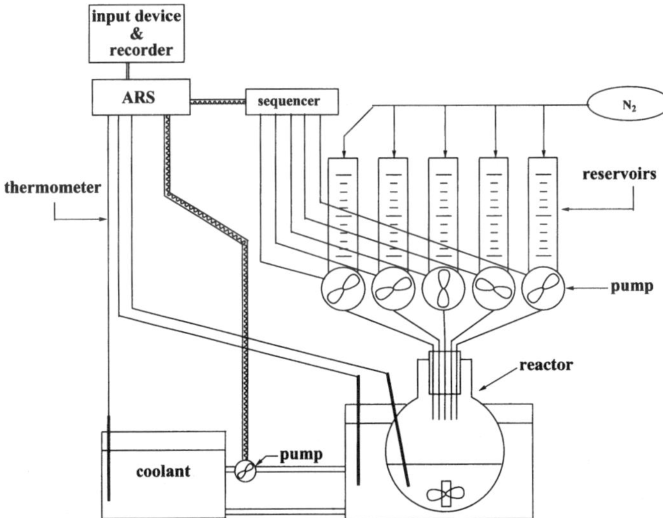

PhCOMe + PhCH(OMe)₂ + **23** + ⌒⌒TMS ⟶ (OTBS ... Ph...CO₂Et, R) + (OMe ... Ph) 95% 100%

Scheme 23

4 AUTOMATED SYNTHESIS

We have made an automated synthesizer which is suited to one-pot processes as mentioned above, in the introduction. Figure 1 shows a diagram of our automated synthesizer, which basically consists of a control unit (ARS [21] and a sequencer), a reactor, reservoirs, volumetric pumps, and a cooling unit. The inside of the apparatus is completely sealed so that an inert atmosphere can be obtained by evacuation followed by refilling with nitrogen or argon. The substrate and reagents are pumped into the reaction flask from the reservoirs. The amount (accurate within ± 0.1 mL) and addition rate of the reagents are controlled by ARS. The reaction temperature (accurate within $\pm 0.1°C$) is monitored by immersing the sensor directly into the reaction solution, and it is controlled by ARS. In addition, this system has an advanced control function that can initiate the next task immediately after the end point of the previous one is detected, so that the processing time is minimized. The addition of reagent is controlled by a sequencer in any order and any time. The reactor has a three-layered structure

Figure 1 Diagram of the automated synthesizer.

(Fig. 2): the outermost layer is the vacuum temperature insulator, cold MeOH circulates in the second layer, and the innermost reaction flask is jacketed by a glass tube through which warm (usually room temperature) MeOH flows. By balancing the currents of both cold and warm MeOH, the reaction temperature can be adjusted very quickly. For instance, the temperature is warmed from $-78°C$ to $20°C$ within 5 min, while cooling from room temperature to $-78°C$ takes only 10 min. This is very important for the sequential reactions which take place at different temperatures.

The reactions using Grignard and organolithium reagents proceed smoothly to prove the successful employment of air-sensitive reactions (Eqs. 4 and 5). The utility is further exemplified by successful use of air-sensitive transition metal catalysts (Eqs. 6–8).

$$\text{PhCHO} \xrightarrow[\text{THF, 30°C, 24h}]{\text{EtMgBr (1.2 equiv)}} \text{Ph-CH(OH)-Et} \quad (4)$$

73%

$$\text{PhCO-OMe} \xrightarrow[\text{THF, -78-0°C, 24h}]{\text{BuLi (2.0 equiv)}} \text{Ph-CO-Bu} \quad (5)$$

79%

$$\text{Ph-B(OH)}_2 \;+\; \text{Cl-Ar-R} \xrightarrow[\text{dioxane, 80°C, 24h}]{\substack{\text{i. NiCl}_2\text{(dppf) (0.1 equiv)} \\ \text{ii. K}_3\text{PO}_4 \text{ (3.0 equiv)} \\ \text{iii. BuLi (0.4 equiv)}}} \text{Ph-Ar-R} \quad (6)$$

$$\text{binaphthyl-(OTf)}_2 \xrightarrow[\text{Et}_2\text{O, 0-25°C, 4h}]{\substack{\text{i. (dppf)NiCl}_2 \text{ (0.05 equiv)} \\ \text{ii. MeMgBr (4.0 equiv)}}} \text{binaphthyl-(CH}_3)_2 \quad (7)$$

93%

$$\text{MeO-C}_6\text{H}_4\text{-H} \xrightarrow[\substack{\text{CH}_3\text{NO}_2, 50°C, \\ 18h}]{\text{Ac}_2\text{O (5 equiv)/Sc(OTf)}_3 \text{ (0.2 equiv)}} \text{MeO-C}_6\text{H}_4\text{-OAc} \quad (8)$$

95%

Scheme 24 demonstrates the sequential reactions in automation. Fourteen operations are programmed for the double alkylations of prenyl sulfone. The reactor is purged with nitrogen and then the process is triggered. After this stage, no further manual operation is needed until the completion of the process. It is apparent from this result that the computer program effectively allows the maintenance,

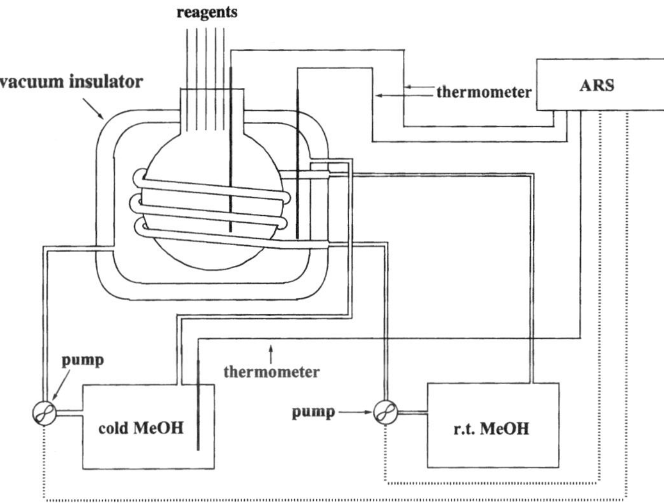

Figure 2 The reactor and cooling unit.

warming, and cooling of the reaction temperature and addition of the reagents in the required order and amount.

Finally, performance of the integrated process is described. The automation of the acetylene synthesis by the Peterson-sulfone elimination protocol (Scheme 11) is depicted in Scheme 25. In this process, 16 operations are performed automatically (Scheme 25) without decrease of the yields.

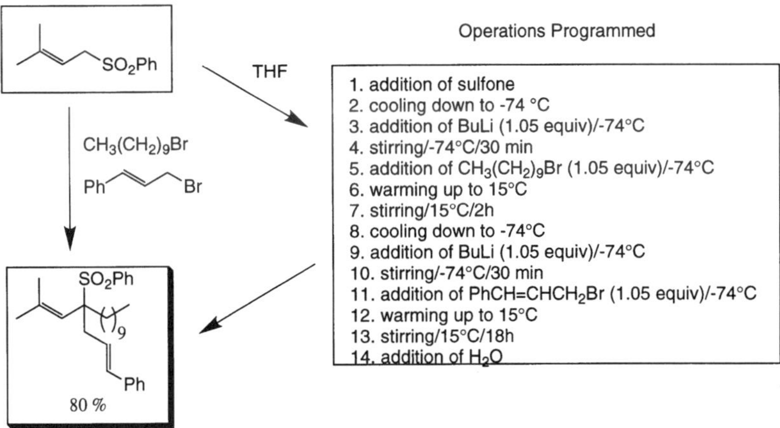

Scheme 24

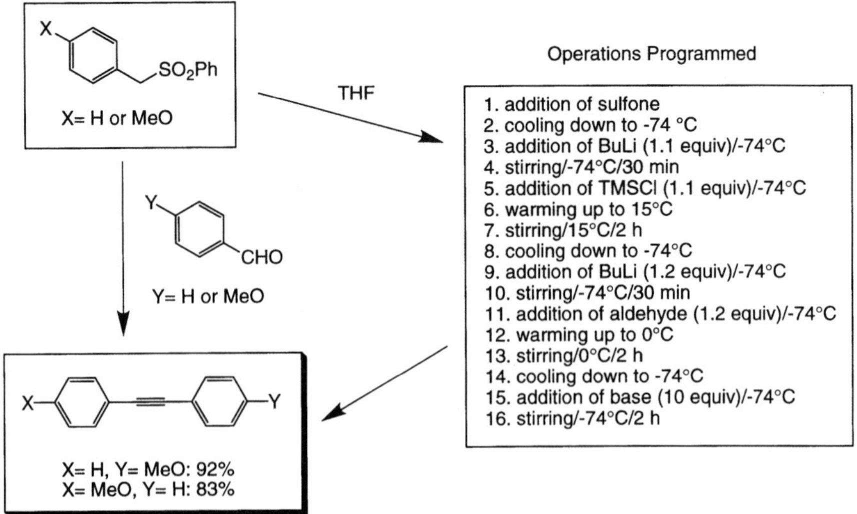

Scheme 25

5   SUMMARY AND FUTURE PROSPECTS

The automation of laboratory operations is of great significance not only for improvements of laboratory work itself but also for commercial production. The demand will be increased for producing a variety of pharmaceuticals and specialty chemicals on a relatively small scale, i.e., only a few hundred kilograms or less per year. Such requirements may be satisfied with the multipurpose lab-scale automated system that is designed to be readily tunable for various types of reactions and a wide scope of reaction conditions. In particular, employment of one-pot reactions heightens the usefulness of this system from both economical and ecological points of view. Nevertheless, few one-pot methodologies have appeared that are applicable to automated synthesis because the reactions that can be employed in these protocols are rather limited.

New concepts for the one-pot reaction, ''integrated chemical process'' and ''parallel recognition,'' have been advanced here. It should be pointed out that normal reactions are employed under normal reaction conditions according to these concepts, so that the limitations encountered in conventional one-pot methods are eliminated. Actually, many integrated chemical processes have been proved to work efficiently on the automated synthesizer. Significantly, the synthetic routes have been determined from the standpoints of process chemistry as well as synthetic chemistry. In this regard, the collaboration between synthetic chemists, process chemists, and engineers from the beginning of the synthetic route design is crucial for realizing the synthetic automation. So far, synthetic chemists have concentrated mostly on planning synthetic routes for benchwork without consideration of what follows. In my opinion, they should pay more attention on the process chemistry.

ACKNOWLEDGMENTS

This work could not have been achieved without the collaboration of my coworkers, whose names are given in the references with my sincere appreciation. The financial support of the Japan Society for the Promotion of Science under the ''Research for the Future'' program is gratefully acknowledged.

REFERENCES

1. For recent reviews see Lindsey, J. S. Chemom. Intell. Lab. Syst. 1992, 17, 15. Sugawara, T., Cork, D. G. Lab. Rob. Autom. 1996, 8, 221. Harre, M., Tilstam, U., Weinmann, H. Org. Process Res. Dev. 1999, 3, 304.
2. For the importance of process design see Repic, O. Principles of Process Research and Chemical Development in the Pharmaceutical Industry. John Wiley, New York, 1998.
3. Posner, G. H. Chem. Rev. 1986, 86, 831. Tietze, L. F., Beifuss, U. Angew. Chem. Int. Ed. Engl. 1993, 32, 131. For special issues on this topic see Chem. Rev. 1996, 96(1). Tetrahedron 1996, 52(35).
4. For relevant studies on integration of multistep processes see Franzone, G., Carle, S., Dorizon, P., Ollivier, J., Salaün, J. Synlett 1996, 1067. Dorizon, P., Ollivier, J., Salaün, J. Synlett 1996, 1071. Haight, A. R., Stuk, T. L., Menzia, J. A., Robbins, T. A. Tetrahedron Lett. 1997, 38, 4191. Piva, O., Comesse, S. Tetrahedron Lett. 1997, 38, 7191. Fleming, F. F., Huang, A., Sharief, V. A., Pu, Y. J. Org. Chem. 1997, 62, 3036. Bednar, P. M., Shaw, J. T., Woerpel, K. A. J. Org. Chem. 1997, 62, 5674. Nakamura, E., Kubota, K., Sakata, G. J. Am. Chem. Soc. 1997, 119, 5457.
5. Mandai, T., Yanagi, T., Araki, K., Morisaki, Y., Kawada, M., Otera, J. J. Am. Chem. Soc. 1984, 106, 3670. Otera, J., Misawa, H., Sugimoto, K. J. Org. Chem. 1986, 51, 3830.
6. Otera, J., Misawa, H., Mandai, T., Onishi, S., Suzuki, S., Fujita, Y. Chem. Lett. 1985, 1883. Otera, J., Misawa, H., Onishi, T., Suzuki, S., Fujita, Y. J. Org. Chem. 1986, 51, 3834.
7. Orita, A., Yamashita, Y., Toh, A., Otera, J. Angew. Chem. Int. Ed. Engl. 1997, 36, 779.
8. Orita, A., Yaruva, J., Otera, J. Angew. Chem. Int. Ed. Engl. 1999, 38, 2267.
9. Müller, M., Kübel, C., Müllen, K. Chem. Eur. J. 1998, 4, 2099.
10. Orita, A., Yoshioka, N., Struwe, P., Braier, A., Beckmann, A., Otera, J. Chem. Eur. J. 1999, 5, 1355.
11. Lagow, R., Kampa, J. J., Wei, H. C., Battle, S. L., Genge, J. W., Laude, D. A., Harper, C. J., Bau, R., Stevens, R. C., How, J. F., Munson, E. Science 1995, 267,

362. Yamaguchi, M., Park, H. J., Hirama, M., Torisu, K., Nakamura, S., Minami, H., Nishihara, H., Hiraoka, T. Bull. Chem. Soc. Jpn. 1994, 67, 1717.

12. Chodkievicz, W., Ann. Chim. (Paris) 1957, 2, 819. Bahlmann, F., Herbst, P., Gleinig, H. Chem. Ber. 1961, 94, 948. Miller, J. A., Zweifel, G. Synthesis 1983, 128. Schulte, K. E., Reisch, J., Hörner, L. Chem. Ber. 1962, 95, 1943. Rubin, Y., Lin, S. S., Knobler, C. B., Anthony, J., Boldi, A. M., Diederich, F. J. Am. Chem. Soc. 1991, 113, 6943.

13. Tour, J. M. Chem. Rev. 1996, 96, 537. Diederich, F., Rubin, Y. Angew. Chem. Int. Ed. Engl. 1992, 31, 1101. Diederich, F. Nature 1994, 369, 199. Moore, J. S. Acc. Chem. Res. 1997, 30, 402.

14. Sonogashira, K., Tohda, Y., Hagihara, N. Tetrahedron Lett. 1975, 446.

15. Simpkins, N. S. Sulfones in Organic Synthesis. Pergamon Press, Oxford, 1993, Chaps. 3 and 9.

16. Mohri, M., Kinoshita, H., Inomata, K., Kotake, H. Chem. Lett. 1985, 451. Mohri, M., Kinoshita, H., Inomata, K., Kotake, H., Takagaki, H., Yamazaki, K. Chem. Lett. 1986, 1177. Inomata, K., Igarashi, S., Mohri, M., Yamamoto, T., Kinoshita, H., Kotake, H. Chem. Lett. 1987, 707.

17. Orita, A., Watanabe, A., Otera, J. Chem. Lett. 1997, 1025. Orita, A., Watanabe, A., Tsuchiya, H., Otera, J. Tetrahedron 1999, 55, 2889.

18. For representative papers on reaction of π-allylpalladium with organozinc nucleophile see Matsushita, H., Negishi, E. J. Am. Chem. Soc. 1981, 103, 2882. Negishi, E., Chatterjee, S., Matsushita, H. Tetrahedron Lett. 1981, 22, 3737. Fiaud, J.-C., Aribi-Zouioueche, L. J. Organomet. Chem. 1985, 295, 383. Hayashi, T., Yamamoto, A., Hagihara, T. J. Org. Chem. 1986, 51, 723. For a review see Knochel, P. In: Transition Metals for Organic Synthesis (Beller, M., Bolm, C., eds.). Wiley-VCH, 1998, Vol. 1, p. 467.

19. Chen, J., Sakamoto, K., Orita, A., Otera, J. J. Org. Chem. 1998, 63, 9739.

20. Chen, J., Otera, J. Angew. Chem. Int. Ed. Engl. 1998, 37, 91. Chen, J., Otera, J. Tetrahedron Lett. 1998, 39, 1767. Chen, J., Sakamoto, K., Orita, A. Otera, J. Tetrahedron 1998, 54, 8411.

21. ARS is an automatic reaction controller commercially available from Sogo Pharmaceutical Co., Ltd., Japan.

9
Optimization of Organic Process Chemistry

John E. Mills
*Johnson & Johnson Pharmaceutical Research & Development, L.L.C.
Spring House, Pennsylvania*

1 GOALS OF PROCESS DEVELOPMENT

The decrease in the cost of computers over the last few decades has resulted in the application of automation in numerous ways to chemical development. Before enumerating the application of automation in this area, it may be useful to provide background on the goals of chemical development in the pharmaceutical industry. Although some of the details may differ from specialty to specialty, the main features of chemical development should be consistent throughout the chemical industry. While the production of kilograms of active pharmaceutical ingredient or bulk drug substance is typically thought to be the major responsibility of a chemical development group, an equally important product from these efforts is information. The combination of data produced from the studies utilizing the drug substance with the data gathered during the definition of the process provides a basis for decisions concerning the economic viability of a potential drug. It is important to understand that in the past many of the advances in automation have impacted the collection, manipulation, storage, retrieval, and dissemination of chemical information more than the way in which the data was generated.

Within the past decade, the methods used in the discovery process have undergone a tremendous change. As indicated in earlier chapters in this book, many companies are relying upon combinatorial chemistry, often using solid-phase reactions, to generate large libraries of compounds that are screened for activity. The screening processes themselves have gone from observations on tens of compounds per week in the 1970s to thousands of compounds per day in the late 1990s. The efficiency in the production of potential new therapeutic agents and advances in the selection of the most promising leads have resulted in pressure for development of those leads to occur in a rapid and efficient manner.

Development chemists are seeking methods to improve their efficiency without losing the elements of creativity that they enjoy in their work. Today, the need to produce more material and information in shorter times is forcing chemists to reevaluate the way that they work and to determine where automation may be applied most effectively. The success in the utilization of automated synthesis achieved by discovery chemists has forced development chemists to ask many of the same questions and address many of the same issues concerning automation in organic chemistry that were identified by Jean-Paul Guette et al. over a decade ago [1].

1.1 Information

The information that is generated and collected during development of a new process or chemical is quite varied. Evaluation of the individual reagents and reactants utilized to effect a given transformation is common to all processes. Calculation of the yield of a given step, determination of the purity of a product, and calculation of the throughput in a reaction are all basic operations in chemical development. Because the last operation(s) in any synthesis is typically purification of the product, the impact of the purity of the raw materials going into that synthesis is sometimes overlooked. The toxicological properties of compounds used in a reaction may require that special precautions be taken during the step. In some instances, the limitations are so severe that a different reaction must be used.

The thermal properties of a reaction can provide valuable information both on the safety of the step and on the course of the reaction. Tools for the evaluation of thermal hazards are available and should be applied early in the development process. It is common practice to use reaction calorimetry to determine the potential for an uncontrollable exotherm before a reaction is scaled above several liters. The observation of a buildup of latent heat in a reaction is indicative of potential problems during scale-up and manufacturing. Until the thermal behavior of a reaction has been determined one may not be aware that a potential problem exists.

There are opportunities for utilizing automation to facilitate collection, analysis, and utilization of data throughout development.

1.2 Investigational Materials

The active pharmaceutical ingredient typically is introduced into a number of processes to determine its suitability for its intended use. In addition to studies intended to establish the compound's biological properties, studies to determine its physical properties such as stability, solubility, and suitability for formulation are performed. Although the primary focus of this chapter is not on the methods used to generate physical and biological properties, some of the automation involved in the determination of physical properties will be included.

1.3 Economics

As indicated above, the products coming from chemical development are both the investigational active pharmaceutical ingredient and the data related to its production. At some point in the development of a new chemical entity, a decision must be made concerning the economic viability of the compound. The availability of the necessary information in a concise, understandable format is a valuable aid in the decision-making process.

2 DEVELOPMENTAL STRATEGY

2.1 Early Development

During early development of a new chemical entity, the strategy is focused on production of sufficient amounts of the compound for physical characterization and suitability testing. In the pharmaceutical industry, physical characterization includes studies designed to identify potential problems associated with polymorphism, stability, and solubility. Suitability testing includes formulation work and safety evaluation studies. A demand in early development is frequently production of sufficient material to answer critical questions within a short time frame regardless of cost. Because of the short time frames, early work on new chemical entities frequently involves evaluation of the known synthesis to determine whether that process can be scaled up directly.

Many of the automated tools that were previously used primarily in discovery are being utilized in chemical development in order to compress development time as much as possible. In early development, the most frequently used automated tools are commercial chemical databases, spectrometers, and chromatography instrumentation. It is becoming very common to see instruments in which separation and identification of compounds are performed in tandem. Gas chromatography/mass spectroscopy (GC/MS) has been augmented with liquid chromatography/mass spectroscopy (LC/MS) and, more recently, with liquid chromatography/nuclear magnetic resonance (LC/NMR) or LC/NMR/MS [2].

Many of the experiments performed at this stage in development are screening experiments. Such experiments are performed to determine which parameters are and are not important in a given reaction. Knowledge of those factors that impact upon the yield and purity of the product of a reaction allows for more rational decisions during the initial scale-up of a synthesis.

In addition, the thermochemistry associated with the reactions and chemicals under study is frequently evaluated very early in the development cycle. Without such data, a large scale-up of an existing reaction may be dangerous. Once a potential problem has been identified, steps can be taken to circumvent the buildup of excess heat in a reaction that could result in an uncontrolled release of gasses. This can be accomplished either by changing the chemistry or by establishing sufficient process controls to ensure that there is never a large amount of latent heat in a reactor from unreacted starting materials. Sometimes the required change in the chemistry is as simple as changing the order of addition of the reactants and reagents.

2.2 Later Development

Once a new compound has shown promise for commercialization, supplies of the material are required for additional testing. It is at this point in development that optimization of the chemical process typically occurs. The focus of the work changes from obtaining material to maintaining sufficient material for ongoing studies and to developing a safe, viable commercial scale synthesis. Statistical methods are frequently employed during this stage of development to provide assurance that one has truly reached the optimum performance of each step, and to obtain a better understanding of the ruggedness of the system.

Initial work is designed to find the location of a maximum in the defined system. Later studies should identify the ''size'' of the ''sweet spot'' in the process parameters.

Validation studies are performed to establish that the chemistry and the process are ''under control.'' It is useful to employ statistical methods during the development of the process since such methods can greatly simplify identification of those parameters that must be maintained within strict limits in order to produce material that passes specifications. If a process has been developed by changing only one reaction variable at a time, one does not necessary know the consequences of simultaneous variations of two or more process parameters. It is possible that either a planned or an unplanned deviation of multiple process parameters from their set point to near their limits can result in material that is outside the specifications set in the validation documentation. If that is the case, the process cannot be considered to be under control.

This stage in development is frequently ideal for the application of automation in the development of a new process. Automated in-line controls make it easier to reproduce a given experimental. The number of experiments that are required within a study makes it easier to justify the time investment needed to automate the work. The repetitious nature of much of the work also provides an incentive to the chemists working on the project to find an ''assistant'' to do the experiments. In addition, most equipment performing automated processes is designed to produce documentation on those processes. Once the experiment is started, the equipment tracks and records specified parameters. The automated logging of reaction parameters frees the experimenter from the drudgery of manually recording the data while allowing for more time to be spent on the analysis of the data produced.

Entry of the data into a process database allows both for the rapid recovery of information regarding a given reaction and for the identification of the full range of experimental conditions that were utilized in development of the process. Well-designed databases will simplify the identification and compilation of subsets of data that aid in justification of the conditions used in the commercial process.

2.3 Postmarketing

Once a compound has been commercialized, the process of fine-tuning its production is typically continued. While the variation in the range of parameters is not as great as that employed in the development process, batch-to-batch variability in critical parameters such as yield, cost, throughput, and purity should be tracked. Commercial processes are typically run under automated control. Statistical methods can be applied to the data produced during normal operation of a process to determine where modifications of the process can be made to improve the output.

3 COMMONLY USED AUTOMATED TOOLS

Currently, the areas in which automation has been most effectively implemented are those related to analytical chemistry, physical chemistry, and spectroscopy. Some of the reasons for the long-standing use of automation in these areas are (1) the repetitious or tedious nature of the work, (2) the accuracy and precision required in the measurements, (3) the large number of samples to be processed, (4) the cost of the equipment, and (5) the consequences of errors in the results [3]. In many cases the automation of data collection, storage, and analysis has become so common that it can easily be overlooked.

3.1 Reactors

Much of the early development of automated reactors centered on the success of solid phase reactions used for the production of peptides. Difficulties encountered in the solution-phase synthesis of polypeptides can, to a large extent, be overcome through the use of solid-phase synthesis. Differences in the reactivity of different coupling steps can be programmed into commercial units. Each step in a linear peptide synthesis may be driven to completion through the use of excess reagents. Racemization can be minimized by the selection of the appropriate coupling reagent and reaction conditions. Reaction by-products from deblocking reactions can easily be separated from the desired product by washing the resin with a suitable solvent. Last, but not least, the product can be isolated in high purity after cleavage from the resin. Modification of the chemistry has allowed this technology to be applied to the production of oligonucleotides [4,5].

The reduction in the amount of material needed for initial screening experiments, together with the desire to produce a large number of compounds in a small amount of time, led medicinal chemists to seek unconventional methods to prepare new compounds. The ability touted by peptide chemists to produce a small amount of pure material with a minimal amount of workup led to the investigation of solid-phase synthesis for the preparation of small molecules. Initial success using specially modified resins indicated that solid-phase synthesis may be a viable tool and helped drive the commercial development of the multireactor systems used in the production of libraries of compounds common in many pharmaceutical companies.

One of the strengths of solid-phase synthesis is also one of its weaknesses. The low density of binding sites on the resin allows for the chemistry to be performed without concern for interactions between the molecules that are being produced. The same property also results in low volume efficiency and potentially higher costs to scale up the process. Despite these problems, scale-up of the production of solid-phase synthesis of antisense oligonuclides to ton quantities has been reported [6].

The success in identifying new chemical leads experienced by medicinal chemists has resulted in the need for development chemists to examine their practices. As indicated above, scale-up of the solid-phase synthesis is typically not a viable option. The need to run larger numbers of solution phase reactions in a shorter time frame led to the request from development chemists for solution phase workstations. The robotics industry has responded by introducing equipment that is based on modifications of the software and hardware used for solid-phase synthesis. Such units are capable of running multiple reactions in which several parameters are independently varied among the reactions. Unlike the ''personal synthesizers'' used by discovery chemists, the development workstations typically are capable of performing fewer reactions, but the reactions are run on a larger scale.

In addition to the multireaction automated workstations described above, single-vessel automated reactors have been commercially available for several years. Such reactors typically have a working volume of 100 mL to 2 L. Commercial units, as delivered, typically do not provide automated reaction sampling. However, users of such equipment have developed systems that allow for automated reaction sampling. FTIR instrumentation is also available that allows for continuous monitoring of the reaction in such equipment. Such in situ monitoring of a reaction circumvents a number of problems associated with removal of an aliquot from a reaction.

Harre et al. have published a review containing information on the commercial availability of a number of automated systems used in chemical process research and development [7].

3.2 Chromatography

Commercial hardware and software are available for most tasks involved in gas liquid chromatography (GLC), high performance liquid chromatography (HPLC), and capillary electrophoresis (CE). In addition to automating routine analyses, advances in this area allow for automated reanalysis of fractions obtained from semipreparative or preparative scale HPLC work. In early development, the use of preparative scale HPLC can make moderate amounts of material available without the need to optimize each reaction step. Such systems are designed to make a series of injections automatically, to detect peaks of interest as they elute, and to send the eluent to specific holding tanks or waste disposal as appropriate.

Commercial systems using nuclear magnetic resonance for identification of eluent peaks have been introduced. Fully automated, continuous, flow-through identification of unknown peaks is not feasible for all samples. The cost of the chromatography is higher than with conventional detectors since deuterated solvents must be used as eluents. The additional cost of solvents is largely offset by the reduction in time necessary to identify impurities, degradation products, and metabolites using such systems.

3.3 Spectroscopy

The high cost of high field NMR spectrometers has been one of the driving forces in the automation of these instruments. Addition of an autosampler and automated shimming routines allows the instrument to be run nearly continuously. The net result is a lower cost per sample. The spectroscopists' time can be spent on interpretation of the spectra rather than on running the sample. In addition, software is available for the production of simulated spectra, thus speeding the interpretation process.

Fourier transform infrared spectroscopy (FTIR) has made a tremendous impact on sample preparation and on the types of experiments and the quality of infrared data that can be routinely obtained. The process from data collection to IR spectrum performed on commercial instruments has been automated to the point where FTIR spectra are as easy to obtain as conventional IRs. Software that uses chemometric techniques to identify chemical samples is available. In addition, the short times required for data collection allow for continuous monitoring of reactions [8]. Once a process has been developed utilizing IR criteria, rugged equipment designed for use in the manufacturing setting can be used to ensure that the chemistry observed on a laboratory scale is reproduced in the plant setting [9].

3.4 Differential Scanning Calorimetry and Thermogravimetric Analysis

Differential scanning calorimetry (DSC) and thermogravimetric analysis (TGA) can be very useful tools in the identification and characterization of changes in the solid state. Such data are especially important when studying solids that can exist in multiple morphologies or as solvates or hydrates. Utilization of DSC in the development of methods to resolve enantiomers through diastereomeric salt formation as well as in the identification of conglomerates and racemic compounds can be very useful. While autochangers are available for such instrumentation, their applicability to certain classes of compounds must be carefully considered. Rupture of a pan seal or decomposition of a compound in an open pan may lead to contamination of the equipment and result in erroneous results.

3.5 Dynamic Vapor Sorption

This technique is used to determine the stability of a solid under various temperature and relative humidity conditions. Standard manual methods for determination of moisture sorption by solids involves storing samples of the solid in humidity chambers and periodically measuring the mass of the samples until equilibrium is achieved. Because of the length of time necessary to achieve a true equilibrium in such studies, the experiments are necessarily long in duration. Such studies readily lend themselves to automation. Changes in temperature, relative humidity, and mass of a sample over a period of time serve as the criteria for additional changes in the experimental conditions [10].

Commercial units are available that utilize a carrier gas with controlled but variable amounts of moisture flowing over a small sample on a microbalance. A record is maintained of the change in the sample mass as a function of time. Once the rate of change drops below a preset limit or when a preset time limit is reached, the relative humidity of the carrier gas is changed. This process of changing the relative humidity and monitoring the approach to equilibrium is repeated until the complete vapor sorption isotherm over the desired humidity range has been produced. Because the sample is never removed from the microbalance, weighing errors and changes due to differences in relative humidity during the weighing are minimized, so that smaller sample sizes can be used and more rapid equilibration occurs [11].

It is not unusual for a compound to be isolated in more than one polymorphic form. Dynamic vapor sorption, in conjunction with DSC, can aid in the selection of the most appropriate polymorph for further development. It has been applied to the quantitation of small amounts of amorphous material in crystalline lactose [12]. It can also provide guidance concerning potential problems during formulation. If a compound is shown in dynamic vapor sorption studies to exist in more than one hydration state, the potential impact of a wet formulation should be considered.

3.6 Reaction Calorimetry

Several companies have commercial units available for evaluation of heat flow during the course of a reaction. The reaction size and degree of automation available on such systems varies considerably between vendors. In general, the reaction can be run either as a fully automated process or as a semiautomated process. In the fully automated process, all reagents and reactants are typically liquids or in solution. Semiautomated work frequently involves the manual addition of a solid to a reaction. Analysis of the reaction data is automated with some manual intervention to define the limits of the reaction(s).

Utilization of this equipment in development allows for greater reproducibility in performing a series of reactions. The technique can identify potentially hazardous processes and help define intrinsically safer reaction conditions [13–15]. It has also been shown to be useful for determining kinetics in complex reactions [16].

3.7 Accelerated Rate Calorimetry

Accelerated rate calorimetry (ARC) is a technique that is applied both to reaction mixtures and to pure solids. When applied to reaction mixtures, it provides information concerning the potential for runaway reactions. It is a very useful tool for rapidly assessing the stability of compounds and determining conditions under which they may be used or stored safely.

3.8 Spreadsheets

While some may disagree with the inclusion of spreadsheets as an automated tool, they are being included here because of the potential saving in time and effort once the sheet is programmed. Furthermore, many chemists are unfamiliar with their use and may discover that these versatile tools are applicable to their jobs.

Spreadsheets are useful tools within chemical development. The use of spreadsheets allows for information concerning materials and costs in a multistep synthesis to be easily tracked. The initial programming of a spreadsheet can be time consuming. However, once programmed, it can be used to address a number of questions such as how much of each raw material is required in a synthesis, how much product can be produced from a given limiting reagent or reactant, or what will be the cost of producing a compound. A well-designed spreadsheet can help to focus development efforts on those steps in a synthesis where the return on invested time can be the greatest. If designed properly, changing a single entry such as the cost of a reagent in the spreadsheet can result in that change being incorporated throughout the sheet. This capability is especially important when one is investigating the impact that a change will have on the overall production of a material.

In addition to tracking cost and materials, spreadsheets track trends in numeric data and generate theoretical data. A theoretical binary phase diagram can be constructed from the heats of fusion and melting points of a pure enantiomer and racemic material. Information obtained from a minimal number of experiments can be used to calculate the theoretical recovery from a crystallization designed to separate enantiomers or diastereoisomers.

3.9 Databases

There are a number of chemically related databases and database management systems currently available. The scope of commercial databases ranges from comprehensive to specialized. The ease with which database management systems handle specific types of chemical information also varies greatly. Chemical drawing software frequently includes the capability to calculate the molecular weight, empirical formula, and theoretical elemental analysis for any drawn structure. The various structure-drawing programs use different atomic weights and different algorithms to calculate molecular weights. Consequently, the molecular weights calculated for given molecular structures using different chemical drawing packages are typically only the same to 0.1 amu. If one plans to use the molecular weight generated automatically within a drawing package, the end use of the calculation should be considered. The accuracy given should certainly be high enough for any calculations on the number of moles used in a given reaction. However, the molecular weight produced by a given drawing package may not be the same as that published by a pharmacopeia.

In-house databases of chemical reactions can serve as valuable repositories of information. Construction of such databases, however, can be a time consuming and expensive process.

4 VALUABLE MATHEMATICAL TOOLS

This section contains a brief overview of mathematical tools that are used in development of chemical processes. They are being included in this book because they are available in a number of software packages. Although the calculations may be performed manually, the commercial products perform much of the tedious and repetitive work, thus automating the use of the method. These packages can facilitate interpretation of the raw data.

Simplex optimization can be used when the optimum values for reaction parameters are needed. The method provides information concerning the shape of the response surface in the vicinity of the experiments used on the approach to the optimum. It does not provide a good model of the entire response surface within the experimental domain.

Design of experiments (DOE) can be used to address a number of questions. One common application is determination of the reaction parameters that are most significant in producing changes in a measured response. A DOE study can be useful as a prerequisite to a simplex optimization in order to minimize the number of factors included in that study. A second application is construction of a model of the response surface within a defined experimental domain. The model is useful when validation of a process is required. Process validation requires that a series of reactions be run in which predefined results are to be obtained. If the obtained results meet the predefined (expected) results, the process is considered to be validated. Knowing the shape of the entire response surface within the experimental domain defined by the validation protocol provides assurance that no unexpected results will be observed during the validation study. A robust chemical process, however, is dependent both upon a knowledge of the response surface in the region in which the process is being run and upon the utilization of proper and adequate engineering controls to ensure that the process variables remain within the specified ranges.

Principal components analysis can be used as a tool to simplify classification of chemicals based upon their physical properties. Through chemical components analysis, a number of different physical properties can be examined for correlations. Those properties that display minimal correlations can then be used as the basis for formulation of a graphical representation of the diversity of all physical properties in one or more dimensions. The graphical representation can then be used in planning screening experiments to aid in the selection of compounds that can be representative of the full diversity of compounds in the original set.

4.1 Simplex Method

The simplex method of process optimization relies upon identification of the impact of a change in the value of multiple factors on a response. Commonly used responses include yield, strength, molecular weight (for polymerization reactions), and purity. Factors that frequently impact on a chemical response include temperature, concentrations, stirring speed, pressure, and pH. Once the factors have been identified, a series of experiments, in which the total number of experiments is one more than the number of factors, is conducted. In this series of experiments, each individual factor is initially set to a different value.

After the initial series of experiments has been performed, the individual sets of experimental conditions are force ranked from most desirable to least desirable based upon the observed response. A new experiment is then designed by finding the "center of mass" (centroid) of all of the experiments excluding the one least desirable experiment. A new set of experimental conditions is defined such that the experimental conditions associated with the least desirable response are reflected through the centroid. The set of experimental conditions associated with the least desirable response is then dropped from further consideration. The new experiment is performed, and the results are evaluated as above. Repeating the procedure typically results in movement toward the desired outcome.

The number of experiments required to reach the optimum outcome is dependent upon a number of factors. A small increment for a given factor may result in a large number of experiments. The starting location on the response surface plays a large role in the number of required experiments. The presence of a local maximum in a response surface may prevent discovery of the global maximum. A large plateau on the response surface may also interfere with the discovery of the global maximum.

Depending upon how critical the work is, it may be necessary to run a second series of simplex experiments to verify that the global maximum has been found. In this case, the conditions used to define the starting simplex should be well removed from those in the preceding series of experiments. If both sets of experiments converge to the same area in the experimental domain, one has greater confidence that the global optimum has been obtained.

Modifications in the execution of the standard simplex method may reduce the number of experiments necessary to locate a maximum in the response surface. Expansion or contraction of the interval between experimental conditions may result in more rapid evolution of the system toward the desired goal.

There are limitations in the use of the simplex method for optimization. Although simplex optimization provides satisfactory results, this method may not be appropriate when calendar time for a project is important to the success of a project. The method is essentially sequential in nature. After the initial set of experiments has been performed, one must always wait for the results of the following experiment before any additional experiments can be designed and performed. If it is critical that the global maximum within the experimental domain be found, then the optimization process may be made somewhat more parallel by running at least two simplex optimizations starting from widely separated regions simultaneously.

A concern in utilizing this method is that one will not necessarily know the shape of the complete response surface at the conclusion of the experiments. The method is successful in defining the slope of the surface along the path to the maximum response, but it is difficult to extrapolate beyond that path. This is not normally a great concern, since the response surface in the vicinity of the manufacturing conditions is the information that is most critical for long-term success.

Finally, the method is sensitive to errors in a response. Repetitions of experimental points are not typically done. An error in a response may lead to the propagation of errors in calculation of the next experiment. When this occurs, it may appear that the optimum has been reached. Additional work may be necessary in order to identify experiments containing the erroneous responses.

Despite the limitations inherent in the method, it can be a valuable tool in investigating difficult development problems [17].

4.2 Design of Experiments (DOE)

Included in this section is a brief discussion of statistics-based tools that are typically used for the exploration of response surfaces, especially in close proximity to an optimum. Factorial design, fractional factorial design, Box–Behnken design, and Plackett–Burman design are a few of the experimental designs used today. All of the design strategies are intended to maximize the information content from a series of experiments. Each has been applied successfully to chemical development problems. The use of each is determined, to some extent, by the experiences of the individual responsible for investigating a given project. More important however, is the type of model that can easily be fitted to a given experimental design. A number of commercial software products are available that automate the DOE process and analysis of the experimental data. Good experimental designs should include repetitions of one or more experiments. The repetitions allow for estimation of the variability (error) in the response. Inclusion of several repetitions of experiments makes DOE much more robust than simplex optimization if an error is made in a single experiment.

The major differences between the experimental designs mentioned above are the number and location of the conditions (points) chosen for the individual experiments. In any experimental design, each individual experiment can be represented by an equation of the form of Eq. (1). In this equation, R represents the response to a given experiment, the β_0 term is a constant, β_i, β_{ip}, and β_{ii} are coefficients related to each of the experimental variables, x_i, x_p, and x_{ii} are the values of each of the individual variables used in the experiment, and ϵ is the error in determining the response of the specific experiment.

$$R = \beta_0 + \Sigma\beta_i x_i + \Sigma\Sigma\beta_{ip} x_i x_p + \Sigma\Sigma\beta_{ii} x_{ii}^2 + \epsilon \qquad (1)$$

In Eq. (1), the terms $\Sigma\beta_i x_i$ are the linear terms, $\Sigma\Sigma\beta_{ip} x_i x_p$ are interaction terms, and $\Sigma\Sigma\beta_{ii} x_{ii}^2$ are quadratic terms. The analysis of the complete set of experiments involves solving the entire set of simultaneous equations to find values of β_0, each β_i, each β_{ip}, each β_{ii}, and ϵ that minimize the difference between the measured and calculated R across all experiments. The terms that can be unambiguously determined are dependent upon the experimental design [18]. Such calculations are performed automatically in commercially available DOE software. For screening purposes, designs that only incorporate linear or linear and interaction terms are typically used. The exploration of response surfaces in preparation for validation studies will normally include quadratic terms.

The general considerations used in DOE will be illustrated using a factorial design in n dimensions (n factors). The range of each factor is determined in advance of the experiments so that a series of experiments can be performed in which each factor can be set independently to either its high or its low value. The full set of experimental conditions defining the selected range is called the experimental domain. In order to obtain all combinations of all factors at either their high or their low values, a total of 2^n experiments must be performed. Once a response under each experimental condition has been obtained, DOE software can be used to fit the data to a linear mathematical model. Addition of a point at the center of the design will allow for estimation of both linear and quadratic terms.

The output from analysis of the data produced using DOE is a mathematical model of the impact that the factors have on the measured response within the boundaries of the experimental domain. It is important to remember that the mathematical model does not necessarily explain why the response is observed. A good model does, however, provide the magnitude of the response to be expected under a given set of reaction conditions. If a maximum is predicted by the model at a point other than one already included in the study, then the model must be tested by running an experiment under those conditions.

In contrast to simplex experiments, DOE studies may be conducted in parallel rather than sequentially. Parallel experiments provide the opportunity to reduce significantly the amount of time necessary to perform a complete series of experiments. If equipment is unavailable for simultaneously performing all experiments in an experimental design, running the experiments sequentially or in blocks is a possibility.

The response surface that is constructed will most accurately describe the actual response surface if the study is conducted over a relatively small range for each variable. The response surface generated using, for example, a temperature range of 0 to 100° C and a pH range of 3 to 12 could provide a much less satisfactory fit across the entire experimental domain than one in which the temperature range was 45 to 55° C and the pH was between 7 and 9.

Studies involving DOE provide much more information than conventional experiments. Unfortunately, it is difficult to obtain useful information until the study is complete. For most chemical reactions involving three or more variables it is not necessary to perform a full factorial design in order to interpret the data. Studies that include all interactions between any two variables frequently provide models that fit the data very well. For example, a complete factorial experiment involving five factors involves a minimum of 2^5 or 32 experiments. Since it is unlikely that there will be any significant interactions among more than two terms, it is possible to perform a partial factorial design (2^{5-2}) for 5 factors that includes as few as eight experiments. Addition of a point at the center of the experimental design allows for detection of any curvature that otherwise may not be detected. Addition of three repetitions provides additional information on the error associated with any response. Thus a total of 12 experiments can provide information on the shape of a response surface and the error associated with any measurement. Data from such an abbreviated study must be interpreted cautiously, since the results are confounded as a consequence of the design.

The data obtained from any given experimental design may indicate that the optimum is not contained within the experimental space. In such cases, it is often possible to extend the study to adjacent experimental areas without starting an entirely new study in the adjacent area. By building a new study utilizing existing data, a model for a larger response surface can be constructed.

The efficient utilization of an automated workstation for the optimization of a synthetic reaction depends upon a good experimental strategy. That strategy should include the use of DOE to determine what variables are truly important in achieving the desired response. Well-designed studies can provide valuable information necessary to design robust processes.

4.3 Principal Components Analysis

Another multivariate analysis tool used in chemical development is principal components analysis (PCA). Although it is used much more frequently within discovery research, there are also proponents of the use of this technique within development.

PCA is a subset of the mathematical tools composing factor analysis. Such tools allow for comparison of different data sets to determine if there are any correlations among the data. As used in chemical development, the goal of PCA is to reduce a large number of physical properties into a small number of independent factors. The tool can be used effectively to compare compounds that serve the same function in a reaction. For example, the properties of a number of compounds can be compared when the compounds are intended for use as the reaction solvent.

One use of PCA is to represent similarities and dissimilarities between numerous physical properties of compounds. This is done by displaying a projection of multidimensional data (the values of the individual properties) into a lower dimension. The projection is selected so that it shows the largest spread of properties available in the lower dimensional view.

The projection of measured physical properties to two dimensions that describe the majority of the variation in the data creates a ''map'' on which compounds with similar properties are found near one another. Compounds selected from different areas, but evenly distributed across the entire map, then represent nearly the full range of diversity of all compounds on which the PCA was conducted. Once selected, the compounds should be evaluated to determine if their functionality is compatible with the reaction chemistry being screened. If the functionality is incompatible, an alternative compound from the same region of the map should be selected. The initial screening experiments are then run under the same conditions with the desired response, e.g., product formation, being monitored as a function of time. The maximum yield and time necessary to achieve it can then serve as measures of the relative suitability of the compound in the reaction. If desired, additional compounds in close proximity to the most suitable compound(s) found in the original study are investigated.

Figure 1 shows a score plot from a principal components analysis of eight physical properties associated with 82 solvents. The figure was produced using published values of the melting point, boiling point, dielectric constant, dipole moment, refractive index, E_T, density, and log P of the individual solvents. The two-dimensional plot of the solvents accounts for 51% of the variation in the physical properties of the entire set of solvents. Addition of a third principal component did not improve the explained variance in the data [19]. The solvents represented in Fig. 1 are listed in Table 1.

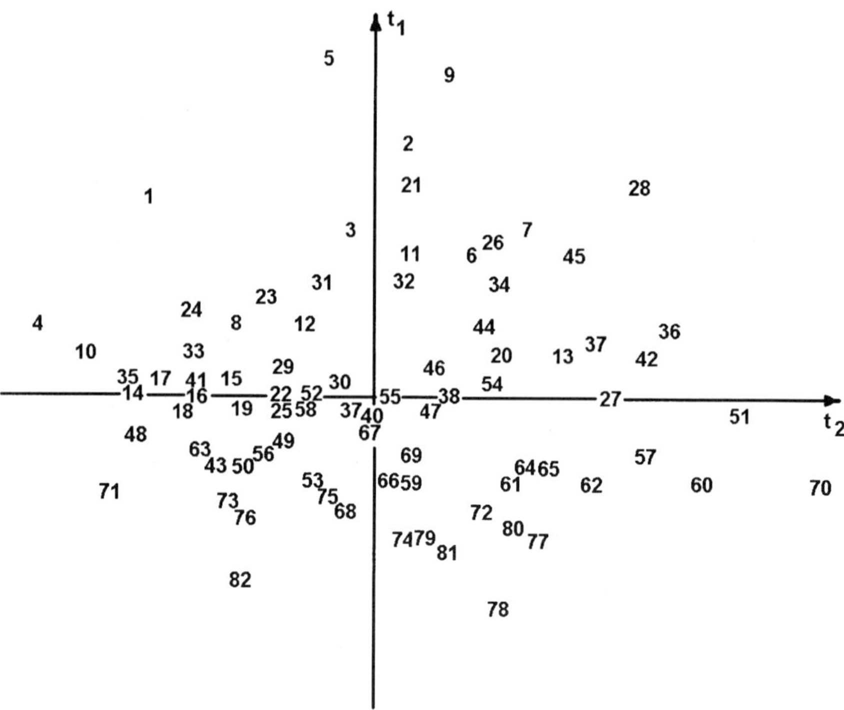

Figure 1 Plot of two principal components for 82 solvents. (Reprinted with permission from *Acta Chemica Scandinavica B39* (1985), pp. 7–91.)

Table 1 Solvents Represented in Fig. 1

Number	Solvent	Number	Solvent
1	Water	42	Acetophenone
2	Formamide	43	Dichloromethane
3	1,2-Ethandiol	44	1,1,3,3-Tetramethylurea
4	Methanol	45	Hexamethylphosphoric triamide
5	N-Methylformamide	46	Cyclohexanone
6	Diethylene glycol	47	Pyridine
7	Triethylene glycol	48	Methyl acetate
8	2-Methoxyethanol	49	4-Methyl-2-pentanone
9	N-Methylacetamide	50	1,1-Dichloroethane
10	Ethanol	51	Quinoline
11	2-Aminoethanol	52	3-Pentanone
12	Acetic acid	53	Chloroform
13	Benzyl alcohol	54	Triethylene glycol dimethyl ether
14	1-Propanol	55	Diethylene glycol dimethyl ether
15	1-Butanol	56	Dimethoxyethane
16	2-Methyl-1-propanol	57	1,2-Dichlorobenzene
17	2-Propanol	58	Ethyl acetate
18	2-Butanol	59	Fluorobenzene
19	3-Methyl-1-butanol	60	Iodobenzene
20	Cyclohexanol	61	Chlorobenzene
21	4-Methyl-1,3-dioxol-2-one	62	Bromobenzene
22	2-Pentanol	63	Tetrahydrofuran
23	Nitromethane	64	Anisole
24	Acetonitrile	65	Phenotole
25	3-Pentanol	66	1,1,1-Trichloroethane
26	Dimethylsulfoxide	67	1,4-Dioxane
27	Aniline	68	Trichloroethylene
28	Sulfolane	69	Piperidine
29	Acetic anhydride	70	Diphenyl ether
30	2-Methyl-2-propanol	71	Diethyl ether
31	N,N-Dimethylformamide	72	Benzene
32	N,N-Dimethylacetamide	73	Diisopropyl ether
33	Propionitrile	74	Toluene
34	1-Methyl-2-pyrolidone	75	Di-n-butyl ether
35	Acetone	76	Triethylamine
36	Nitrobenzene	77	1,3,5-Trimethylbenzene
37	Benzonitrile	78	Carbon disulfide
38	1,2-Diaminoethane	79	Carbon tetrachloride
39	1,2-Dichloroethane	80	Tetrachloroethylene
40	2-Methyl-2-butanol	81	Cyclohexane
41	2-Butanone	82	n-Hexane

Using the above approach, the most effective solvents can be found for each step in a multistep synthesis. Alternatively, it may be possible to select a single solvent that provides satisfactory results in many of the steps of a multistep synthesis. Once the solvent(s) is (are) selected, each of the steps can be individually optimized in the solvent.

Carlson [20] proposed a number of strategies for the selection of solvents, reagents, and catalysts based upon the results of principal components analysis of the physical properties of the materials.

Principal components analysis has also been applied to the evaluation of screening experiments in which multiple responses are measured [21].

5 EXAMPLES IN PROCESS DEVELOPMENT

The following sections provide specific examples of many of the applications of automation in chemical development. Examples are chosen to be illustrative, not comprehensive. The examples may include the use of more than one automated tool to address a specific issue.

5.1 Reactors

Automated reactors used in development may be classified into three main categories: small reactors used for data production, reactors used for production of both data and moderate amounts of chemicals, and reactors used primarily for production of chemicals. Each of these categories may be divided further into solid-state chemistry and solution-phase chemistry. All of the automated reactors require the operator to enter some instructions concerning the parameters to be used during the reaction(s). The detail required is dependent upon the software driving the system.

5.1.1 Small Automated Reactors

Small reactors are typically designed for screening experiments or initial optimization studies. The reaction volume may range from a few microliters to about 250 mL. The reaction vessel itself is often a vial, test tube, or modified test tube. Agitation in such vessels can be provided by placing the reactor on an agitator, by bubbling a gas through the reaction mixture, or by magnetic or mechanical stirring. Temperature control in such vessels is provided by a gas, by a circulating liquid, or by contact with a metal reaction block held at the desired temperature. Addition of reagents is accomplished by a syringe drive through a cannula.

The user interface on such equipment may consist of a limited set of instructions such as transfer (from one location to another), add (a specified amount of a solvent), set the temperature, wait (for a given amount of time or until a condition is met), and sample. Although the instruction set appears to be limiting, execution of different combinations of instructions allows for considerable flexibility in defining an experiment or series of experiments.

A wide variety of reactions have been studied using such equipment. Boettger [22] published an early example of the use of such equipment in a palladium catalyzed coupling reaction. Armitage et al. [23] reported success in palladium-catalyzed coupling reactions, a Michael addition reaction, screening of Lewis acids in a displacement reaction, desilylation reactions, alkylation of isatin, and bromination of an indoline derivative. Corkan et al. [24] reported results of a study in which a direct comparison was made between simplex optimization and a 4 × 4 factorial design experiment for the optimization of the synthesis of porphyrins.

Such reactors have the advantages of the use of small amounts of starting materials, production of relatively small amounts of waste, and the simultaneous performance of numerous experiments (parallel experimentation). Disadvantages of such experiments include larger relative errors on the delivery of small amounts of reagents, less precise temperature control, and the need to develop rapid, robust analytical methods.

Depending upon the exact setup and the chemistry being evaluated, the analysis of a reaction mixture may be the major factor in designing a series of experiments. This is especially the case if a quenched sample of the reaction is not stable. For example, utilizing equipment in which a total of 12 simultaneous reactions may be run, an analytical method that requires 5 min to perform limits the sampling rate from a given reaction to once per h. If a reaction is to be monitored throughout the course of the reaction and the reaction requires several h for completion, one sample per h does not pose a major problem. However, an hourly sampling rate for a reaction that is complete in 3 h or less poses significant data loss when 12 reactions are run simultaneously. In this case it is beneficial to decrease the number of reactions run simultaneously and run the total of 12 reactions in a series of four to six blocks.

Sample delivery and workup on such systems may impact upon the analysis method. If the equipment is capable of delivering a precise volume of the reaction mixture into a precisely controlled volume of quench solution, it may not be necessary to include an internal standard in the reaction mixture. External standards may be satisfactory in such work. If the volume of reaction mixture or the quench volume or both cannot be controlled precisely, an internal standard is required for quantitative work.

Commercially available units are sold by a number of companies [7]. Due to the number of units currently available and the differences in the capabilities of the commercial units it is recommended that the companies be contacted directly.

5.1.2   Reactors Used for Production of Both Data and Moderate Amounts of Chemicals

Reactors with working volumes of between 250 mL and 5 L have been used as multipurpose equipment. Because of the space required for the reactor, associated pumps, and other control components, such equipment typically utilizes only one reaction vessel at a time. Reaction vessels are often glass and are configured to resemble miniature production scale equipment. Agitation is provided by an overhead mechanical stirrer. Temperature control is provided by a liquid circulating through a jacket, or a combination of a thermostated jacket and an internal heater. Reagents may be added as solids, as neat liquids, or as solutions through metering pumps.

These reactors provide a much greater variety of control devices than the smaller reactor systems. Temperature, pH, and pressure control, as well as simultaneous addition of multiple reagents, are possible. Additions can be based upon mass or upon volume. Because of the larger number of control elements, the instruction set used for programming the equipment may contain a greater variety of control elements. Such control elements typically include conditional statements. The use of such equipment in the preparation of butylmagnesium chloride and its subsequent reaction with acetaldehyde has been reported [25].

An advantage of these larger automated reactors is the capability of using either the jacket or the internal temperature to control the reaction temperature. Because of the scale on which the reactions are run, it may be possible to generate sufficient quantities of product to support other developmental work during the course of optimization studies. The larger reaction scale also provides the possibility of continuous nondestructive monitoring through Fourier transform infrared spectroscopy, ultraviolet spectroscopy, or even calorimetry. Reaction volumes are large enough that removal of samples for chromatographic analysis should not impact the yield. Larger samples can be obtained for analysis. This can simplify analysis, since errors in the volume of the sample and quench solutions can be minimized. Smaller relative sampling errors can allow the use of external standards. Care must be taken in obtaining samples from sensitive reactions, since opening the reactor may introduce moisture or air.

The disadvantages of such reactors include the relatively large amounts of reagents, reactants, and solvents necessary to run individual reactions. Because only one reactor is typically available on such equipment, DOE studies reactions must be run sequentially. The total clock time necessary to complete a series of reactions can be much longer than a similar study utilizing small automated reactors. A runaway reaction in the larger scale equipment has the potential for greater environmental impact. Consequently, the settings for an emergency shutdown should be carefully considered before running such experiments.

Two commercial units that include the capability of performing calorimetry are widely available. ASI's RC-1 unit is available in numerous configurations with reaction vessels designed for reaction volumes of 100 mL to 25 L and reaction pressures ranging from atmospheric to 350 bar. HEL's Auto-Lab allows for calorimetry to be performed on a 200 mL to 20 L scale and at pressures up to 200 bar. A third system, the LABMax automated reactor, is similar to the RC-1 calorimeter but does not include the capability of measuring heat flow in a reaction.

5.1.3 Reactors Used Primarily for Production of Chemicals

Reactors included in this category are specially constructed workstations that are intended to produce small to intermediate amounts of material with minimal operator intervention. The equipment, in many cases, is not commercially available. Details of a number of components have been reported in the literature [26]. The equipment fills a need for the production of material on a scale necessary for further development.

Takeda Chemical Industries, Ltd., has developed a number of reactor systems that are used primarily for the automated production of a moderate amount of a series of compounds with a minimal amount of operator intervention. Two factors contribute to achieving this level of automation. First, the chemistry involved in preparation of each member of the series of compounds is very similar. Thus a ''master synthesis'' is developed for one member of the series. Details on the operations used to produce the material are then used as the basis for the production of all other members of the series. Multiple steps such as synthesis, purification, and isolation can be performed in a specific workstation. Second, the workstation is equipped with specific sensors that are interfaced to a computer with access to a database that is used to determine when unacceptable conditions exist for continuation of the programmed sequence. Identification of unacceptable conditions results in an alarm to summon the operator, who intervenes by correcting the condition and manually advances the sequence to a point where automated control is again possible. Utilization of such equipment allows for much of the work required to produce a series of compounds to be done automatically while ensuring that extraordinary conditions are addressed by a qualified human whenever necessary [27,28].

5.2 Spreadsheets

The examples provided for the use of spreadsheets are intended to provide a compromise between the functionality provided by the spreadsheet and the time necessary to program it. Once programmed for a specific purpose, the spreadsheet can repeat the same series of calculations using various values. Therefore they are most useful when it is known that variations on the same calculations will be needed repeatedly.

5.2.1 Calculation of Materials and Costs

In its most general form, a spreadsheet for a multistep synthesis contains information on the amount and cost of the reagents used in each step and on the yield of each step. Linking the amount of an intermediate produced in one step of a reaction sequence to the amount of that intermediate used in the next step of the sequence allows for the overall material needs to be calculated.

Once programmed, the spreadsheet can be used for a number of purposes. In addition to finding the total cost of raw materials and the amount of each raw material used in the sequence, the impact of improvements in yield in a given step can be determined. The impact on total cost of changes in the cost of a given raw material can be determined. In addition, most spreadsheets allow for calculation of a specific cell based upon changes to another cell on the spreadsheet. This capability can be used, for example, to determine how much of the product can be produced from an intermediate or raw material currently in stock. Although programming such spreadsheets is straightforward, the space necessary for a detailed explanation of the programming for a spreadsheet used for a multistep synthesis is nearly as great as the spreadsheet itself. A detailed explanation of the use of the spreadsheet is comparable in size. Consequently, a specific example of this application will not be provided.

5.2.2 Binary Phase Diagrams and Estimation of Resolution Efficiency

On a preparative scale, enantiomers are typically separated either by preferential crystallization of one enantiomer (entrainment) or by preferential crystallization of a single diastereomeric salt (classical resolution). Using a spreadsheet to produce the theoretical phase diagram of the system can facilitate development of either process. Preparation of Spreadsheet 1 requires only knowledge of the Schröder–Van Laar equation [29,30] and the Prigogine–Defay equation [34]. Application of the spreadsheet to a specific resolution also requires the use of DSC to determine the heats of fusion and the melting points of the racemate and a pure enantiomer of the compound.

	A	B	C	D	E	F
1	**Compound:**		o-chlorophenylhydracrylic acid			
2	$\mathbf{\Delta H_a^f}$		$\mathbf{\Delta H_r^f}$	$\mathbf{T_a^f}$	$\mathbf{T_r^f}$	**R**
3	24.7		20.1	393	359	**0.008315**
4	**Chart Min.**	355		**Chart Max.**	395	

Spreadsheet 1   Data entry area of a spreadsheet used to prepare a theoretical phase diagram from racemate and a single pure enantiomer.

	A	B	C	D	E	F	G
1	**Salt:**	α-phenylethylammonium mandelates					
2	$\mathbf{T_{eu}}$	$\mathbf{T_{50}}$	$\mathbf{Q_{eu}}$	$\mathbf{Q_{fr}}$		**C**	$\mathbf{X_{eu}}$
3	381	436	19.1	15.7		*0.63*	*0.12*
4							
5	**R**					**Efficiency**	
6	**0.0083145**	**kJ/mol K**				**S**	
7						*0.86*	

Spreadsheet 2 Data display area of a spreadsheet used to calculate the theoretical efficiency of a resolution based upon DSC data obtained from a 50:50 mixture of diastereomeric salts.

Spreadsheet 2 can be used when an equimolar mixture of two diastereomeric salts are known to be present in a solid sample. In this case, the heat of fusion of the eutectic and its melting point and the heat of fusion and final melting temperature of the sample are determined by DSC. The spreadsheet will provide a value for the eutectic composition.

From Pure Components.   Several defined terms will be used throughout the following discussion. A racemate will be used for a 1:1 mixture of both enantiomers of a chiral compound without consideration of the physical state of that mixture. A racemic compound indicates a 1:1 mixture of both enantiomers present in a well-defined crystalline lattice. In racemic compounds, both enantiomers are present in a single crystal. A conglomerate is a mixture of crystals in which each crystal contains a single enantiomer. A racemic conglomerate is therefore a 1:1 mixture by weight of crystals, each of which contains a single enantiomer. A solid solution is a mixture of enantiomers in which the proportion of the two enantiomers has no set value between individual crystals [35] (Fig. 2).

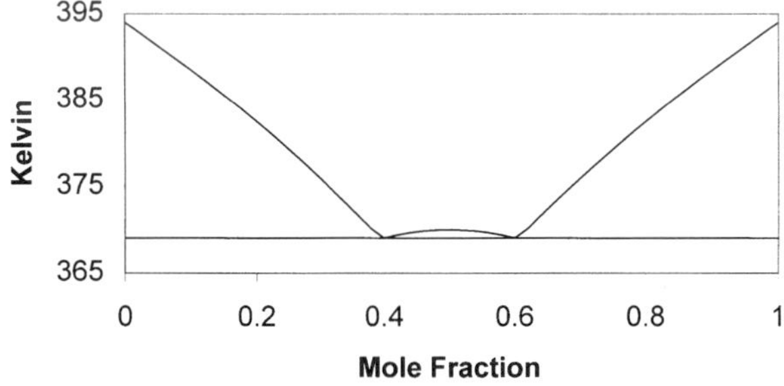

Figure 2 Calculated phase diagram for *m*-fluoromandelic acid displaying two eutectic compositions.

Racemates are considered first, since they are the examples that are the most straightforward to program and to understand. The theoretical binary phase diagrams that are produced using the spreadsheet will have one of two general shapes. The shape will resemble either that of a V or that of a stylized W. In both cases, the lowest melting points are the eutectic. The composition of the eutectic can be read from the x-axis.

Of the two types of diagrams, the V is preferred, since it indicates that the material exists as a conglomerate. For compounds that form crystalline conglomerates, there is the possibility of achieving a resolution without an enantiomerically enriched auxiliary agent. Although technically difficult, entrainment can be cost- and time-effective for the separation of enantiomers showing such behavior.

The W indicates that attempts to resolve the two enantiomers through crystallization of racemate will be unsuccessful. When this type of phase diagram is observed, it indicates that the racemate exists as a true racemic compound. Intermolecular forces between $(+)$- and $(-)$-enantiomers are stronger than those between enantiomers with the same optical rotation. The consequence is that the first solid that crystallizes from solutions produced from racemate is the racemic compound. For molecules that form racemic compounds, resolution to a single pure enantiomer may be achieved if an enantioselective synthesis has produced material with a composition outside the region between the two eutectic compositions.

Inspection of the theoretical phase diagrams provides rapid feedback concerning the preferred method for resolution. If a V-shaped diagram is observed, one should consider the possibility of fractional crystallization. When a W-shaped diagram is observed, it is useful to consider the possibilities of formation of mineral acid salts or formation of a pair of diastereomeric salts with an enantiomerically enriched resolving agent. Both of these options can be easily explored using automated systems.

Spreadsheet 1 shows the normal view of the data entry portion of an Excel spreadsheet designed to calculate the theoretical phase diagram for mixtures of two enantiomers.

The numbers displayed on the spreadsheet and the calculated phase diagram (Fig. 3) are for o-chlorophenylmandelic acid [33]. Figure 2 can be prepared by entering the following values in the appropriate cells in Spreadsheet 1: $\Delta H_a^f = 24.3$, $\Delta H_r^f = 24.7$, $T_a^f = 394$, and $T_r^f = 370$. ΔH_a^f (Cell A3) is the enthalpy of fusion of a pure enantiomer in kJoule mole^{-1}. ΔH_r^f(Cell C3) is the enthalpy of fusion of the racemate in kJoule mole^{-1}. T_a^f (Cell D3) is the final melting point in kelvins of a pure enantiomer. T_r^f (Cell E3) is the final melting point in kelvins of the racemate.

To use the spreadsheet, place the name of the compound in Cell B2 and values of the enthalpy of fusion and melting points in the appropriate cells. Significant deviations between an observed melting point and the theoretical melting point are observed when solid solutions are formed.

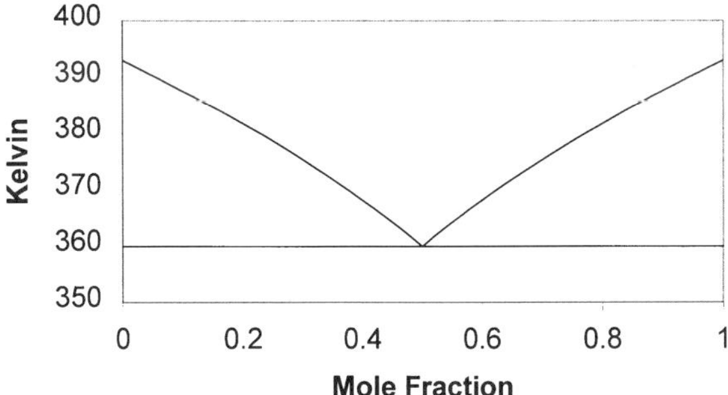

Figure 3 Calculated phase diagram for o-chloromandelic acid displaying a single eutectic.

The directions provided in Table 2 are applicable to Excel. The integer (INT), maximum (MAX), minimum (MIN), and natural log (LN) functions used for this spreadsheet should be available in other commercially available spreadsheets. The syntax of the calculations may vary. In the column labeled Cell location, A8 . . . A32 indicates that all cells from A8 to A32, inclusive, should be filled as indicated in the adjacent column. The text included between quotation marks under the Cell contents heading should be entered as indicated without the quotation marks. Texts not included between quotation marks are specific instructions that explain how to fill the given cells or produce the graph (Table 2).

From 50:50 mixture of diastereomeric salts. The following spreadsheet is based upon the paper by Kozma et al. [34] in which it is reported that the efficiency of a resolution using diastereomeric salt formation can be estimated from the DSC curve obtained from a 50:50 mixture of the two diastereomeric salts. The method may be used when the 50:50 mixture of salts is a solid. Since the pure diastereomeric salts need not be individually prepared, considerable time may be saved in screening numerous resolving agents. Details on the mathematics used in the calculations can be found in the original article. In order to use the spreadsheet, the required values for four variables, T_{eu}, Q_{eu}, T_{50}, and Q_{eu}, must be determined.

The DSC sample is prepared starting from racemic material. One molar equivalent of a pure enantiomer of the resolving agent is added in a suitable solvent and the resulting mixture of salts is concentrated to dryness. If the residue is a solid, the DSC curve is obtained. Using standard methods, the curve is analyzed as shown for the melting curve of a 50:50 mixture of (R,S)- and (R,R)-α-phenylethylammonium mandelate to determine the melting point of the eutectic (T_{eu}) in kelvins, the heat necessary to melt the eutectic (Q_{eu}) in kJoule/mole, the

Table 2 Instructions for Preparation of the Spreadsheet Used to Prepare a Theoretical Phase Diagram from DSC Data Obtained from Racemate and a Single Pure Enantiomer

Cell location	Cell contents
A1	"Compound:"
A2	"ΔH_a^f"
A4	"Chart Min."
B4	"=5*(INT(E6/5))"
B1	This cell is used to enter the name of the compound under study
B6	"Schöder–Van Laar"
C2	"ΔH_r^f"
C6	"Prigogine–Defay"
D2	"T_a^f"
D4	"Chart Max."
D6	"Larger"
E2	"T_r^f"
E4	"=5*(INT(E7/5))+5"
E6	"=MIN(D7:D57)"
F2	"R"
F3	"0.008315"
G3	"kJ/mol K"
A7	"1"
A8 . . . A32	Increment the value in each cell immediately above the cell by -0.02
A33	"0.52"
A34 . . . A57	Increment the value in each cell immediately above the cell by $+0.02$
B7	"=\$A\$3*\$D\$3/(\$A\$3-\$D\$3*\$F\$3*LN(A7))"
B8 . . . B57	Copy the formula in cell B7 into each of these cells
C8	"=(2*\$C\$3*\$E\$3)/(2*\$C\$3-\$E\$3*\$F\$3*LN((4*A8)*(1-A8)))"
C9 . . . C56	Copy the formula in cell C8 into each of these cells
D7	"=IF(B7>C7,B7,C7)"
D8 . . . D57	Copy the formula in cell D7 into each of these cells
E7	"=MAX(D7:D57)"
F7	"0.00"
F8 . . . F57	Increment the value in each cell immediately above the cell by $+0.02$
G7	"=D7"
G8 . . . G57	Copy the formula in cell G7 into each of these cells
G7 . . . G57	Format these cells to display 1 numeral to the right of the decimal point
A6 . . . E57	If desired, the value in these cells can be hidden by setting the text color to that of the background
H7 . . . H57	Observed melting points from mixtures of known composition may be added and plotted for comparison to the theoretical curve
Graph	Select Cells F7 . . . H57 as cells to be graphed. Use a scatter plot to plot the values in Column F as the x-axis values and the values in Columns G and H as the y-coordinate values. Add a title and scale the axes as needed.

final melting point of the solid (T_{50}) in kelvins, and the heat necessary to melt the excess of the remaining solid (Q_{50}) in kJoule/mole. Insertion of the obtained values into the appropriate cells of the spreadsheet produces a calculated efficiency (S) for the resolution using that resolving agent.

Repeating the above process with a series of resolving agents results in calculated efficiencies for the series of compounds. Comparison of the S values can facilitate selection of the best resolving agent. One must keep in mind, however, that it may be possible to achieve a more efficient resolution using one of the resolving agents that produces only a single crystalline diastereomeric salt. The method also fails (1) if any diastereomeric salt decomposes before or during melting, (2) if the diastereomeric salts form a solid solution, or (3) if either salt crystallizes as a hydrate or solvate (Fig. 4).

Spreadsheet 2 shows the normal view of the data entry and results portion of an Excel spreadsheet designed to calculate the theoretical efficiency of a resolution. Text appearing in bold face is programmed into the spreadsheet. The

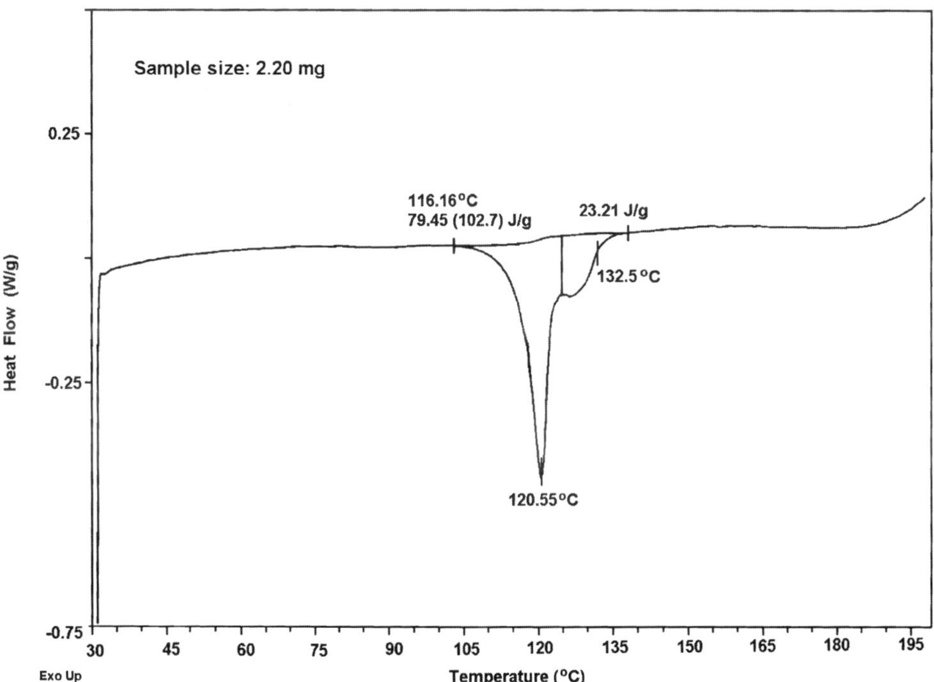

Figure 4   DSC data for 50:50 mixture of n and p salts of 3-methoxy-α-methylbenzylammonium mandelate.

Table 3 Instructions for Preparation of the Spreadsheet Used to Calculate the Theoretical Efficiency of a Resolution Based upon DSC Data Obtained from a 50:50 Mixture of Diastereomeric Salts

Cell location	Cell contents
A1	"Salt:"
A2	"T_{eu}"
A5	"R"
A6	"0.0083145"
B2	"T_{50}"
B6	"kJ/mol K"
C2	"Q_{eu}"
D2	"Q_{fr}"
F2	"C"
F3	"=(D3/A6)*((1/A3)−(1/B3))"
F5	"Efficiency"
F6	"S"
F7	"=((1−2*G3)/(1−G3))"
G2	"X_{eu}"
G3	"=D10"
A10	"0.00"
A11 . . . A60	Increment the value in the cell that is immediately above by +0.02
B11	"=−(((1−(2*A11))/(2−2*A11))*LN(2*A11))"
B12 . . . B59	Copy the formula in Cell B11 into each of these cells
C10	"=MIN(C11:C60)"
C11	"=ABS(B11−F3)"
C12 . . . C60	Copy the formula in Cell C11 into each of these cells
D10	"=SUM(D11:D60)"
D11	"=IF(C10=C11,A11,"")"
D12 . . . D60	Copy the formula in Cell D11 into each of these cells
A3 . . . D3	Format these four cells with a different background color or protect all cells on the sheet except these to allow changes in the values of only these four cells
F7	Format this cell with a different color to indicate that this cell contains the results of the calculations
A10 . . . D60	If desired, the value in these cells can be hidden by setting the text color to that of the background

data in cells A3 through D3 and in cell B1 is data that can be entered to test that the spreadsheet is functioning correctly. The italicized numbers in Cells F3, G3, and F7 should be the calculated values produced by the spreadsheet and rounded to two decimal places. Table 3 provides instructions for setting up the spreadsheet.

5.3 Solvent or Reagent Selection

A strategy similar to that reported by Carlson [19] can be useful in the selection of solvents. Utilizing Fig. 1, a number of points should be selected that are approximately equally spaced from one another and from different regions of the diagram. The solvents corresponding to the points are identified and a judgment is made by the chemist on the compatibility of the solvent with the reaction being studied. For example, an alcohol solvent would be inappropriate in a Grignard reaction. If a solvent is incompatible with the reaction, then a compatible solvent represented by a nearby point is selected.

Once a set of several solvents have been selected, the reaction is set up and run under the same conditions with one solvent used in each reaction. Excluding the scale, the reaction should be run under conditions that are considered reasonable on the anticipated final scale. The response to the reaction is determined. In screening experiments, commonly used responses are yield, raw materials cost, and purity. It is very desirable to determine the response as a function of time. Reactions that proceed at different rates or those in which the product may decompose may give misleading results if sampled at only one time point.

The desirability of each solvent is force ranked based upon the response in that solvent. The solvent with the best response should be selected as the focus for additional experiments. It may be useful to run a second set of screening experiments with different solvents. Depending upon the number of experiments that can be performed in the allotted time, the best one or two solvents are selected, and several points in close proximity to either of those solvents can be selected as above. The reaction may be run again using the new set of solvents and the responses are determined. The solvent with the best overall response should then be selected.

Note that this strategy will not necessarily provide the solvent that is the best overall solvent for the reaction. Different solvents may stabilize any transition states on the reaction pathway to different extents. In order to determine the global best solvent, the reaction should be optimized in each solvent. In setting up the original screening experiments, conditions were selected close to those desired for large-scale production. An assumption is made that the reaction that produces the best response will already be closer to its optimum conditions than a reaction run in any other solvent. A second assumption is that full optimization in the selected solvent will be easier than for any other solvent. While the first assumption may be correct, the second assumption is not necessarily valid. Additionally, thought should be given to the reactions preceding and following the reaction under current consideration. Considerable cost savings can be realized industrially if an intermediate can be carried through several steps in the same solvent without isolation. The experimenter should rely upon past experience as well as upon the current results in selecting the solvent to be studied in optimization studies.

Similar considerations can be made in selection of reagents used to effect a given transformation. Published correlations for Lewis acids are available [35]. At present such correlations for other reagents such as bases, halogenating reagents, coupling reagents, etc. are not readily available. Commercial databases and the literature should be used to obtain information about the relative utility of a series of reagents for a given transformation. Screening experiments, often utilizing automated equipment to facilitate data collection, should be designed with the desired final reaction conditions in mind.

The reagent(s) that produces the most desirable results is typically used in optimization studies. Here also, the specific reagent should be considered with regard to the overall process and to the ease of isolation of the desired product. Despite the desire to automate process development, it is not currently feasible to rely entirely upon automation to develop the best overall process for production of a chemical.

5.4 High-Throughput Optimization of Reaction Parameters

At the R. W. Johnson Pharmaceutical Research Institute (RWJPRI), high-throughput optimization of reaction parameters (HTORP) is performed utilizing robotic workstations. These workstations consist of two specialized robotic platforms. The first platform, Bohdan's Automated Process Development Workstation, performs the synthetic operations. The second platform, a Gilson Combinatorial Chromatography System, performs sample analysis. The personal computers driving these units are connected to one another in order to coordinate sample delivery and information tracking between the two platforms. Whenever a sample is transferred from the Bohdan unit to the Gilson unit, a sample identifier is stored on both computers. Figure 5 shows a typical setup of the two units.

Transfers and solvent delivery on the synthesis unit are made through either a three-channel cannula or a slurry-handling cannula. Both cannulas are mounted on the same robot arm. The temperatures of the reactors are individually controlled. All reactors are placed in a reaction block that shares a common inert atmosphere. General information concerning the capabilities and limitations of the synthesis unit are provided in Table 4.

Samples are removed from the reaction using the slurry-handling cannula. During a set of reactions, samples may be delivered either directly to the analytical unit through a filtered narrow-bore teflon tube or to a sample rack that replaces one of the reactant racks. When the reactions are completely in solution, the preferred delivery is directly to the analytical unit. As currently configured, the maximum dilution factor of about 600 is available for samples delivered directly

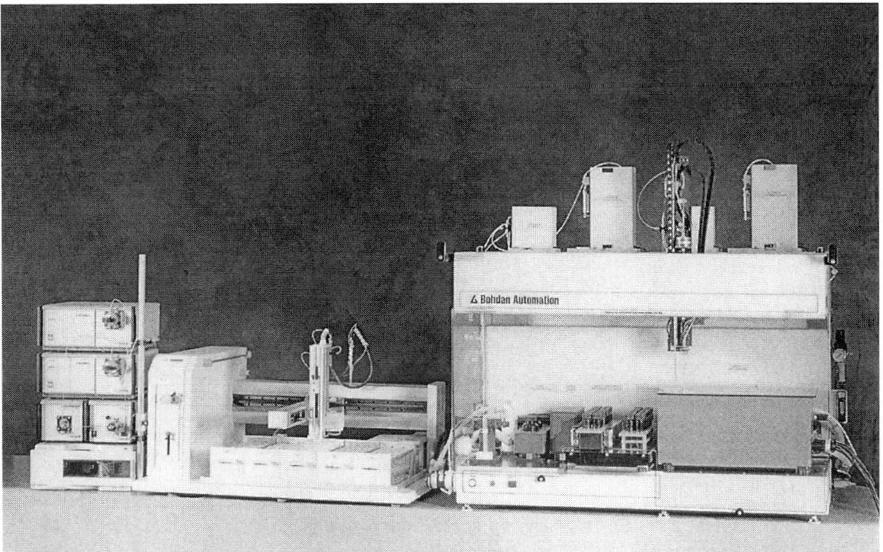

Figure 5 Photograph of the workstation used for HTORP studies. (Photo courtesy of Bohdan Automation, Inc.)

to the analytical unit. Sample agitation on the Gilson unit is through an aspiration and dispense sequence into a given tube. The maximum stroke volume during the mixing routine is 5 mL. Samples that contain solids should be delivered to the sample rack in order to avoid clogging the transfer line. Additional workup of these samples can then be performed on the analytical workstation.

Although a single analysis may take several minutes, samples can be transferred from the synthesis workstation to the analytical workstation at any time that the arm of the analytical robot is free. During an analysis, the arm on the analytical robot is only busy during the transfer of the sample from the test tube to the injector and during the wash of the cannula.

Development of rapid, efficient analytical methods is crucial to the effective use of the equipment. When all twelve reactors are in use, an analytical method that requires more than 5 min will not allow more than one sample per hour to be obtained without formation of a backlog of samples. In those instances when the quenched samples are known to be stable, long analysis times are not problematic. Even when the product is known to be chemically stable in the quench solution, incorrect analyses can be obtained. Crystallization from the quenched reaction mixture has been observed on occasions when the samples have stood for extended periods of time.

Table 4 Information on Limitations and Features of the Synthetic Workstation

Feature	Associated data
Simultaneous reactions	12 (maximum)
Reaction volume	25–30 mL (maximum)
Maximum reagent volume	ca. 120 mL in each of 4 containers ca. 30 mL in each of 32 reagent vials
Samples per reaction	ca. 30; the number is dependent upon the configuration and number of simultaneous reactions
Pressure	Atmospheric
Transfer range	0.1 mL to 10 mL in single step
Accuracy of transfer	ca. 0.05 mL, small dependence upon solvent
Liquid delivery rate	0.017 to 2.5 mL/min
Reaction atmosphere	Air, nitrogen, or argon
Solvents	12 different reservoirs (maximum) with possibility to make additional solvents available from reagent racks. Each solvent can have transfer rates individually set.
Reaction temperature range	-20 to 140°C
Reaction heating/cooling rate	3°C/min (maximum) 0.1°C/min (minimum)
Reaction temperature accuracy	ca. ±0.5°C (at constant temperature)
Reflux	Allowed—the block temperature should be set near reflux temperature to avoid distillation of the solvent
Reactors	Standard 25 $\times$ 150 mm screw top test tubes
Reaction time	$>$ 24 hours allowed
Agitation	Magnetic stirring bars, 3 zones, each 100 to 1000 rpm, all vessels in each zone are stirred at same rate, the reaction block is a single zone
Waste volume	5 gallons

Short analytical columns are useful in reducing the time for a single analysis while achieving acceptable resolution. One must be aware of the potential impact on the chromatography of overloading such columns. Because of the smaller capacity of these columns, skewed peaks and loss of resolution may be observed over the course of a reaction as relative concentrations of different chemicals change. Higher dilutions of the reaction mixture help address these problems, but quantitation of small peaks becomes more difficult at higher dilutions. Short, wide-bore columns can provide a good compromise between the speed provided by a short column and the capacity allowed by longer analytical columns.

An alternative approach to sample throughput with longer analytical methods is the selection of sampling times that vary over the course of the reaction and delay of the start of reactions to minimize sample backlog on the analytical instrument. Since most reactions display kinetics that are exponential, it can be beneficial to obtain samples more frequently near the beginning of a reaction than near the end.

Internal standards are used in all reactions from which quantitative data are desired. The internal standard is necessary since there are errors made in the volume of the sample removed from the reaction mixture. In addition, for those reaction samples transferred directly to the analysis unit in an undiluted form, some liquid always remains in the transfer line. Consequently there is an error on the total volume of solvent used in the dilution. Similar considerations apply to neat samples delivered to the sample rack.

The original software driving the synthesis workstation included a simple scheduler. All of the steps involved in an individual reaction were required to reach completion prior to starting any step in a subsequent reaction. The scheduler checked to ensure that reactions were started at times such that conflicts between two different reactions' need for the robot arm for sampling were prevented. In addition, the scheduler did not account for the actual time necessary to perform a given operation. Consequently the individual reactions were, to a great extent, performed sequentially. Times specified for samples to be withdrawn from the reaction mixture were measured from the completion of a preceding task rather than from the start of a reaction. When numerous samples were withdrawn from a reaction, large deviations from the desired sampling times were observed.

The workstation software has recently been modified to address concerns caused by the scheduler. The improved scheduler allows several reactions to be run much more nearly in parallel. Implementation of more parallel reaction processing has also required that the sample cannulas be washed both prior to and immediately following sample transfers. The addition of a wash following a transfer ensures that residues are washed from the cannula with a compatible solvent. The additional wash also minimizes the potential for crystallization of a solid in the cannula.

Programming the synthesis unit to run a series of experiments is a multistep process. The synthesis to be performed is examined to determine what unit operations can be performed as prereaction steps, what unit operations should be included as reaction steps, and what sampling times should be included in the study. The concentration and total volumes of reagent and reactant solutions required in the study are calculated and decisions are made concerning the locations those solutions will occupy on the workstation deck.

A reaction setup file is created that contains those steps that typically begin with the first operation that contains a parameter to be varied among the separate reactions. The final step in the reaction setup file is typically the removal of the last sample from the reaction. The synthesis workstation software requires a reaction setup file to be generated externally and then imported. Error checking on the imported file is conducted as the file is imported. When an unrecognized parameter is encountered, an error message is generated and the entire content of the file is rejected. An Excel spreadsheet has been programmed with data validation controlled using vocabulary lists in order to circumvent this problem.

The prereaction steps such as initial solvent addition, setting the initial temperature, turning the stirrer on, placing the reactors under an inert gas, and any initial transfers of reactants or reagents are programmed. The reactors to be used in a given study are specified and the reaction file is imported, and the point at which that file is to assume control of the sequence is specified. Any operations that should be performed after execution of the reaction setup file is complete are specified. A file used by the scheduler is then generated.

The scheduler allows the priority of any or all reactions to be specified. Once those priorities are provided, the scheduler analyzes the tasks and generates a list of optional procedures from which the chemist can select a preferred sequence. Each optional sequence is assigned a relative figure of merit that can be used to compare the relative efficiency of any sequences in terms of the total amount of time that is required to complete the entire study. Once the preferred sequence has been selected, a list of the actions to be performed as specified in the selected sequence is generated.

The reactants, reagents, and reactors are loaded onto the workstation. Required solvents are checked to make sure that sufficient solvent is present, and lines leading from those reservoirs are primed. The reaction sequence can then be run. Should unanticipated problems arise during the execution of the reaction sequences, the software allows independent termination of specific reactions. The types of chemistry that have been examined at RWJPRI utilizing this equipment include oxidations, hydrolyses, acylations, Friedel–Craft chemistry, condensations, solvent selection, and formation of organic salts.

6 CONCLUSION

6.1 Automation

The need to rapidly develop cost-effective, safe, reproducible chemical syntheses is leading to numerous applications of automation by developmental chemists and their partners. Spectroscopists and analytical chemists were leaders in the application of automation to solving problems associated with the need to improve productivity without increasing headcount. In many instances, automation has been successfully applied to specific needs related to narrowly defined problems. It has been possible to develop automation that is easily learned to address repetitive and time consuming tasks such as NMR data acquisition and injection of samples onto a variety of chromatographic equipment. Many commercial vendors of chromatographic systems produce software that is capable of selecting specific sets of data for further analysis and compilation into summary reports.

Much effort has gone into the development of automated reactors for use in development of organic processes. The need for a generalized automated reactor system intended to study a wide variety of reactions, workup the reaction, and analyze the product is still unmet. The reason that this goal remains elusive is perhaps the great variety of conditions that are encountered during the steps, from addition of reagents to a reactor until a sample is submitted for analysis. Despite the advances that have been made in the development of expert systems, it is difficult to anticipate all problems that may be encountered in an experiment and the best solutions to address them. Until such automated systems become easy to program and until those programs can be easily modified during a run, generalized reactor systems will be impractical.

A number of mathematics packages are available that automate the production of various experimental designs. Such packages may be designed for end users who are not necessarily familiar with the details of the calculations that are being performed but who are familiar with the benefits of well-designed studies. Investigators who are seeking software to aid in the planning and interpretation of experiments should look for software that also provides guidance when a study does not produce readily interpretable results. An indication that there is lack of fit to a selected model can provide the stimulus needed to seek an expert who can help identify and address those factors causing the problem.

In addition to automation that is designed for very specific purposes, software is readily available that is intended to be a useful multipurpose tool. Spreadsheet software was once considered to be a tool useful only for accountants. The availability of graphs within spreadsheets as well as a number of included data manipulation capabilities makes them very useful for a variety of purposes within chemical development. A modest knowledge of spreadsheets allows one to program a spreadsheet to calculate total raw materials costs and requirements for a multistep synthesis. The same spreadsheet can then be used to determine the impact of changes in yield or cost of raw materials or to determine the amount of product to be expected if certain materials are in limited supply. With a greater knowledge of the functions available within the spreadsheet software, it is possible to use DSC data to produce theoretical graphs of phase diagrams and use those graphs to aid in planning experiments for the resolution of enantiomers. Based upon the cost of this software and its many potential uses, it is possibly the most economical automation tool currently available.

6.2 Robotics

When one thinks of automation, robotics readily comes to mind. Most of the chemistry workstations currently available utilize one or more robots. In some instances a single robot fitted with a variety of attachments is utilized to perform a variety of tasks. In other instances multiple robots working independently but under coordinated control from a central command unit are utilized. In the later case, each robot performs a specialized task. In both cases, components should be selected to provide temporal compatibility. The systems that work most efficiently are those in which one component does not have to wait on another in order to perform its task [36].

Automation trends in chemical synthesis appear to be toward construction of specialized robot modules. With such systems, each module can be constructed and programmed to do its job efficiently. A disadvantage to such a system is sample tracking from station to station. An error in identification of a sample as it is passed from one location to the next can lead to completely erroneous results. It therefore becomes critical to maintain an accurate log of the samples as they move onto and off of each module and to pass that information to a repository for future access.

The scheduler coordinating all tasks is a critical component of the entire system. A well-designed scheduler provides the ability to perform tasks associated with multiple reactions in parallel or, if time constrains prevent truly parallel work, with a minimal delay between initiation of each individual reaction. A scheduler that allows for modification of experimental conditions while a reaction is in progress should allow for greater flexibility in a system. Experience with such systems will determine whether the enhanced flexibility must come at the cost of longer and more detailed training on the system.

6.3 Future Developments

As speed-to-market, cost, and reliability continue to grow in importance in a global economy, it is anticipated that automation will continue to gain in importance in the development of new chemical processes. It is anticipated that advances in software and hardware will aid in the development of systems that will address current issues in the use of automation in chemical development.

It is easy to imagine the ultimate automated process development workstation, one in which a variety of open-loop and closed-loop experiments can be easily programmed. Its equipment would include a combination of hardware and software designed to run a wide variety of chemical reactions. Its equipment would monitor the course of the reactions it is running and terminate those in which the product has reached a predetermined level. A workstation programmed with algorithms would provide the ability to anticipate materials requirements before they become critical and inform the chemist/operator to replenish those supplies that are needed. It would contain an expert system that could diagnose problems encountered during the process and would take actions to correct them or to bring the reaction to a set of safe conditions, if necessary, and inform the operator of the status of the process. The equipment would be capable of handling the preparation of reagents, running synthetic steps, isolation of the product, and cleaning itself in preparation for the next reaction.

Although a number of individual items listed in the above description of the ultimate process development workstation have been reported, no automated process development equipment incorporates them all. One should consider whether such generalized equipment is really practical and necessary. An overlooked cost in flexibility of an automated system is often an increase in the complexity of the user interface. As more options are added to a system, the time required to program the desired set of options into a process increases. Equipment designed to do many jobs satisfactorily may not be capable of doing any of them well [37].

It may be more reasonable to view the ultimate process development workstation as a coordinated group of equipment. In such a system each component is designed to do a specific job well. With a modular system the chemist could select those components that are most suitable for each required operation. A central unit would be responsible for coordinating the activities associated with a given reaction. When desired, data associated with the reaction and its samples would automatically be compiled into a meaningful report. Data from a series of related reactions would be automatically processed, and a summary report would be generated. A database of the work would automatically be updated so the knowledge and experience gained in development of one process could be easily accessed to determine if it could be applied to others.

It must be stressed that the development and utilization of such systems will not necessarily result in the need for fewer chemists. It does mean that the nature of the work that is done by chemists can change. Time currently spent performing manipulations may be spent in designing experiments to produce data that is truly important in achieving goals set for projects. Time will be spent evaluating and interpreting the greater amounts of data that can be generated automatically.

The use of automation in chemical development will have a great impact on the way that chemistry is studied. The use of NMR and solid-phase synthesis has revolutionized the way in which discovery research in chemistry is performed. It is reasonable to expect that advances in automation and its application in chemical development will have a similar impact on moving new chemical entities from discovery into production.

REFERENCES

1. Guette, J.-P., Crenne, N., Bulliot, H, Desmurs, J.-R., Igersheim, F. Pure Appl. Chem. 1988, 60(11), 1669–1678.

2. Clayton, E., Preece, S., Taylor, S., Wilson, L., Wright, B. Adv. Mass Spectrom. 1998, 14 C104540/1-C104540/10 (CAN 130:130041).

3. Lindsey, J. S. Chemom. Intell. Lab. Sys. 1992, 17(1), 15–45.

4. Caruthers, M. H. Science (Washington. D. C.) 1985, 230(4723), 281–285.

5. Yeung, A. T., Miller, C. G. Analytical Biochemistry 1990, 187, 66–75.

6. Sanghvi, Y. S. Past, present and future of automated manufacture of antisense oligonucleotides. Presented at The Evolution of a Revolution: Laboratory Automation in Chemical Process R&D, Scientific Update. The Chester Moat House, Chester, U.K., 17 Nov. 1998.

7. Harre, M., Tilstam, U., Weinmann, H. Org. Process Res. Dev. 1999, 3, 304–318.

8. Vacque, V., Dupuy, N., Sombret, B., Huvenne, J. P., Legrand, P. J. Mol. Struct. 1996, 384(2–3), 165–174.

9. Dozeman, G. J., Fiore, P. J., Puls, T. P., Walker, J. C. Org. Process Res. Dev. 1997, 1(2), 137–148.

10. Bergen, M. S. Int. J. Pharm. 1994, 103, 103–114.

11. Levoguer, C. L., Williams, D. R. Application Note 101, Surface Measurement Systems Ltd.

12. Buckton, G., Darcy, P. Int. J. Pharm. 1995, 123, 265–271.

13. Lerena, P., Wehner, W., Weber, H., Stoessel, F. Thermochem. Acta 1996, 289, 127–142.

14. am Ende, D. J., Clifford, P. J., Northup, D. L. Thermochem. Acta 1996, 289, 143–154.

15. Rowe, S. M. Thermochem. Acta 1996, 289, 167–175.

16. LeBlond, C., Wang, J., Larsen, R. D., Orella, C. J., Forman, A. L., Landau, R. N., Laquidara, J., Sowa, J. R., Jr., Blackmond, D. G., Sun, Y.-K. Thermochem. Acta 1996, 289, 189–207.

17. For a more detailed discussion of simplex optimization see Walters, F. H., Parker, L. R., Jr., Morgan, S. L., Deming, S. N. Sequential Simplex Optimization—A Technique for improving Quality and Productivity in Research Development, and Manufacturing. CRC Press, Boca Raton, FL, 1991.

18. Carlson, R. Data Handling in Science and Technology—Volume 8, Design and Optimization in Organic Synthesis. Elsevier, Amsterdam, The Netherlands, 1992.

19. Carlson, R., Lunstedt, T., Albano, C. Acta Chem. Scand., Ser. B 1985, B39(2), 79–91.

20. Carlson, R. Chem. Scr. 1987, 27, 545–552 and Ref. 3, 5, 14, 15, 17, and 20–24 therein.

21. Carlson, R., Nordahl, A., Barth, T., Myklebust, R. Chemom. Intell. Lab. Sys. 1992, 12, 237–255.

22. Boettger, S. D. Lab. Rob. Autom. 1992, 4, 169–181.

23. Armitage, M. A., Smith, G. E., Veal, K. T Org. Process Res. Dev. 1999, 3, 189–195.

24. Corkan, L. A., Plouvier, J.-C., Lindsey, J. S. Chemom. Intell. Lab. Sys. 1992, 17, 95–105.

25. Porte, C., Mouzayek Kous, E., Delacroix, A. Lab. Rob. Autom. 1994, 6, 119–129.

26. Porte, C., Canatas, A., Delacroix, A. Lab. Rob. Autom. 1995, 7, 197–204.

27. Cork, D. G., Kato, S., Sugawara, T. Lab. Rob. Autom. 1995, 7, 301–308.

28. Cork, D. G. Laboratory automation systems and workstations for organic chemistry research. Presentation at The Evolution of a Revolution: Laboratory Automation in Chemical Process R&D, Scientific Update. The Chester Moat House, Chester, U.K., 17 Nov. 1998.
29. Schroeder, I. Z. Phys. Chem. 1893, 11, 449.
30. van Laar, J. J. Arch. Nederl. 1903, 264.
31. Prigigine, I., Defay, R. Chemical Thermodynamics. Longman and Green, London, 1967, p. 375.
32. Jacques, J., Collet, A., Wilen, S. H. Enantiomers, Racemates, and Resolutions. Krieger: Malabar, FL, 1991, pp. 3–5.
33. Collet, A., Jaques, J. Bull. Soc. Chim. Fr. 1973, 12(2), 3330–3334.
34. Kozma, D., Pokol, G., Acs, M. J. Chem. Soc., Perkin Trans. 2 1992, 435–439. For a related paper that explains how to treat DSC data from an unknown mixture of two diastereomeric salts see Ebbers, E., Ariaans, G. J. A., Zwanenburg, B., Bruggink, A. Tetrahedron: Asymmetry 1998, 9, 2745–2753. A separate spreadsheet (instructions are not provided) needs to be prepared for the work described in the Ebbers paper.
35. Carlson, R., Lundstedt, T., Nordahl, A., Prochazka, M. P. Acta Chem. Scand., Ser. B 1986, 40, 522–533.
36. Lindsey, J. S., Corkan, L. A. Chemom. Intell. Lab. Sys. 1993, 21, 139–150.
37. Corkan, A.,Lindsey, J. S. Adv. Lab. Autom. Rob. 1990, 6, 477–497.

10
Automated Calorimetry in Process Development

Nick Evens
Avecia Pharmaceuticals, Grangemouth, Stirlingshire, Scotland

1 INTRODUCTION

This chapter describes the applicability of automated calorimetry to each of the different phases of process development from route selection through optimization to siting and manufacturing support. Commercially available systems are discussed, and a series of case studies is presented that highlight the key benefits that automated calorimetry equipment can deliver. In summary, these are

> Increased speed of development
> Higher quality/more robust processes
> Cost saving
> Increased process safety
> Job enrichment

Ever contracting time scales for the discovery, development, and launch of new chemical entities (NCEs) have necessitated the development of higher yielding, more robust, and more environmentally friendly processes in shorter times.

The use of automation in drug discovery, for example in library generation and high-throughput screening (HTS) has been amply demonstrated over the past decade or so. The use of automation for chemical process development is now beginning to mushroom in the same way. There are two key reasons [1] why automation is being used.

1. Speed. For a blockbuster drug, every day after launch can equate to several million dollars worth of sales. Hence pharmaceutical companies are keen

to contract drug development timescales as much as possible. For the process development chemist this means developing the process as quickly as possible. Using automation, many more experiments can be performed in a given time.

2. Cost. The development of processes that deliver appropriate quality products in high yield and at low cost are key, especially as pharmaceutical products mature and face generic competition.

One of the ways in which calorimetry can impact on cost is improved process safety, with increased personal safety being the main benefit. Personal injury claims, prosecution, removal of license to operate, and loss of sales can all result from a poor process safety performance.

The safety implications of cost saving process changes such as waste minimization and yield/productivity improvement all need to be assessed prior to implementation, and here too, automated calorimetry equipment has a vital role in delivering cost benefits.

Automation has other indirect effects on cost. Experimentalists spend less time manually controling their preps and more time thinking about their results. The ability to do more preps means that it is much easier to examine breakout options, which might have been left untouched during manual development.

Finally, a detailed appreciation of process robustness gives the opportunity to minimize supervision, hence lowering production costs, and also a lower batch failure rate.

These cost savings mean that the payback period on seemingly expensive calorimetry equipment may be measured in months, or even weeks.

2   The Role of Automation in Process Development

We have developed a simple model to explain the different stages in the development and siting of chemical processes. In particular, we were interested in using this model to examine how the different development phases could be accelerated through the use of automation.

2.1   Stage 1: Route Selection

The objective of this stage is to provide an initial view, both internal and external, on competitive position. In addition to actual route selection, other activities include crude reagent/solvent screening, sample delivery, and cost view.

Traditionally, this phase of development has used manually operated single reactions using segmental reaction screening coupled with labor-intensive product isolation. The word segmental is used to describe different and not necessarily connected areas of experimental space.

Currently we would expect to perform this activity in parallel in a multireactor automated system with in situ analysis. The Anachem SK233 [2] is an example of automated equipment ideally suited to this phase of development. In-line HPLC monitoring of reactions is a major step forward, but it can still lead to bottlenecks. Often in this phase of development, analytical methods are not fully defined, and the use of a real-time measurement technique such as calorimetry can give a major benefit in terms of speed. An ideal scenario would be monitoring a reaction using calorimetry, and on completion doing only one off-line analysis to confirm yield and quality.

2.2   Stage 2: Optimization

Here the objective is the development of a competitive process and activities center around reagent selection, determining the optimum operating conditions and minimization of SHE (safety, health, and environment) impact. As above, this phase was traditionally carried out in manually operated single reactors and involved stepwise optimization and labor intensive product isolation. We now carry out this stage with multireactor systems such as the HEL Auto-MATE [3]. Statistical experimental design techniques are used to define optimum reaction conditions.

2.3   Stage 3: Siting

The objective is the development of a robust process fitted to the manufacturing assets. Specific activities are focused on definition of the robustness envelope, mimicking plant conditions in the lab, and resolution of SHE issues. We carry this phase of development out initially at small scale on the auto-MATE, where calorimetry has an important role in the early identification of reaction hazards, and at larger scale, for process proving runs, using the HEL Simular.

2.4   Stage 4: Manufacturing Support

The objective here is to maintain process effectiveness and ongoing cost reduction. Specific activities are the resolution of problems encountered during initial manufacture, monitoring of process performance, and establishment of process improvements. Traditionally, siting suboptimal processes led to protracted shift cover and reactive lab work to solve plant problems. Execution of these problem solving activities meant that cost reduction was often marginalized, leading to a potential erosion of competitive position.

   We now expect that the development of more robust processes in the preceding phases will mean that cost reduction activities can commence much sooner, therefore maintaining or enhancing competitive position.

3 APPROACHES TO REACTION AUTOMATION

Broadly speaking, there are two approaches to the optimization of chemical reactions. The first involves carrying out a very large number of relatively low-definition preps. This type of experimentation is ideal for solvent and reagent screening but often gives relatively low confidence for scale-up due to factors such as non-ideal reactor geometry, relatively poor temperature control, and agitation. Commercial systems vary from the fully automated Zymark [4] to simple manual multiple reaction stations such as STEM racks [5].

The second approach involves performing fewer high-definition preps. They are typified by high-grade reaction control (temperature and agitation, for example) and may have additional features such as feedback loops (e.g., reagent addition controlled to maintain isothermal conditions or pH) and calorimetry. The HEL [3] calorimeters (Auto-MATE and Simular), together with the Mettler-Toledo [6] RC1e exemplify this second group.

These approaches are complementary—neither is a universal panacea.

An assessment of the number of preps per unit time versus the relevance of the results to scale-up is shown in Table 1.

At the bottom of the league, both in efficiency and in relevance to scale-up, is the traditional round-bottom flask with magnetic stirring. Although suffering from poor temperature control and agitation, together with a geometry that is far removed from any plant reactor, this setup retains a place in many process development laboratories. STEM Blocks in isolation are probably of low relevance for actual siting studies, though they are invaluable at the earlier stage of reagent screening. However, in the Anachem SK233, where the STEM reaction

Table 1 Benefit Analysis of Different Process Development Systems

		Low	Medium	High
			Number of preps per unit time	
	High	Mettler-Toledo RC1 HEL Simular HEL Auto-Lab Metler-Toledo LabMax	HEL Auto-MATE	?
Relevance of results to scale-up	Medium	Large jacketed vessels—mechanical stirring	Anachem SK233 Bohdan Nautilus/ Surveyor Bohdan PDW	Zymark
	Low	Small round-bottom flask—magnetic stirring	STEM Block	

block is PC controlled and combined with an on-line HPLC, the relevance to scale-up is greater because of the automated process control and analytical capability. The SK233 system was probably the earliest system of its type, but there are now other systems with similar characteristics, such as the Nautilus and Surveyor from Argonaut Technologies [18], chem-SCAN from HEL, the Process Development Workstation from Bohdan [6], and the Zymark BenchMate, to name but a few. The larger Zymark systems can perform very large numbers of reactions and hence explore a larger region of experimental space.

Larger scale jacketed vessels with mechanical agitation can provide more accurate scale-up information, and provision of a bottom runoff can accelerate clean-downs, reducing downtime. The provision of PC control and data logging, as exemplified by the LabMax from Mettler-Toledo and the Auto-Lab systems from HEL, reduces operator supervision requirements and dramatically improves efficiency.

Isothermal calorimeters are basically PC controled jacketed vessels that have the ability to measure reaction power output. They are the tool *par excellence* of the process development chemist. The Mettler-Toledo RC1 and the HEL Simular are extremely accurate calorimeters with a very high level of control. The HEL auto-Mate, while still able to generate good quality calorimetry data, has a great advantage in being able to do four (or more) reactions at one time.

There do not appear currently to be any systems that would fit into the high-throughput/high-scale-up relevance box. The requirements of such systems will be addressed in Section 11.9.

4 CALORIMETRY

4.1 Process Safety

Calorimetry involves the study of the thermochemistry of individual compounds and chemical reactions. It provides vital information for the safe scale-up of chemical processes. The amount and quality of information required for safe scale-up is often underestimated. A typical type of plant operation is the maintenance of a given temperature during a chemical reaction. The underlying chemistry is often complex and it is difficult if not impossible to predict accurately reaction exotherms without experimentation. The key to this type of operation is to balance the exotherm, i.e., the rate of heat production, with the cooling capacity of the vessel. Extrapolation from standard lab preps can be at best misleading. A typical reaction carried out on a 1 liter scale might be expected to give out up to 100 watts of heat. The heat loss to the surroundings could be 200 W, i.e., double the power output. Hence, in this case, the reaction may be wrongly assumed to be endothermic. Because of the surface-area-to-volume relationship [20], the heat loss per kg of batch from typical chemical reactors, with volumes

Table 2 Comparison of Cooling and Heat-Loss Rates for Various Sizes of Reaction Vessels

Reactor volume	Heat loss per kg of sample, W	Time for 1 K loss at 80°C
10 mL	350	11 s
200 mL	50	17 s
2.5 m^3	3	21 min
25 m^3	0.03	233 min

of 25 or even 2.5 m^3 are much less than those from lab scale reactors, as shown in the Table 2 [21,22].

A large uncooled reactor can self-heat under almost adiabatic conditions, i.e., zero heat loss, and even simply stirring such a system can lead to a temperature increase. This is what makes knowledge of reaction thermochemistry so vital for scale-up purposes.

Data from the Health and Safety Executive [21] (HSE UK) (Table 3) shows that a lack of knowledge of reaction chemistry is one of the major causes of reactor incidents.

The other major causes of incidents are closely related:

Temperature control problems—was the temperature sensitivity of the reaction understood?

Agitation problems—was the effect of agitation failure on the reaction considered? Did the system layer? Was any reagent addition tripped on agitator failure?

Mischarging of reactants or catalyst—were the potential hazards understood? Had the effect of mischarges or the wrong order of addition been assessed?

Table 3 Causes of Batch Reactor Accidents in the UK (1962–1987)

Cause	Contribution (%)
Mischarging of reactants or catalysts	21
Lack of knowledge of reaction chemistry	20
Temperature control problems	19
Poor maintenance	15
Agitation problems	10
Problems with material quality	9
Operator error	6

To put the above numbers into perspective, on average, there is a thermal runaway incident reported to the HSE [21] in the UK every 2 weeks.

If the hazards of a process are not understood, an uncontrolled exotherm may cause a dramatic temperature rise in the batch. This in itself could lead to overpressurization if the batch boils. More sinister however, is the possibility of this temperature rise allowing the batch to access an exothermic decomposition. This would be expected to give rise to a further temperature rise with boiling and/or gassing. It is the latter that can lead to rapid overpressurization, vessel failure, widespread damage to the plant and the environment and, potentially, loss of life.

Thankfully, most chemical reaction hazards are identified before the process reaches the production plant. However, the identification of reaction hazards, through a detailed investigation of reaction calorimetry, is often carried out late in the process development cycle. This can lead to production delays and even lost business, if processes need to be reengineered to fit the plant or vice versa. The use of calorimetry, at an early stage, to understand and quantify reaction hazards should be fundamental to any process development program.

There are two basic types of calorimeter, adiabatic and isothermal. Adiabatic calorimeters give information on temperature and pressure increases during a reaction runaway situation. Isothermal calorimeters give information of heats and rates of reaction during normal operation.

4.2 Adiabatic Calorimetry

For the siting chemist, whose job involves safely implementing a new process on plant, adiabatic calorimetry is just as important as isothermal calorimetry. Isothermal calorimetry will determine if the plant cooling capacity can cope with the process heat output under the normal operating conditions. In contrast, adiabatic calorimetry will determine if there is a dangerous exothermic decomposition lurking just outside the normal operating conditions. Differential scanning calorimetry (DSC) is another technique that can be used to estimate thermal stability, but because of factors such as small sample size and lack of agitation, the results are often inaccurate and imprecise. To overcome this, a 100°C safety margin is often set between the normal operating temperature and the DSC exotherm onset temperature. The use of automated adiabatic calorimeters such as the PHI-TEC2 [3] give a much more accurate determination of potential instability problems. Screening devices such as the TS^U [3] give performance at moderate cost relatively close to that of systems like the PHI-TEC2. These devices, unlike DSC, allow the determination of both pressure and temperature rises, accurate onset temperatures, and decomposition kinetics. This information is critical in defining safe operating conditions and relief sizing.

4.3 Isothermal Calorimetry

An isothermal calorimeter basically consists of a stirred reactor under PC control, with automatic data acquisition and automated addition of liquids, gases, and possibly solids. Close control and real-time monitoring of reaction parameters such as pH, temperature, pressure, reflux rates, and gas evolution are all more or less standard. The basic function of an isothermal calorimeter is to measure the power output of a given reaction versus time.

4.4 Methods of Determining Power Output

A brief description of the theory behind the determination of the calorimetry data will now be given, although it should be emphasized that this is all performed automatically by the PC and no actual number crunching is required. The following data can be determined—maximum power output, rate of power output, total power output, and evidence of accumulation, together with data more familiar to the bench chemist such as percentage conversion to product.

There are two main methods of determining calorimetric data.

4.4.1 Heat Flow Calorimetry

In this method, the reactor jacket provides temperature control—in order to maintain isothermal conditions, the temperature of the circulator oil is reduced in response to an exotherm and increased in response to an endotherm. The power output is calculated from the heat transfer rate from the batch to the coolant. This requires calibration at the start and end of the reaction as it varies with batch volume, viscosity, and composition. It is a time-consuming process, often taking 2 to 3 hours, and even then is subject to inaccuracy, because of the need to extrapolate throughout the reaction period. The power output Q_r from a reaction is given by the following relationships:

$$Q_r = Q_{rem} + Q_l + Q_{accum} + Q_{dose} = UA \cdot \Delta T$$

where

$$
\begin{aligned}
Q_{rem} &= \text{the power removed (or added) by the oil in the vessel jacket} \\
Q_l &= \text{the power loss from the vessel walls} \\
Q_{accum} &= \text{the power taken up by the reactants and the reactor} \\
Q_{dose} &= \text{the power taken up by the feed material} \\
U &= \text{overall heat transfer coefficient} \\
A &= \text{wetted area of reactor} \\
\Delta T &= \text{jacket/batch temperature differential}
\end{aligned}
$$

Several other useful relationships can be derived:

The overall heat of reaction, $\Delta H_r = \int Q_r \cdot dt$

The conversion, $\chi_t = \dfrac{\int Q_r \cdot dt}{\Delta H_r}$

The concentration of a given reactant, $Ct = [1 - \chi_t]C_0$

where
C_0 = initial concentration of the reagent

The power output can be used as a reliable measure of the extent of reaction, this gives the calorimeter tremendous power as a general process development tool, especially in those cases where traditional analytical methods are unavailable.

4.4.2 Power Compensation Calorimetry (CMC—Combined Method of Calorimetry)

In this method, a calibrated heater inside the vessel working against an external cooling jacket maintains isothermal conditions. The reactor temperature is denoted by T_r, the jacket temperature by T_j, and the temperature differential across the jacket inlet/outlet by ΔT_{oil}. At the start, prior to the addition of the feed reagent, there is no power output from the reaction, i.e., $Q_r = 0$. Hence the baseline power $Q_{p(0)}$ required to maintain constant reactor temperature T_r is equal to the sum of the power loss through the oil in the jacket and that through the vessel walls. It is given by

$$Q_{p(0)} = Q_{rem} + Q_l$$

where

$Q_{p(0)}$ = baseline power
Q_{rem} = power removed (or added) by the oil in the jacket
Q_l = power loss from the vessel walls

Once the feed is started and if the reaction exotherms, the instantaneous power from the heater Q_p is equal to the difference between the baseline power $(Q_{p(0)})$ and the heat of reaction Q_r, i.e.,

$$Q_p = Q_{p(0)} - Q_r$$

The key benefit of this approach is that no calibration (UA) is needed, and the reaction power output is determined from the baseline power and the instantaneous power output of the heater. Baseline drift is an issue, but the baseline power input is proportional to the temperature differential across the oil in the jacket. This fact allows the real-time determination of the actual reaction power output. This combination of power compensation calorimetry with baseline adjustment through the oil temperature differential is called the combined method of calorimetry (CMC) approach. One disadvantage is that in highly viscous systems, charring may occur on the element. This may be overcome, either by increasing the agitation rate or by using a heater with a larger surface area.

In order to monitor calorimetry reliably, reactions are best carried out isothermally, although nonisothermal operation is possible. The vessels are most effectively operated in the semibatch mode (in which all the nonreactive components are charged to the vessel). After an appropriate equilibration period (generally 30 min), the limiting reagent and solvent (if required) are charged via an automated pump. A heater inside the reactor, working against an external cooling jacket, provides control. If the reaction exotherms, power input from the heater is reduced to maintain the same temperature. In contrast, if the reaction endotherms, power input from the heater is increased. The calorimetric data is obtained from a plot of power input versus time.

Heat-flow and power-compensation calorimetry can give different results, basically, as a result of the different ways in which the baseline is calculated. In fact the heat-flow method may underestimate the power output, but for most systems, the heat-flow and power-compensation methods will give the same results. The main difference is the relative ease of use of power compensation.

4.4.3 Calorimetry at Reflux

Reactions are often performed at reflux, and the determination of calorimetry data is of course complicated by heat loss through the latent heat of vaporization of the solvent. One option would be to perform the calorimetry just below the reflux temperature. An alternative procedure is to perform a heat balance on the condenser. This involves setting up a fixed temperature differential between the jacket and the batch, i.e., a constant reflux rate. Any changes in the heat flow across the condenser as measured by the difference in coolant inflow and outflow temperatures are therefore due to the reaction. This method is affected by changes in UA or overall heat transfer coefficient, and as above in the CMC method, baseline interpolation is carried out using the difference in the input and output temperatures of the oil in the vessel jacket. The power output Q_r is given by

$$Q_r = C_{pc} \, M_c \, (\Delta T_r - \Delta T_b)$$

where

$\quad\quad C_{pc} \;=\;$ coolant flow-rate through condenser
$\quad\quad M_c \;\;=\;$ specific heat capacity of coolant
$\quad\quad \Delta T_r \;=\;$ temperature differential between condenser input/output during reaction
$\quad\quad \Delta T_b \;=\;$ temperature differential between condenser input/output prior to reaction

5　DESCRIPTIONS OF COMMERCIALLY AVAILABLE CALORIMETERS

5.1　Hazard Evaluation Laboratories (HEL)

There are basically two types of HEL isothermal calorimeter:

1.　Auto-MATE—consists of blocks of four mechanically stirred reactors with a working volume of 50–75 mL
2.　Simular—a single larger (nominally 500 mL) version of the above

Essentially, the auto-MATE gives the benefit of running up to four (or more) simultaneous reactions, whereas the Simular can be used to run one reaction. However, the quality of the calorimetric data from the latter is better due to the more favorable surface-area-to-volume ratio.

5.1.1　HEL auto-MATE

The HEL auto-MATE (Figure 1) consists of four reaction vessels arranged in a linear configuration. The vessel lid has six ports: the central one is fitted with a mechanical stirrer motor, which has interchangeable stirrers; of the others one holds a thermocouple and the third a Hastelloy® heating element. Of the remaining two ports, one is a B14 ground glass joint and one has a septum screw

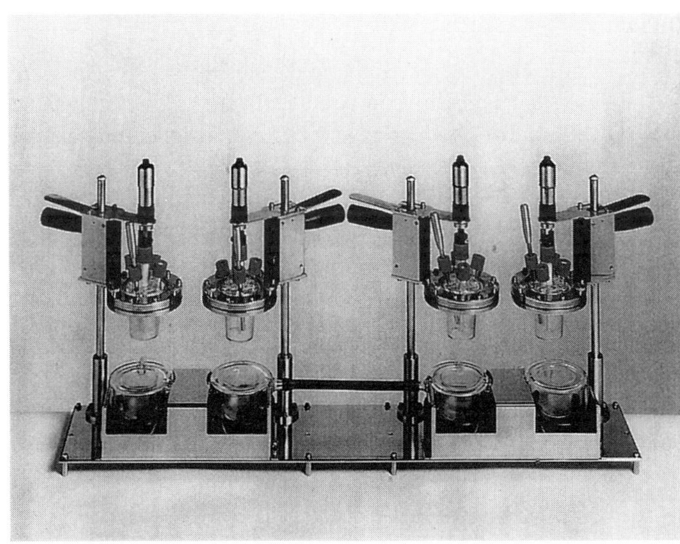

Figure 1　HEL auto-MATE.

cap for liquid feeding. Each pot has a syringe pump for liquid feeds. The reaction vessels sit in a base unit, which functions as an oil circulator jacket. The jacket temperature is controlled by a suitable heater/chiller unit such as a Julabo FP50 unit, which has an upper limit of 200°C, although other circulators can be used that go up to 300°C.

Calorimetry is available on any or all of the reactors. A wide range of conditions can be tolerated (-40 to 300°C); atmospheric pressure up to at least 100 bar using metal versions. The reactors are typically fabricated from glass but can be made in a wide range of materials such as stainless steel, Hastelloy, and PTFE.

Temperature control is excellent ($+/-$ 0.2°C), and each pot can be set to a different temperature, subject to a maximum of 30°C between pots.

Each vessel is mechanically agitated, and various agitator designs are available.

Feed-rates for reagent addition are easy to set, but experimentation is sometimes needed to determine the optimum rate, since if the addition rate is too high and the reactions power output is greater than the heater power, the conditions become nonisothermal. However, the system does have an autoshutdown system that switches in if, for example, the pot temperature exceeds a set maximum.

A key advantage of this system is the ability to feed all four (or more) reactors simultaneously.

5.1.3 HEL Simular

The Simular (Figure 2) is a larger scale isothermal calorimeter. Reactors are available in glass, stainless steel, and alloys in volumes from 0.2 to 20 liters and pressures up to 300 bar. Specialist geometries and custom-designed vessels are available to simulate real plant conditions.

Operating Temperature: The cooling and heating circulation unit is capable of operating from -80 to 350°C, with control of temperatures to within 0.1°C.

Liquid Feed: Liquids can be fed, at controlled rates, into or out of the reactor vessel using a pump and balance combination at pressures up to 200 bar.

Solid Feed: Precise dosing of powders is possible using a screw type conveyor.

Gas Feed: The system can be adapted to handle the feed of gases into the reactor, with pressure or flow-rate measurement and control. This allows the study of reactions such as hydrogenations.

Reflux Operation: A glass condenser system can be added for reflux operation, with or without calorimetry options. Both total and partial refluxes are possi-

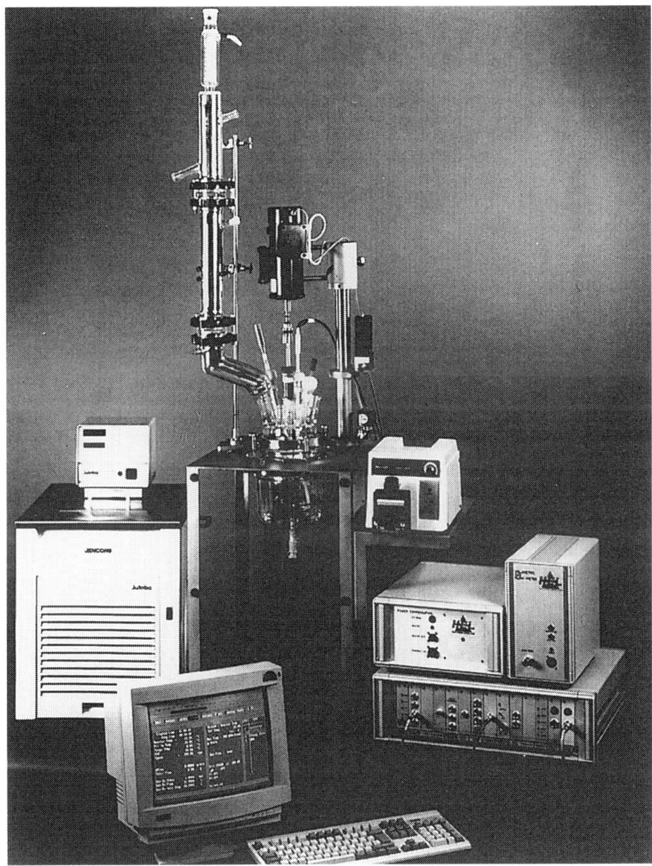

Figure 2 HEL Simular.

ble, as is reflux under pressures of up to 200 bar. An example of a high-pressure version of the Simular is shown in Figure 3.

Software and Electronics: Dedicated software with full safety features as standard.

FTIR Probes: Spectroscopic probes including FTIR can be used in situ to both monitor and control the reaction. This enables real-time analysis and control of the system based on the values monitored from selected IR peaks.

Conductivity/pH: The signals from a variety of sensors that can be connected to the system can be monitored and used to control parameters such as pH, conductivity, etc., using metered addition for example. Special probes can be supplied to cope with high pressure or hostile environments.

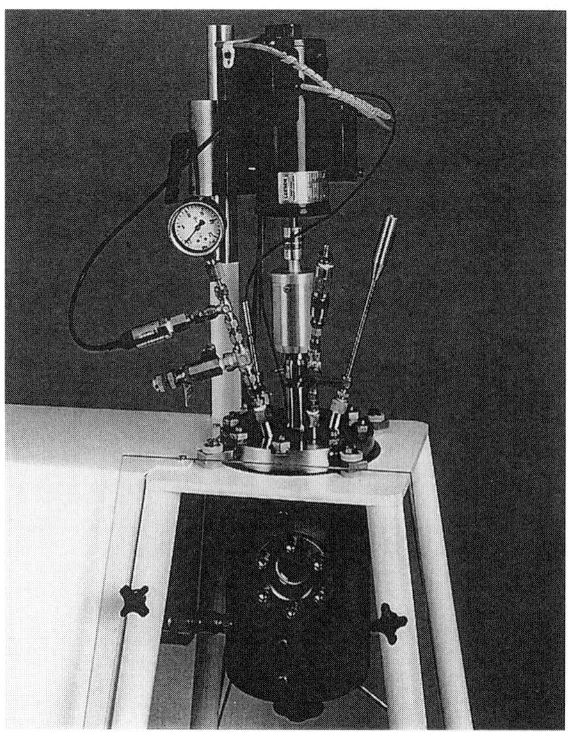

Figure 3 High-pressure Simular.

5.2 Mettler Toledo RC1e Isothermal Calorimeter

The standard calorimeter supplied by Mettler Toledo [6] is the RC1e. The RC1e is an automated lab reactor and an isothermal calorimeter. It has a nominal 2 liter volume but can be operated down to 0.5 liter. It has automatic data acquisition and real time calorimetry display. There are various models encompassing a range of temperatures from −70 to +300°C (jacket temperature). The standard material of construction is Duran® glass. It has a glass head, which can accommodate pH and temperature sensors, a distillation kit, and/or an FT-IR probe. A distillation/reflux kit is available that allows calorimetry at reflux. Distillate weights can also be tracked. Other vessels types are available. For example, the SV01 is a small glass vessel with a bottom runoff and a workable volume of between 80 and 100 mL. It operates between −70 and +230°C.

Autoclaves such as the HP350 may be used for the study of high-pressure reactions. The latter vessel operates at up 350 bar (HP350), with a workable volume of 0.2–1 liter. Materials of construction include Hastelloy® and 316 stainless steel.

5.3 Software

Both HEL and Mettler-Toledo have proprietary software that controls the equipment and data acquisition/processing. In the HEL systems, a package called Win-ISO controls vessel temperature, agitation, and feed rates. It also records data (for example, power output, agitator power, pH) for the vessels *vs.* time. This data is displayed in real time during the experiment and also logged in a data file. Once an experiment is complete, a separate package, WinCALC, is used to perform energy calculations and plot graphs of the data generated.

6 CALORIMETRY TRACES EXPLAINED

The basic utility of the auto-MATE can be illustrated by examining the aqueous hydrolysis of acetic anhydride. The reaction proceeds readily at ambient temperatures with the formation of acetic acid as outlined in Scheme 1.

The reaction is exothermic and therefore results in a decrease in the power supplied to the internal heater to maintain isothermal conditions. By examining the power output from the reaction, it is possible to determine the maximum heat output rate (which will be proportional to the maximum cooling duty required on scale-up) and the total heat release from the reaction. A typical isothermal calorimetry trace for the dosing of acetic anhydride into water is shown in Fig. 4.

As the experiment is initiated the reactors and their contents need to be raised to their set point values. To achieve this as quickly as possible the heater is turned on at a near-to-maximum value (just over 21 W). On achieving the desired reactor temperatures, the power then falls and stabilizes at approximately 7 W (since the heater is now only required to make up the temperature differential between the oil jacket and the desired reactor temperature). When the addition of acetic anhydride starts after approximately 40 minutes the heater power is observed to fall further as it compensates for the energy that is liberated by the reaction, thus maintaining isothermal conditions. On integration of the power–time curve by selection of two suitable baseline regions before and after the reaction period, it is possible to evaluate the heat of reaction. For the reaction

$$(CH_3CO)_2O \ + \ H_2O \ \longrightarrow \ 2CH_3CO_2H \qquad (1)$$

Scheme 1 Hydrolysis of acetic anhydride.

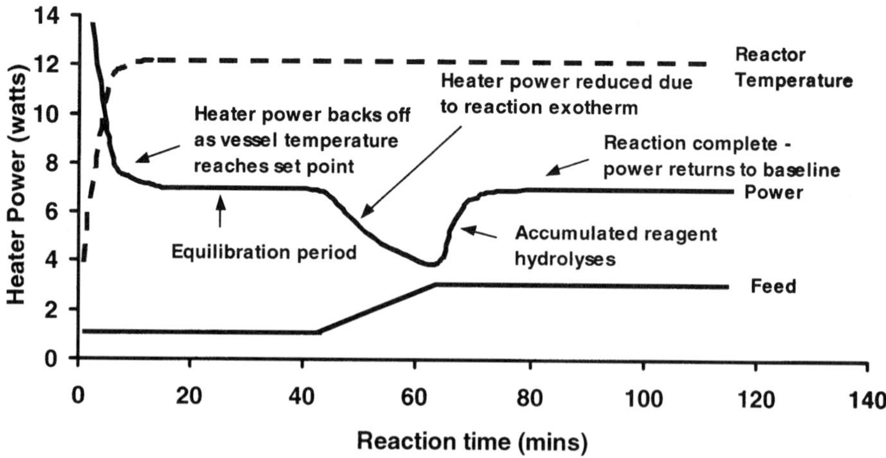

Figure 4 Isothermal calorimetry for the hydrolysis of acetic anhydride in water.

between water and acetic anhydride, this is typically found to be in the region of -60 kJ (per mole of acetic anhydride) [9].

In addition to providing important data regarding the magnitude of heat release during the reaction, the shape of the heater power profile also provides useful information regarding the global kinetics. It is observed that on termination of the acetic anhydride feed the heater power returns gradually to a new baseline value. This is indicative of a degree of accumulation of unreacted acetic anhydride. This unreacted material then gradually hydrolyzes, giving the observed power profile. This information is especially important when you want to scale up your reaction within the confines of your existing plant capacity. These kinetic effects are more clearly shown in Figs. 5 and 6. In Fig. 5 the feed rate of acetic anhydride is kept constant at 0.125 g/min and the isothermal reaction temperature increased from 25 to 35°C. Examination of the heat release profile under these conditions shows that as the reaction temperature is raised so the rate of heat release increases and the reaction terminates more quickly.

In Fig. 6 the effect of feed rate on the heat release power profile is investigated and again a similar effect is observed with the heat release rate becoming greater as the feed rate is increased.

Clearly, it is important to determine the reproducibility of the calorimetry data obtained on this scale and to compare it to data obtained from larger bench scale reactors, especially as this data can be used to give vital information on process scale-up. Our initial calibration experiments (acetic anhydride hydrolysis in water) gave the following data, which was in close agreement with literature

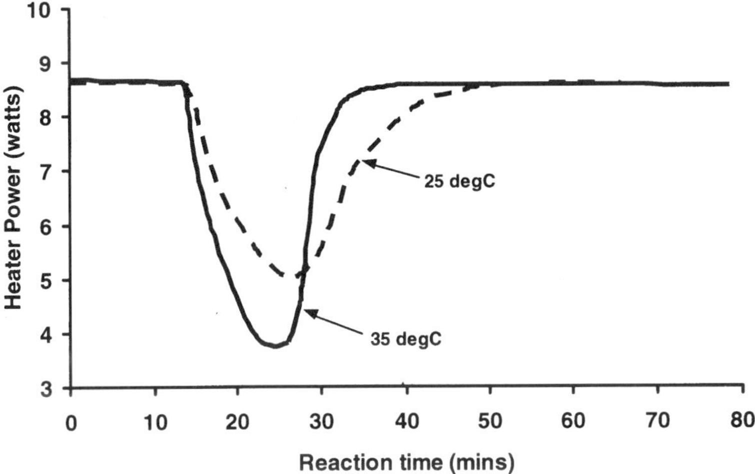

Figure 5 Variation in the power profile for the reaction between water and acetic anhydride with temperature.

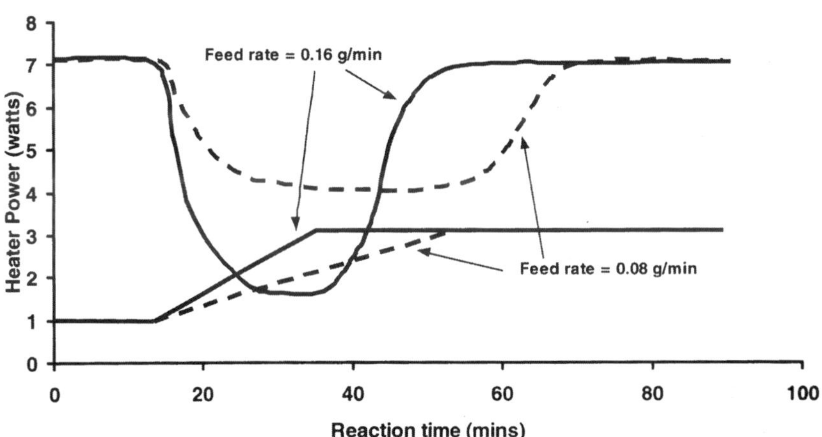

Figure 6 Variation in the power profile for the reaction between water and acetic anhydride with feed rate.

Table 4 Reproducibility of Calorimetry Data from Auto-
MATE and Simular Systems

Run number	HEL Auto-MATE $\Delta H/KJmol^{-1}$	HEL Simular $\Delta H/KJmol^{-1}$
1	65.5	56.8
2	61.2	55.9
3	57.6	55.3
4	59.3	—
Average	60.9	56.5
Standard deviation	3.40	0.75

values and showed a 15% error across the four vessels. Although dependent on the size of the exotherm relative to the plant cooling capacity, a 15% error may not be acceptable for detailed hazard assessment and process siting purposes. However, we believe that being able to run four reactions simultaneously coupled with the availability of semiquantitative calorimetry data is an extremely effective development tool. The actual results are shown below (Table 4). For comparison, the same calibration experiments were performed in the Simular and showed that the reproducibility of the calorimetry data in the Simular was excellent.

Clearly, as with all calorimetry experiments, the uncertainty in the enthalpy determination will be directly related to the heat output during the reaction, with strongly exothermic reactions providing smaller relative errors than less exothermic reactions. In principle the calorimetry information recorded on this scale is not intended to be used for direct scale-up and should be confirmed, once a process has been optimized, with a bench scale calorimeter. In practice, however, this caution is only important for reactions with low heat outputs where the error can be much higher.

7 SAMPLING AND ANALYSIS

In many situations, the measurement of reaction power output gives a real-time measure of the extent of reaction, superseding to some extent the need for in-process sampling and analysis—a confirmatory end-of-reaction sample should be all that is required. For those situations where more frequent sampling is required, HEL [3] has developed a sipper system that can retain up to six samples for later analysis.

Additionally, the auto-MATE can be located in the bed of an XYZ robot which can then be used for sampling the reactors. The ability to take multiple samples and simultaneously determine the calorimetric profile of a reaction represents a major step forward.

8　CASE STUDIES

This section demonstrates the principles of isothermal calorimetry by reference to a series of real examples. These examples vary in complexity from simple esterification reactions through to sequential anion generation/electrophilic quench reactions. It is particularly instructive to note that in the four separate examples of nitro-group reduction that are shown, the calorimetry of each of them reveals quite different features of the reactions. Due to commercial sensitivity of certain of these examples, full chemical structures are not given, but the tremendous power of automated calorimetry should be clear nonetheless.

Case Study 1:　Anion Formation/Acylation

This example involves the deprotonation of a heterocycle with an organometallic reagent followed by C-acylation with an acyl chloride. Both these transformations may be followed by calorimetry, as shown in Fig. 7.

In the initial deprotonation reaction, there is a large initial power spike, which falls away after 1.1 equivalents of organometallic have been added. It does not, however, fall away to zero, as there is an exothermic interaction of the reagent with the solvent, which we established separately. Here, although calorimetry provides an indication of the course of the reaction, it could be dangerous to

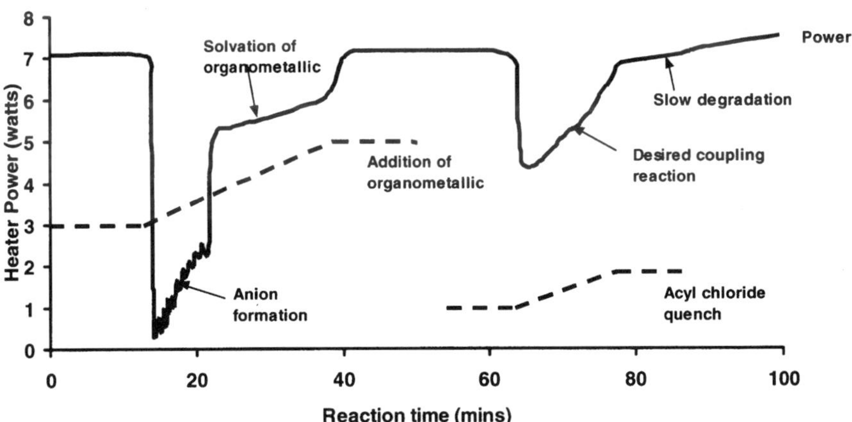

Figure 7　Calorimetric profile of deprotonation and acylation reactions.

assume in the first instance that the observed exotherm was due to the desired deprotonation event. It could, for example, have been due to interaction of the organometallic with residual water. However, after confirming the presence of the product in an initial trial reaction, the calorimetry was used as a reliable check for completion of deprotonation in subsequent reactions. After the organometallic reagent has been added, the batch is aged for 1 hour and then acyl chloride is added. There is a largely feed controlled exotherm, which gradually dies away during the reaction. There is a temperature spike approximately two-thirds of the way through the addition, which is associated with a precipitation event. After the addition is complete, there is a slight exotherm, which dies away over about 1 hour. This is associated with reaction of excess organometallic with the ketonic product and the acyl chloride.

This is a key example demonstrating the benefit of calorimetry in those reactions where analytical protocols are not fully defined. Anion formation reactions could in principle, be monitored in situ by FTIR or off-line by a D_2O quench. Calorimetry gives a viable alternative to these techniques.

Case Study 2:   Safe Scale-Up of an Oligomerization Reaction

Description

An oligomerization process was being developed [10] that involved the dosing of caustic into an organic feed. The caustic acts essentially as a catalyst and normally forms a two-phase mixture with the organic substrate. Due to relatively slow mass transfer across the phase boundary, it accumulates in the reactor and occasionally leads to a large output of heat accompanied by a sharp temperature spike.

The objective was to understand the mechanism producing the temperature spikes and propose a safe mode of operation.

Experiment 1:   Base-Case Scenario

Typical data is shown in Fig. 8. The temperature remains under control at 25°C, as caustic feed is commenced and the heat output is essentially constant. After dosing for 50 to 60 minutes, a sudden burst of heat leads to a corresponding temperature spike. It was observed that the sharp rise in temperature corresponded with a significant change observed visually in the reactor vessel. This was a sudden mixing of organic/aqueous layers and the formation of a single homogeneous layer. This sudden improvement in the mixing would lead to a sharp increase in the reaction rate and hence in the power output giving an accompanying temperature spike. Control is however rapidly achieved and the experiment resumes until feed is finally stopped. Note that in this case, it is the reaction power output which is being plotted against time, unlike in earlier examples, where the heater-input power profile is being plotted.

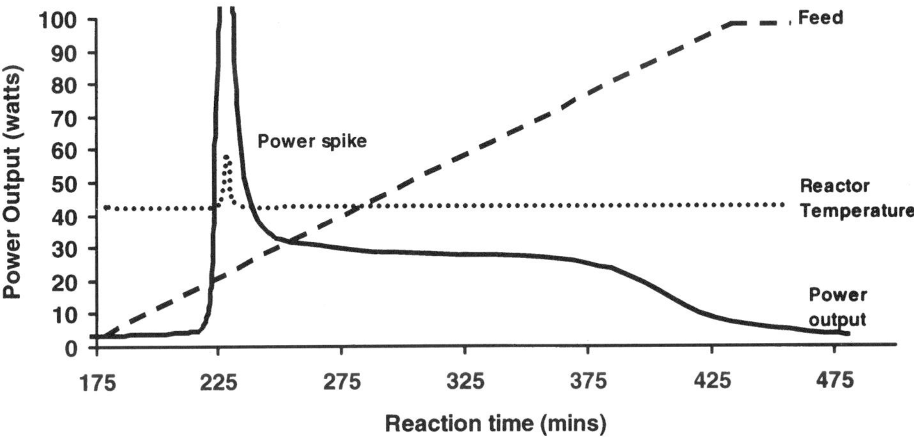

Figure 8　Calorimetry profile of base-case polymerization reaction.

Experiment 2:　Product Recycle

The first test was repeated at the same temperature and with the same caustic feed rate, but in this case a finite amount of the previous batch was initially left in the reactor vessel. The objective was to see if this would improve the mixing between the caustic and the organic phases, thus avoiding accumulation and a subsequent burst of energy. The data for this run are shown in Fig. 9.

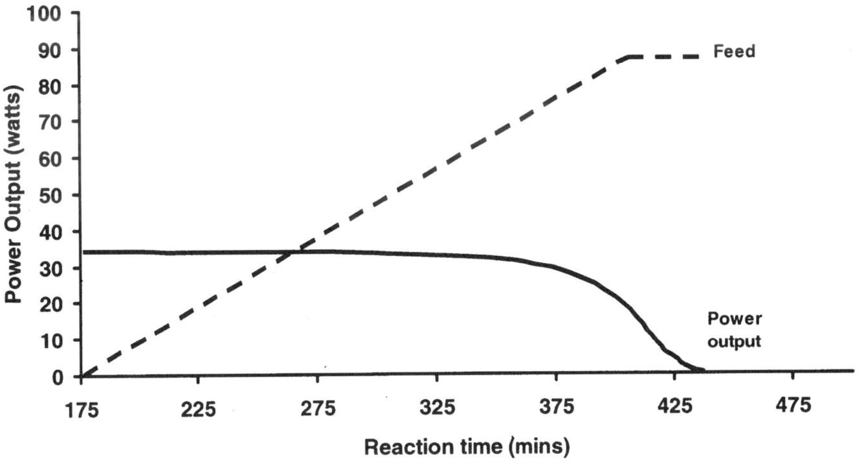

Figure 9　Calorimetry profile of polymerization reaction with partial recycle.

In this case, there was an immediate heat release when dosing was commenced, and power output remains steady throughout the addition. The heat spike was therefore eliminated and the process was able to be safely scaled up on the basis of the second run.

Case Study 3:   Catalysis of an Esterification Reaction

Description

An esterification reaction was carried out [11] at 40°C in a stirred vessel by adding an organic anhydride to an alcohol. The reaction was mildly exothermic and rather slow. The anhydride was added over a period of 100 minutes, but the reaction was slow, taking nearly 12 hours for completion.

The objective was to test the effectiveness of an acid catalyst in reducing the batch time and whether the cooling system on a full scale reactor would be able to cope with the increased power output.

Experiment 1:   Base Case

The traditional method of esterification was evaluated using power compensation calorimetry and the results are shown in Fig. 10. The heat output rate (measured in W) was seen to increase as anhydride was added. This peaked at ~7 W and then decreased gradually after the feed was stopped. It took ~600 minutes after the end of dosing before the power output returned to zero. This confirmed existing plant experience concerning the long batch time. The calorimetry profile

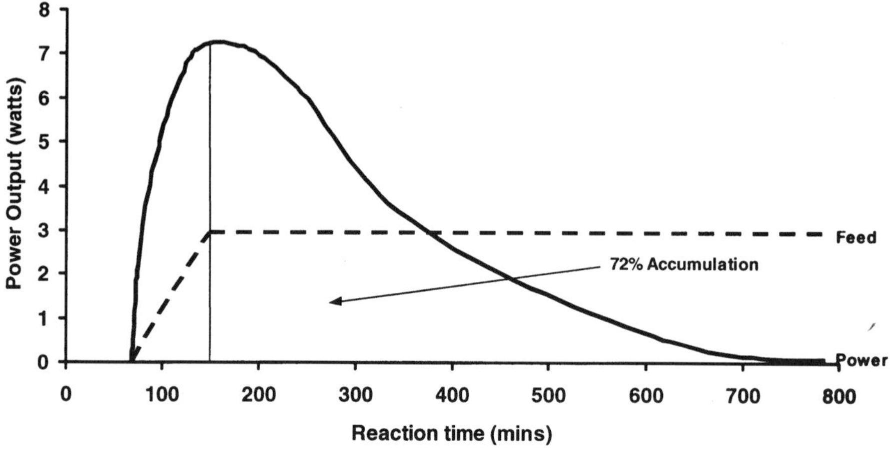

Figure 10   Calorimetry profile of base-case esterification reaction.

suggests an autocatalytic reaction. In this case, acid liberated leads to protonation of the anhydride, with a concomitant acceleration in the rate of attack by the incoming nucleophile.

Experiment 2:   Effect of Catalyst

When anhydride was added to the alcohol, containing catalyst, at the same rate and at the same temperature, the heat output rate was much larger (Fig. 11). The heat output peaked at ∼27 W and after dosing was stopped, returned to zero after ∼150 minutes.

Conclusions

1.   Cycle Time. The two experiments show that a substantive reduction in cycle time was possible. The catalyzed reaction had a cycle time of about 4 hours compared to 11.5 hours for the noncatalyzed process.

2.   Cooling Requirements. In order to achieve this cycle time improvement on scale-up, the reactor cooling system needs to be able to cope with the increased power output. The same amount of heat is being generated in the catalyzed system, but in a shorter period of time. In this particular case, for a 2m3 vessel, the cooling duties are 14 and 44 kW for the uncatalyzed and catalyzed systems, respectively.

3.   Safety Issues. At first sight, the noncatalyzed reaction might appear to be safer than the catalyzed version. The former is slower and has a lower maximum power output.

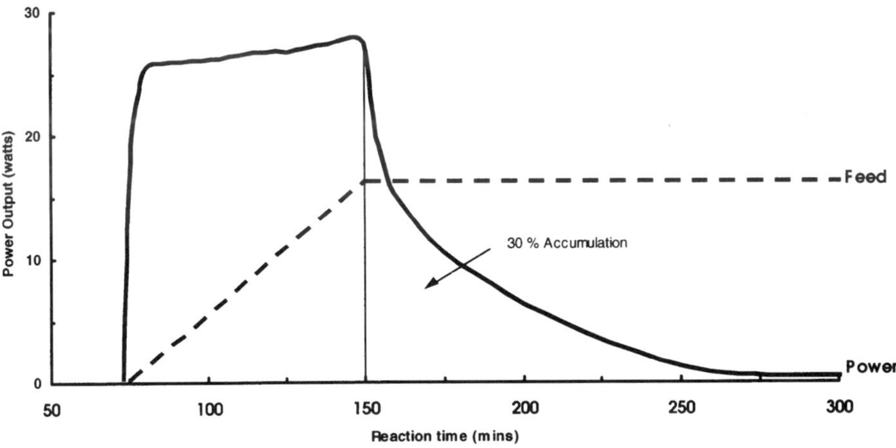

Figure 11   Calorimetry profile of catalyzed esterification reaction.

However, during the addition of the anhydride in the noncatalyzed case, only about 30% of the total heat is given out, i.e., there is about 70% accumulation. Hence if cooling is lost straight after the addition, then there is the possibility of a thermal runaway. On the other hand, in the catalyzed case, the power output curve shows a very much lower degree of accumulation (approximately 30%). Thus if cooling is lost during the addition, then by stopping the addition, a maximum of 30% accumulation is the worst-case scenario.

Case Study 4:   Improvements in Nitration Process Economics

Description

Nitration of an aromatic substrate [13] was traditionally carried out by the addition of two equivalents of nitric acid to the starting material. The yield was known to be relatively poor, and there were a number of other issues threatening the viability of the process. The objective was to increase yield and simplify the product work-up procedures.

Experiment 1:   Base Case

The addition of concentrated nitric acid to the aromatic substrate in the base process was carried out over a period of nearly 6 hours. The power output as a function of time is shown in Fig. 12. Clearly, the heat release rate remains constant at ~8 W for about 4.5 hours and then begins to decrease, finally levelling

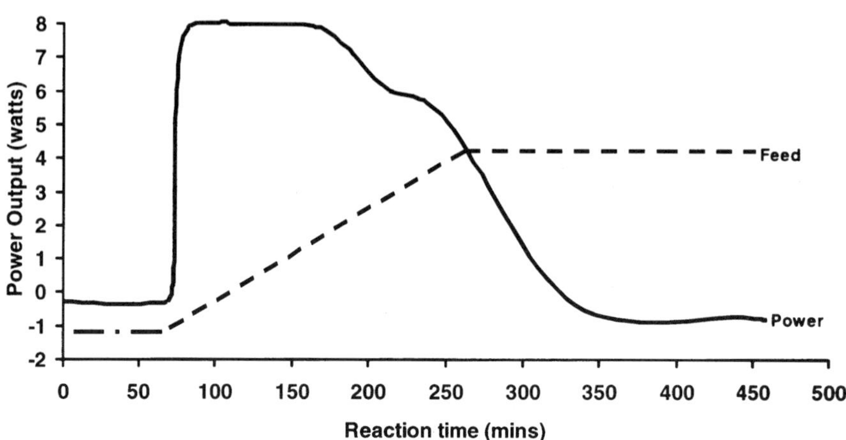

Figure 12   Calorimetry profile of nitration using 2 equivalents of nitric acid.

off after nearly 8 hours. Analysis of the product shows poor selectivity, and extensive work-up is required to recover a pure material.

The calorimetry data showed a significant change in the heat output profile after ~1.2 equivalents of nitric acid had been added. This observation suggested that the quantity of acid could perhaps be reduced significantly, say to a maximum of 1.4 equivalents.

Experiment: Effect of Reduced Nitric Acid Charge

The operation was repeated at the same temperature but with a much reduced nitric acid charge. The results (Fig. 13) show that the power output drops to zero soon after dosing is stopped. The maximum steady-state heat output is ~8 W as in the first experiment.

Benefits of Reduced Charge

Analysis of the product showed that the yield was much better than in the first run. Since the reaction is virtually feed controlled and there is no increase in power output, the cycle time could be dramatically reduced.

Reduction in the amount of excess nitric acid also simplified downstream processing. Typically, optimization of electrophile charge in aromatic substitution reactions leads to a reduction in the levels of di- and trisubstituted by-products.

The quench or drown-out step, where excess acid is either neutralized or

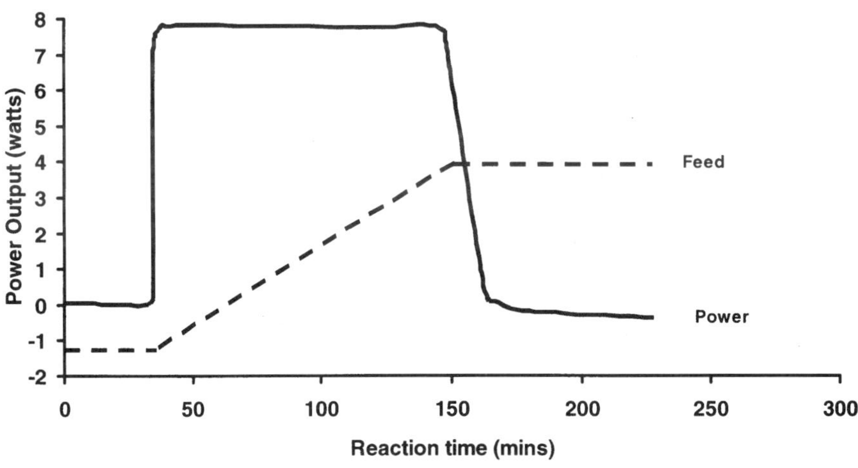

Figure 13   Calorimetry profile of nitration using 1.2 equivalents nitric acid.

simply diluted, will now require less water, hence improving process productivity and generating less waste.

Case Study 5:   Use of the High Pressure Auto-MATE for Nitroarene Hydrogenation [17]

An investigation into the hydrogenation of 2,4-dinitrotoluene to 2,4-diaminotoluene over a palladium-on-carbon catalyst showed [14] that the reaction follows a Langmuir–Hinshelwood kinetic model with a noncompetitive adsorption of the organic species and hydrogen on the active sites with the formation of a 2-(hydroxyamino)-4-nitrotoluene intermediate. It has been shown that both the specific activity and selectivity of the reaction is dependent on the Pd particle size and loading. A high-pressure auto-MATE unit has been used to follow the heat release profile of this hydrogenation reaction. Both of the nitro groups present on the ring are hydrogenated but in a sequential manner, consistent with the proposed mechanism [14]. From the shape of the heat release profile (Fig. 14) it is apparent that hydrogenation of the first nitro group occurs much quicker than that of the second. This is identified by the sharp fall in the power profile over approximately ten minutes on the addition of hydrogen to the system followed by the shoulder corresponding to the reaction of the second grouping over several hours. On determining the total enthalpy release for this reaction for both nitro groupings, we obtain a value in the range of 250 to 300 kJmol^{-1}. This is in good correlation with data from larger calorimeters (1 liter and above) that have been found to give similar values.

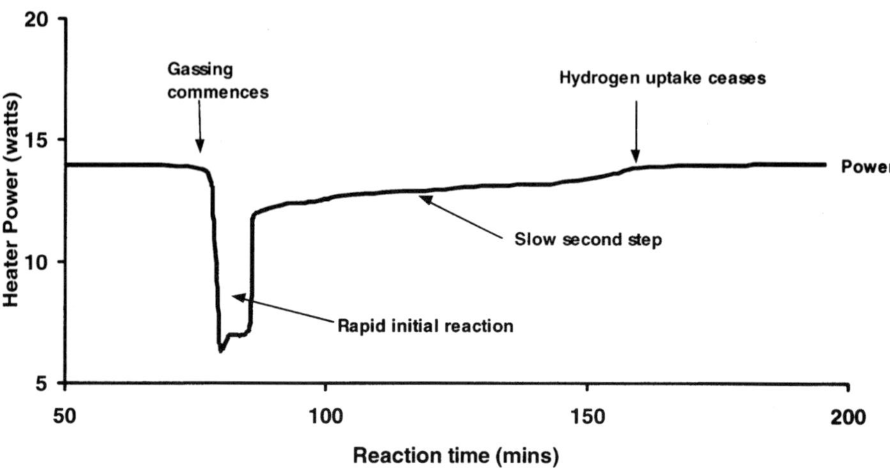

Figure 14   Power compensation data for the hydrogenation of 2,4-dinitrotoluene.

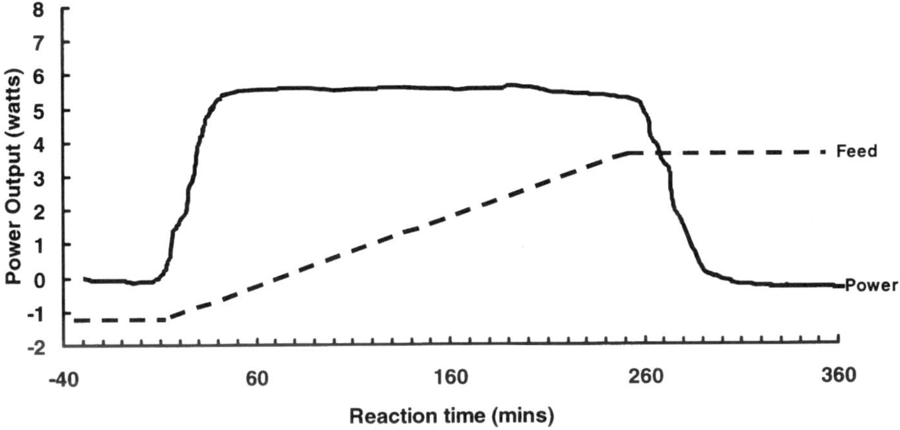

Figure 15 Reduction of nitro-arene using granular iron.

Case Study 6: Reduction of Nitroarene Using Iron/Acid

In this example [16] we were seeking to change the grade of iron used in a reduction process. Historically, we had used granular iron from one supplier and were assessing other suppliers for security of supply reasons. The reaction was run by addition of acid to a slurry of iron in the solvent/nitrocompound mixture. The initial process using granular iron was feed controlled, i.e., as soon as the acid was added the exotherm started. The power output (Fig. 15) was virtually constant throughout the addition at 5 W. When the addition was complete, the exotherm rapidly died away. There was small degree of accumulation.

In contrast, when powdered iron was used (Fig. 16) a similar rapid onset

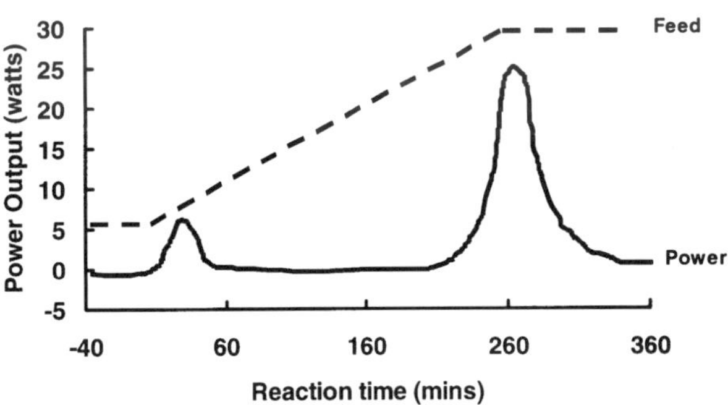

Figure 16 Reduction of nitro-arene using unactivated powdered iron.

of the exotherm was seen, with the maximum power output at about 8 W, although in this case it appeared to rapidly die away. Suddenly, after about 220 minutes, a further large exotherm was seen, with a maximum power output of at least five times that of the granular iron example—this would almost certainly have given major problems on plant, had the cooling capacity been sized on the granular iron example.

Note the difference in scale of the two graphs—the total power output, i.e., the area under the curve, should be the same in both cases. The difference in behavior of the two types of iron was due to the fact that the granular iron, although it had a relatively small surface area, was oxide free and reacted immediately. In contrast, the powdered iron had a small oxide-free area, which gave rise to the initial exotherm. The bulk of the material however had a substantive oxide film. This was slowly degraded throughout the reaction, by mechanical effects and acid digestion. Once the oxide film had been removed, the high surface area of the iron led to an extremely vigorous reaction. In order to control the exotherm in subsequent reactions, we decided to remove the oxide film by acid pretreatment of the iron.

Case Study 7:   Nitroarene Reduction with Hydrazine

This example [16] highlights several interesting points. Firstly, as shown in Fig. 17 the reaction is feed controlled, and the gassing rate largely mirrors the rate of reaction: the increase in gassing rate toward the end of the addition is due to a phase change within the system leading to increased nucleation. Note that the gassing stops soon after the feed—there is minimal accumulation. It is therefore easy to control the gassing rate on plant by altering the feed rate. The apparent

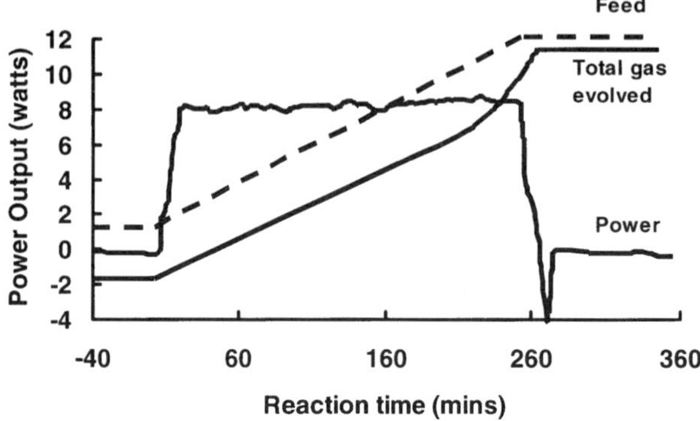

Figure 17   Reduction of nitro-arene with hydrazine.

endotherm at the end of the addition is due to overcompensation. Basically, the system detects the cessation of the exotherm, and the power to the heater increases to maintain isothermal conditions. However, the heater overshoots slightly, giving the apparent endotherm.

Case Study 8:　Transfer Hydrogenation of a Nitroarene

This reaction [9] involves the transfer hydrogenation of an aromatic nitro to an amine, using cyclohexene as hydrogen donor.

Calorimetry (Fig. 18) showed a long induction period in the reaction and alerted us to the possibility of a "runaway reaction" occurring, due to the almost complete buildup of reagent. However, we were able to ascertain that even in a worst-case scenario where the reactants accumulated completely before then reacting instantaneously, the heat evolved (16.8 KJ/mole cyclohexene) would not be enough to cause a hazardous situation. Hence this is essentially an intrinsically safe reaction.

Case Study 9:　Sulphonylation of an Alcohol

In our initial process [9], the sulphonylation was carried out using triethylamine as base in tetrahydrofuran. Calorimetry (Fig. 19) showed a feed controlled reaction, i.e., no accumulation. The heat of reaction was established (107 KJ/mol) and compared with the known cooling capacity of the plant, allowing us accurately to predict and simulate plant addition times in the lab. These expected times, together with worst-case scenarios, were then factored into a SED matrix, along with hold times, temperatures, and reagent charges to allow us to define clearly the reaction robustness envelope.

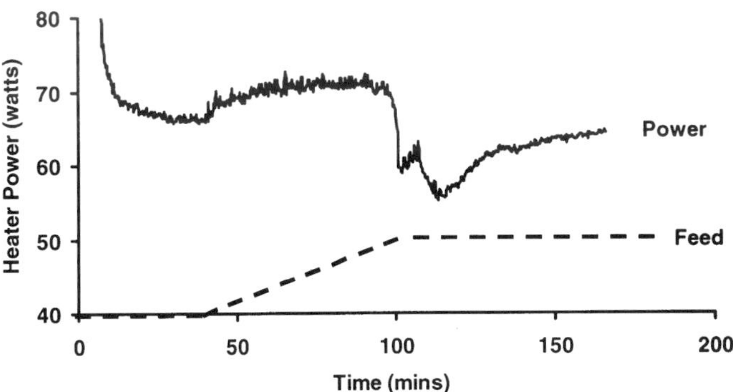

Figure 18　Calorimetry profile of transfer hydrogenation reaction.

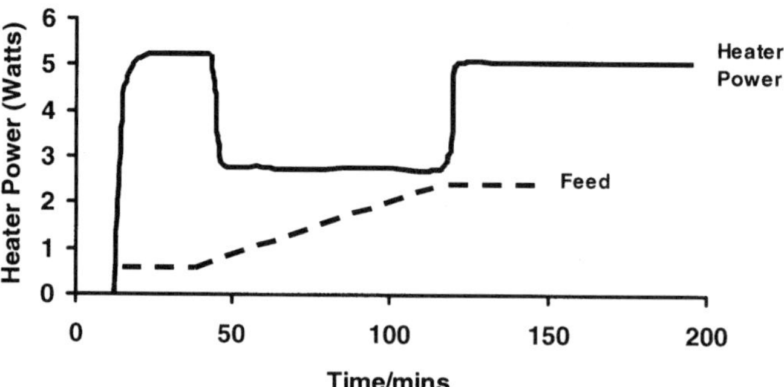

Figure 19 Calorimetry profile of sulphonylation reaction.

A key benefit of using the auto-Mate was that the PC controlled syringe pumps allowed facile replication of plant feed times at the two different levels for the experimental design (e.g., 3 mL added over 8 and 12 hours). This degree of control would have been virtually impossible using a manually controlled addition.

Although our initial manufacture was carried out using tetrahydrofuran as solvent, this meant that we had to drown the batch out into water and extract the product into excess crystallization solvent. The excess solvent was then removed by distillation prior to crystallisation. A mass balance revealed that we were losing yield in two ways. Firstly, there were significant product losses in the spent aqueous after extraction, due to the effect of the tetrahydrofuran. Second, there was substantial product decomposition during the distillation. Although this latter loss had been predicted, and deemed safe by adiabatic calorimetry studies, it was still costly in terms of lost product. We therefore had to reengineer the process completely and subsequently carried out the reaction in the crystallization solvent. Calorimetry showed that the reaction profile in the new solvent was almost identical to that in the original one. Removal of the triethylamine hydrochloride by hot filtration avoided the aqueous losses, and using the minimum volume of crystallization solvent for the reaction avoided the distillation and associated product loss.

9 FUTURE PERSPECTIVES

Currently, calorimeters are largely the domain of process safety specialists. In the near future, the use of isothermal calorimeters linked to an array of techniques such as GC/LC/ReactIR/UV probes/Turbidity meters should become more commonplace in process development laboratories. These techniques could be applied to cascades of reactors to simulate actual plant processes.

The major challenge remaining is to increase the throughput of reaction calorimeters. The HEL auto-MATE has gone part of the way to doing this, but in order to make another step change in effectiveness, a method for allowing continuous operation of multiple reactors is required. This could involve having a series of reactors, probably between 4 and 8, all with automated bottom runoff valves, a waste reservoir or product collection vessels, and a flushing/clean-down system. The key issue here is ensuring that interbatch clean-downs are effective. Obviously this is more acute in smaller vessels with a less favorable surface-area-to-volume ratio, but presumably by choice of appropriate solvent and boil-out protocols, adequate clean-down could be achieved.

Looking further into the future, an Expert System could be integrated. This could involve the chemist defining an initial set of experiments and desired response parameters. The system would then be capable of carrying out and analyzing the reactions, comparing the data with the desired outcomes, and automatically defining and initiating a new set of experiments.

ACKNOWLEDGMENTS

The assistance of Peter Duggan (Avecia Hazards Group), Chris Lampard and Scott Phimister (Avecia Pharmaceuticals), Harjinder Mander and Jasbir Singh (Hazard Evaluation Labs), and Ken Stares (Metler-Toledo) in the gathering and preparation of material for this chapter is very gratefully acknowledged.

REFERENCES

1. N. Evens. The use of automation in process development. In: W. Hoyle, ed. Automated Synthetic Methods for Speciality Chemicals, ISBN 0-85404-825-1, 1999.
2. Supplier: Anachem Ltd., Charles Street, Luton, Bedfordshire LU2 0EB U.K.
3. Supplier: Hazard Evaluation Laboratories Ltd., 50 Moxon Street, Barnet, Hertfordshire EN5 5TS. WEB: *WWW.HELGROUP.CO.UK*.
4. Supplier: Zymark Ltd., The Genesis Centre, Science Park, South Birchwood, Warrington, Cheshire WA3 7BH.
5. Supplier: STEM Corporation, Woodrolfe Road, Tollesbury, Essex CM9 8SJ U.K.
6. Supplier: Mettler-Toledo Bohdan, Inc., 562 Bunner Court, Vernon Hills, IL 60610. Web: www.bohdan.com.
7. Mettler-Toledo Gmbh, Sonnenbergstrasse 74, CH-8603 Schwerzenbach, Switzerland. *WEB:WWW.MT.COM*.
8. J. Singh. Thermochimica Acta 1993, *226*, 211–220.

9. J. Sempere, R. Nomen, R. Serra, P. Cardillo, J. Loss. Prev. Process Ind. 1997, *10*, 55–62.

10. C. Lampard. Avecia LifeScience Molecules, private communication.

11. HEL Application Note No. 98003.

12. HEL Application Note No. 98005.

13. HEL Application Note No. 98002.

14. HEL Application Note No. 98004.

15. G. Neri, M. G. Musolino, C. Milone, G. Galvagno. Industrial Eng Chem. Research 1995, *34*, 2226–2231.

16. P. J. Duggan, Avecia Pharmaceuticals, private communication.

17. G. Bradley, Avecia Pharmaceuticals, private communication.

18. J. Singh, C. Simms. Application of small multiple reactor/calorimeter systems in rapid process scale-up and development, private communication.

19. Supplier: Argonaut Technologies, 887 Industrial Boulevard, Suite G, San Carlos, CA94070, USA. WWW.Argotech.com.

20. J. Singh, Chemical Engineering, May 1997, 92–95.

21. J. Baron, R. Rogers, eds. Chemical Reaction Hazards—A Guide. Inst. of Chemical Engineers, ISBN 085295 2848, 1993.

22. Rogers, R. L. Fact finding and basic data part 1: Hazardous properties of substances. IUPAC Conference—Safety in Chemical Production, Basle, 1991.

11

Parallel Automation for Chemical Analysis

David T. Rossi
Pfizer Global Research and Development
Ann Arbor, Michigan

1 THE NEED FOR AUTOMATED CHEMICAL ANALYSIS

As described in the preceding chapters, the automation capabilities for synthetic chemistry are steadily improving (Chaps. 1, 2, 3, 6, 8, and 9), allowing the creation of larger numbers of synthetic products in less time. This dramatic improvement in productivity has in turn created a strong demand for higher throughput purification and analysis systems to purify and characterize the resulting products. Many of the features of purification systems have been described in Chaps. 4 and 5. This chapter will focus on systems with high-throughput automation capabilities for characterization of synthetic or formulated products.

As molecular diversity initiatives [1,2] and the introduction of combinatorial synthetic processes [3–5] have increased the number of synthetic compounds by orders of magnitude in recent years, chemical analysis has been challenged to catch up, and a great deal of attention has been given to systems for rapid automation of the required analytical chemistry. Although automated systems for serial (or sequential) characterization of synthetic compounds have been available in various forms for more that two decades [6,7], it has become evident in the last five or six years that serial analysis is not conducive to the high-throughput analytical needs of combinatorial chemistry [8]. For this reason, most of the recently introduced automation systems for chemical analysis are parallel systems. The features and performance characteristics of these parallel systems are described below.

2 AUTOMATION APPROACHES IN CHEMICAL ANALYSIS

2.1 Serial Automation Processes

A *serial* (or *sequential*) *automation process* for chemical analysis is one where each of the processes of a chemical determination is executed in turn, before proceeding to the next experiment. If, for example, the moisture content of a process stream is to be monitored, then the sampling would precede the measurement step, which would precede data reduction and reporting. The general characteristic of a serial automation process is that a given determination reaches completion before the next determination begins, although with some serial automation systems, the second determination may be started before the first is completed [9]. A schematic representation of a generalized on-line automated serial process involving analytical sampling, separation, and detection is given in Fig. 1.

Many early automation systems used a single powerful robotic arm operating in a serial processing automation architecture. As Fig. 2 shows, the central arm could access many specialized instrument and apparatus positions. Although these specialized apparatus helped improve throughput, the overall bottleneck for system throughput was still the processing speed of the central arm. Serial automation could also be configured in an *on-line* arrangement, where tandem processes in the chemical determination are performed by collection from a flowing stream. Such an on-line system could be used to acquire a sample, process it, and perform the associated chemical measurement step, as in the case of on-line microdialysis/LC/MS [10] (Fig. 3) or on-line solid-phase extraction/LC/MS/MS [11].

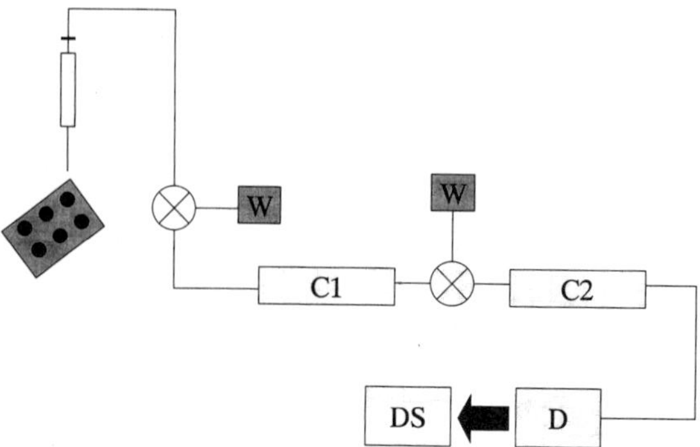

Figure 1 Schematic representation of a generalized serial automation system, including on-line sampling, optional sample concentration/preparation (C1), optional sample separation step (C2), detection (D), and data system (DS) for acquisition and control. Here, W indicates waste.

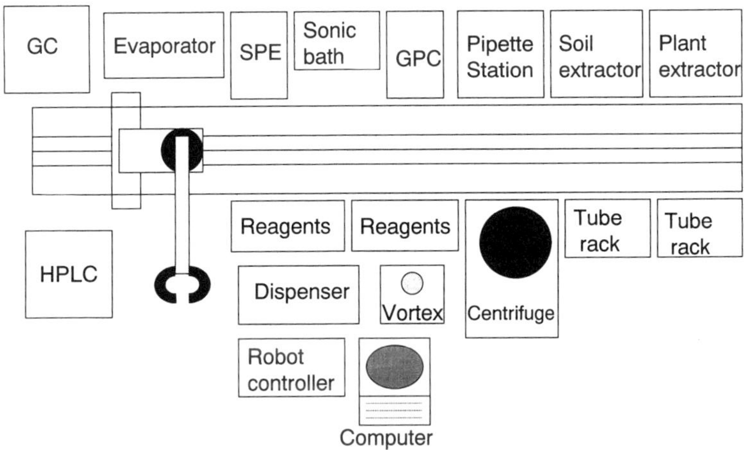

Figure 2 Representative layout of a serial automation system based on a single powerful robotic arm. The arm in this example is reticulated and is capable of traversing along a 6-foot monorail to access a wide array of laboratory apparatus and analytical instrumentation. Despite the complexity of this automation system, the overall throughput of the system is generally limited by the serial processing of the robotic arm.

2.2 Parallel Automation Processes

A *parallel automation process* for chemical analysis is, simply, a process where more than one automated chemical manipulation is performed simultaneously. These parallel manipulations can encompass some or all of the following discrete analytical chemistry operations: experiment initiation, sampling (obtaining the samples), sample preparation, component separation, and analyte detection and data reduction/reporting. Although possible with today's state of the art, it is unusual for all of these chemical determination components to be included in a parallel automation system. Usually, only the rate-limiting components of the determination are included. For example, if the sample preparation step is especially time-consuming, it can be useful to automate it with a multichannel liquid handling workstation.

An assumption made with parallel automation systems is that a number of experiments or samples are available simultaneously, thereby allowing for batch-parallel manipulation or *batch processing*. A schematic representation of a parallel automation process, given in Fig. 4, shows the steps involved in microscale liquid–liquid extraction on a 96-well liquid handling workstation. The procedure is implemented on the six-position stage of a Tomtec Quadra-96 liquid handling workstation. Samples plus standards are placed in a 96-well tube plate, which is located in position 2 of the workstation stage, and to this plate are added an

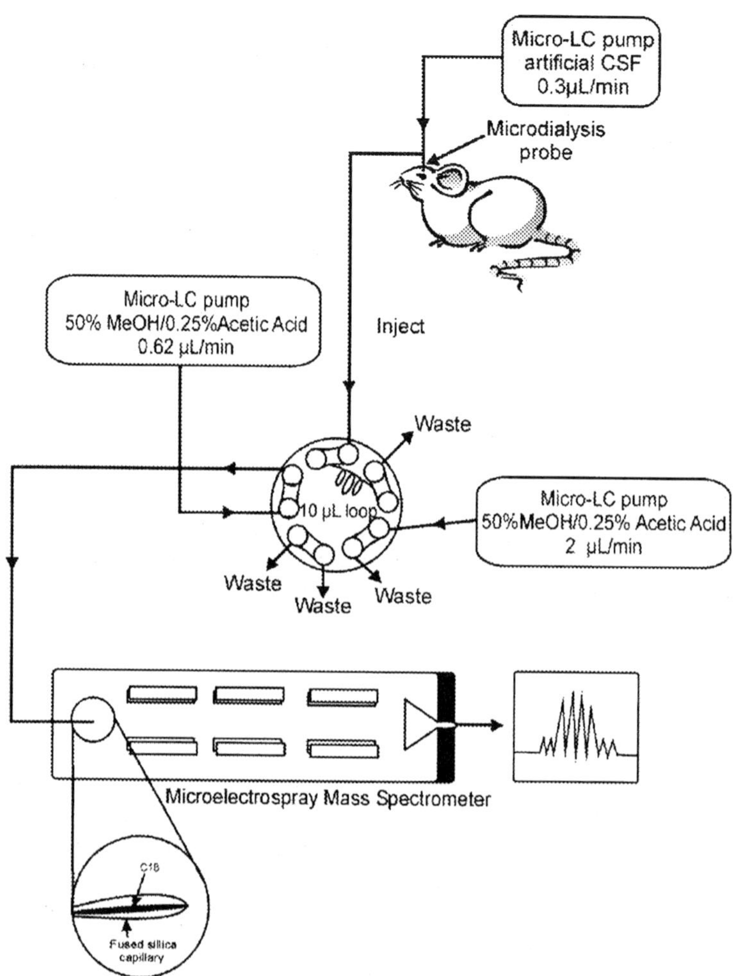

Figure 3 Experimental setup for analysis of substance P metabolism and its metabolic fragments. Substance P was infused into the rat striatum through a microdialysis probe at 0.3 µL min^{-1}, and the probe was used both to deliver and to collect metabolic products. The microdialysate was sampled into a 10-port valve, injected into a fused-silica C_{18} column, and washed with 2% methanol–0.25% acetic acid. The peptides were eluted from the column with 50% methanol-acetic acid at 0.82 µL/min directly into the microelectrospray source on the mass spectrometer by switching the valve back.

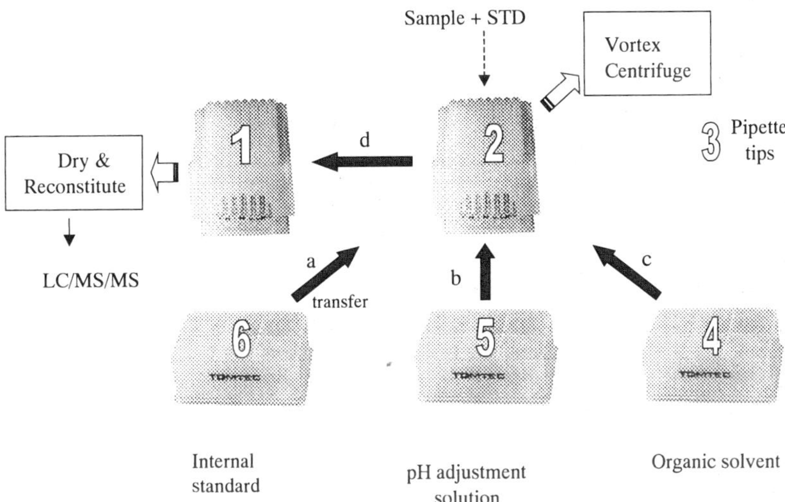

Figure 4 Conceptualized semiautomated 96-well liquid–liquid extraction procedure using a 96-well liquid handling workstation. Refer to the text for a stepwise description of the operation.

internal standard (position 6), a buffer (position 5), and an organic solvent (position 4), using the parallel liquid handling capabilities of the 96-well head. After off-line vortexing and centrifugation, a phase separation allows the organic solvent extract to be transferred to a clean plate located in stage position 1. These extracts are then evaporated to dryness and reconstituted without having ever left the 96-well format. Such 96- or 384-well liquid handling workstations are good examples of parallel automation.

2.3 Hybrid Automation Systems

Hybrid automation systems for chemical analysis combine features of both serial and parallel systems. Some processes are conducted in serial, some are conducted in parallel, and all are integrated into a single system. An example of a hybrid system, given in Fig. 5, shows a central robotic arm (XP) which services various arm peripherals, and a 144-port vacuum manifold, used for solid-phase extraction method development [13]. The system was designed to develop automated methods that could then be transferred to an automated system of similar architecture. The system and associated processes are, for the most part, serial, except for the conditioning and development of the solid-phase extraction columns, which are run in parallel. Although a fully parallel system would have been desirable, it

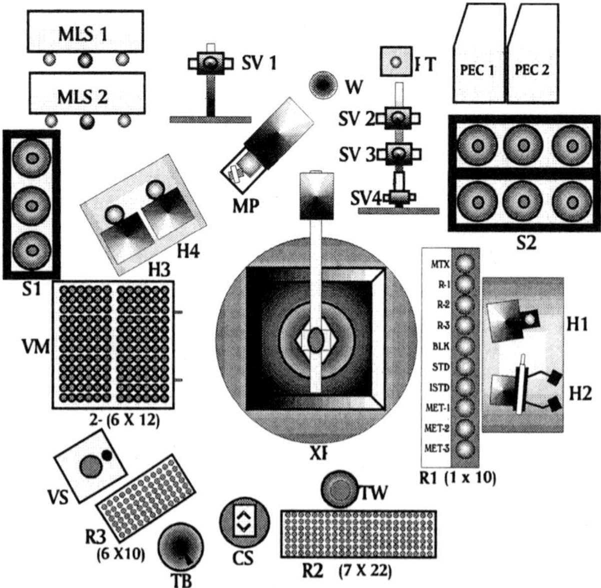

Figure 5 Schematic representation of a robotic solid-phase extraction method development system, including MLS1 or MLS2 = master lab station, SV = switching valve,
W = waste, R1, R2, or R3 = rack. No. 1, 2, or 3 S1 or S2 = solvent reservoir H1, H2,
H3, H4 = hand No. 1, 2, 3, or 4 CS = capper station TW = tip waste, VM = vacuum
manifold, XP = zymate XP robotic arm, PEC1 or PEC2 = power and event controller
1 or 2, PT = pressure transducer, VS = vortex station, TB = tip blower, MP = metering
pump.

was beyond the capabilities of commercially available systems at that time. Other
commercially available hybrid systems for solid-phase extraction represent tran-
sitions between earlier serial systems and more recently developed parallel systems.

3 FEATURES AND ADVANTAGES OF PARALLEL
   AUTOMATION SYSTEMS

Table 1 is a listing of those features and advantages of parallel automation systems for chemical analysis that directly impact their usefulness and value. These
features are discussed below.

   Speed. The biggest single advantage of parallel automation systems is speed.

Table 1 Features and Advantages of Parallel Automation Systems for Chemical Analysis

Feature	Advantage (+) Disadvantage (−)	Rationale
Speed	+++	Parallel systems are intrinsically much faster that serial systems.
Complexity	−	Parallel systems can be more complex, requiring a separate channel for each experiment. Integrated hardware can reduce complexity.
Hardware	+	Multiple channels are required. Integrated hardware is possible.
Software	+/−	Computer control is required for either serial or parallel.
Expandability	+	Parallel systems can be expanded from 2 to 4, 8, 16, 32, 64, 96, 384, or 1536 channels. Odd numbers are also possible.
Unattended operation	+/−	Parallel systems have greater difficulty achieving unattended operation and are more likely to require operator intervention. Once unattended operation is achieved, they can show high reliability.
Error recovery	++	When present in liquid handling systems, error detection is based on conductivity measurements. Error recovery routines allow processes to continue.
Logistics	+/−	All samples or experiments must be available simultaneously.
Cost	+/−	Full-blown parallel automation systems can be more costly to build. Because they are fast, there is usually a payback.

Because many experiments/samples are conducted or processed simultaneously, parallel systems have the potential to reduce experiment time by a factor of N, for at least a part of an analytical determination. Here, N is the number of channels implemented in the parallel system, so for a 96-well workstation, the processing time can be shortened by ~96-fold relative to the serial experiment. This time saving assumes that there is no major increase or decrease in *service time* for each channel, a characteristic of both hardware and software implementation. Sometimes serial systems can make up a portion of this time disadvantage, especially if there is a downstream bottleneck. For example, if samples are being processed automatically before HPLC injection, in a serial system one sample can be chromatographically or electrophoretically separated while the next one is extracted. For other systems, if 96 samples are being separated in parallel, then a $96\times$ increase in speed is sustained. An example of this approach is given later in this chapter.

Complexity. Parallel automation systems are generally more complex because they require a separate channel for each experiment that is run in parallel. The complexity of such a system would increase by a factor of N (the number of channels) to the point where certain hardware features are integrated into a single dedicated component. When integrated multichannel hardware components are created and added into a parallel automation system, complexity decreases. As an example, for some liquid handling workstations (Chap. 2), this could be a 96-channel head that is used for simultaneous transfer. The 96-well head replaces the 4, 8, or 12 automated syringes of other comparable, but less highly parallel, workstations.

Hardware. As described above, it is desirable that specialized hardware be developed and integrated into the system. Under the best and most developed circumstances, this hardware will help achieve greater simplicity in a parallel architecture.

Redundancy of hardware is typical in a parallel system. Redundancy of hardware means that more that one device performing an identical function is present in a system. In a 4-syringe workstation, each of the syringes performs the same function. In a 96-channel workstation, each of the 96-channels should be identical, except for spatial positioning of the tip in the format. This redundancy of hardware is required for the system to handle multiple experiment channels simultaneously.

Software. Control software for a parallel system is specialized but may be no more complicated than the corresponding serial system. While control software for a serial system is focused on manipulating a single sample, container, or syringe, control software for a parallel system is focused on the servicing of an array (possibly an entire rack) of containers or samples.

Expandability. Although parallel systems generally contain dedicated hardware and software to support simultaneous operation, and expansion is generally not an option, in principle a parallel system can be expanded from 2 to 4, 8, or 16 channels by adding more hardware. Expansion as a 2^n series seems to be favored, in order to accommodate the exponential growth in the number of samples, such as in large combinatorial libraries.

Unattended operation. If a serial system were responsible for a large number of samples or experiments, a highly desirable capability would be unattended overnight operation. This is really a necessary feature because serial systems would generally require an overnight timeframe to conduct the necessary operations, depending on the length of the experiments. A drawback for serial systems, then, would be the reliability needed to operate flawlessly overnight, or a reliable error recovery routine. Because of their relative speed, parallel systems would be required to operate reliably for shorter periods of time. Error recovery, while desirable, is less necessary because of the shorter duty cycle requirements. Also, because of the increased speed, semiautomated operation with a higher number of operator interventions is manageable.

Error recovery. Error recovery is required for fully automated parallel systems, so that after error occurrence, the system can continue with the automated process. Error recovery is based on an error detection test. An example of an error routine is as follows: In the transfer of nonhomogenous fluids, a liquid transfer is subject to clogging. This clogging and subsequent lack of fluid in the transfer tip can be detected by a resistance measurement. If electrical resistance occurs at a location in the tip where liquid should be present, then the control software assumes that the tip has clogged. If this error is detected, then the control software will purge the tip and retry the transfer. After a predetermined number of unsuccessful transfer attempts (usually two) on this channel, the control software will instruct the hardware to skip the sample, thereby allowing for error recovery.

Logistics. While serial automation can initiate one experiment or sample in sequence, parallel automation requires the simultaneous availability of all experiments at the same time. This requirement can place certain restrictions on the experiment, depending on what experiment is being automated. If sequential sampling is to be adapted to automation, serial processing is more appropriate. With this strategy, it is often possible to obtain real-time processing and readout of the samples. If, however, an entire set of experiments or samples can be made available simultaneously, then parallel automation would be the architecture of choice. With this approach, sample stability issues can be minimized.

Cost. Cost can be categorized into system cost and operational cost. In principle, system cost will generally be higher for parallel automation because hardware is needed for each separate channel. Hardware cost is rarely N-fold higher than for serial automation because specialized hardware is often developed. In practice, because the extent of automation is often greater for serial systems, these systems are typically more expensive. A driving force for parallel systems is that their cost can be recouped during the operation phase, because greater time saving is usually possible.

4 WHEN DOES PARALLEL AUTOMATION MAKE SENSE?

It is apparent that automation is closely linked to the sample load and, to some extent, the sample generation process. It is most appropriate to choose parallel-automation architecture in an analytical situation where a number of samples are available simultaneously or when a set of experiments must be sampled or processed simultaneously. That is to say, if *off-line batch processing* were necessary, parallel automation would be appropriate. In contrast, if time-dependent sampling of a single experiment or a few samples were in order, such as in-process control sampling, then serial automation would be appropriate. Also, if one-time-use components or difficult-to-create components, such as affinity chromatography columns [13] or immobilized automated membrane (IAM) columns [14], must be employed, then it could be prudent to select a serial on-line automation system. If, however, *postprocess characterization* of a batch of samples is required, then parallel automation processing is most appropriate.

5 POST-COMBINATORIAL-SYNTHESIS COMPOUND CHARACTERIZATION

One important example of postprocess automation is the chemical analysis associated with the characterization of large combinatorial libraries. This type of work assumes that a large number of combinatorial chemistry library samples are available in a readily accessible format, such as 96-well plates or larger. Samples representing library components are constantly being synthesized and require simple, rapid characterization such as molecular weight determination, purity assessment, or molecular-weight-guided fractionation [3]. The high sample numbers that combinatorial synthesis provides makes automated high-throughput approaches highly desirable.

One recent automation approach has been to use liquid chromatography–mass spectrometry (LC/MS) in a parallel way (Fig. 6). In this arrangement, a high-throughput autosampler (AS) services four to eight HPLC systems, and a mass spectrometer is used as the detector. The mass spectrometer is the most expensive component in the system, but a single mass spectrometer can service all the HPLC channels. Only one HPLC channel is routed to the mass spectrometer at a given time through the use of a flow-stream multiplexer, which allows only one channel to reach the mass spectrometer at a time by rapidly switching (50

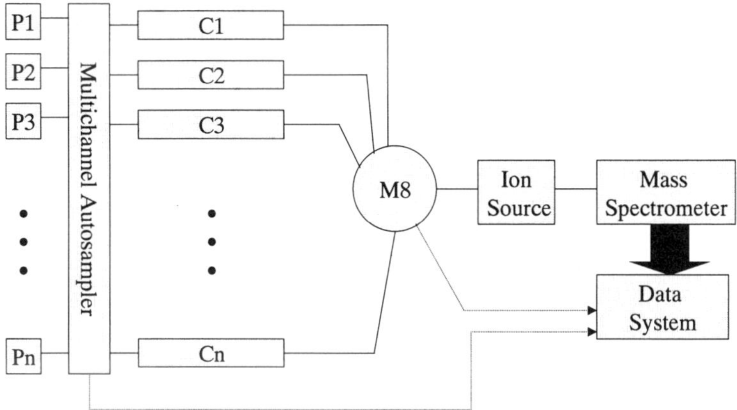

Figure 6 Parallel chromatographic systems used to increase the throughput of liquid chromatography–mass spectrometry (LC/MS). In this arrangement, a high-throughput autosampler (AS) services up to n HPLC systems. As the mass spectrometer is the most expensive component in the system, a flow-stream switching device is used to multiplex all n chromatographic effluents into a single channel.

ms) between channels. Once set, the flow from a selected channel will be selected for 0.5 to 2.0 s before switching to the next channel. In this way, molecular weight information can be collected on up to eight library components simultaneously. The minimum amount of chromatography provided with each channel allows some degree of separation between individual sample components, such as degradation products and unreacted starting materials or other reagents. Although any single chromatographic channel is sparsely sampled, each channel receives adequate sampling for qualitative purposes and the overall throughput is increased relative to a single-channel system.

An additional piece of automation that can be added to this system is molecular weight- or identity-driven fractionation. With this approach, if the identity of a predicted product is confirmed by molecular weight, a fraction collector is automatically triggered, allowing collection of the sample component of interest. The unwanted related substances are routed to waste. Several variations on this theme have been reported, with the gradual tendency to move to higher degrees of parallelism and a greater extent of automation [15–17].

A second example of parallel automation involves the use of an automated matrix-assisted laser desorption ionization (MALDI) sample preparation system for molecular weight characterization of combinatorial libraries of peptides. This characterization approach is normally conducted in a serial automation mode that uses consecutive sample introduction. Samples are applied onto a multiposition plate that is placed in the vacuum region of the MALDI instrument through a vacuum lock. As the sample plate is indexed, the ionization laser will perform a soft ionize on each spot in turn. This technique provides an important alternative for high-throughput compound characterization for compounds that are not easily characterized by atmospheric pressure chemical ionization techniques such as electrospray or atmospheric pressure chemical ionization.

Parallel automation has been used to prepare the samples associated with these MALDI plates [18]. This is accomplished through the application of several individual layers to the plate. First, the plate is conditioned by spraying a thin seed layer of the matrix (methanolic ferulic acid solution). A specially designed sprayer called an oscillating capillary nebulizer (OCN) is used for this purpose. After intervention, an operator transfers the target to an automated workstation (Symbiot I), and then calibrant and samples are spotted (1 to 10 µL) at discrete positions on the target. Although the workstation has a single-tip, it could be upgraded to a parallel-sample handling workstation without a change in the concept. In the last sample preparation step, a matrix over-layer is applied by spraying the target plate again. Only the inner 64 positions of the MALDI target (originally a 10×10 grid) are utilized to prevent fringe electric field effects from interfering with the accurate mass determinations of the test compounds. Using this approach, molecular weights of approximately 1000 Da can be determined accurately, to the third decimal place, as the average deviation in molecular weight is 3.5 ppm.

6   POSTSYNTHESIS IN VITRO DRUG TESTING

Drug metabolite isolation. On-the-fly MS and MS/MS instruments have been coupled to a 96-well fraction collection system to enable on-line characterization and isolation of drug metabolites from biological samples, based on molecular and production information [19]. The instrument system for this work is illustrated in Fig. 7. After an initial assessment of the chromatogram by UV absorbance, two parallel processes are implemented with mass-spectral acquisition and fraction collection for subsequent NMR spectral characterization. A splitter divides the overall chromatographic flow (4 mL/min) between the two processes. Only a minor portion (20 µL/min) of the flow is routed to the mass spectrometer, where it is combined with a makeup flow (100 µL/min) prior to electrospray ion-source introduction. The makeup flow consists of water/acetonitrile/formic acid (50:50:0.1) and is used to promote ionization and decrease ion suppression resulting from the sample matrix. The bulk (95%) of the flow is sent to the fraction collector, so that adequate sample is available for proton NMR.

The mass spectrometer contained an ion-trap mass analyzer that can provide data rich in qualitative information. Fractions are collected at 15 s intervals,

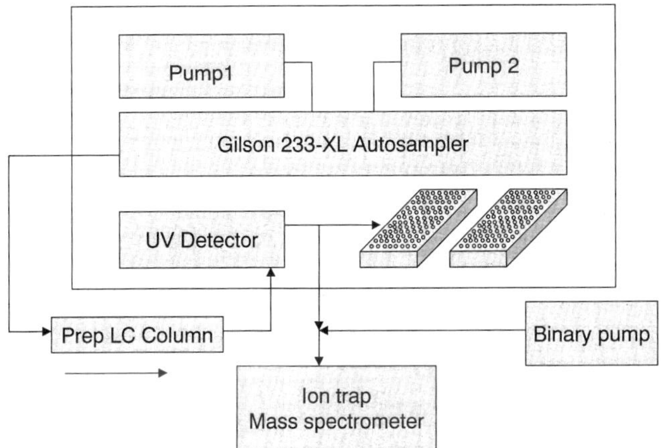

Figure 7 Schematic diagram of a preparative LC/MS/MS linked fraction collection system.

corresponding to specific retention windows on either UV or total-ion chromatograms. Although not completely described, the paper suggests only passive fraction and mass spectral data collection instead of chromatographic peak-directed collection. Although this system has at least one parallel operation, samples are processed serially, making this a hybrid automation system.

Automated dissolution testing. A second area for application of parallel automation is in vitro drug-product dissolution testing. In this type of testing, it is assumed that the rate and pH at which tablets or capsules dissolve in the alimentary canal can affect the rate and extent of drug absorption [20] and ultimately the overall exposure of the drug. Dissolution testing is thought to model the in vivo dissolution of a drug dosage form, hence the release of the drug. It is a widely used pharmaceutical testing method, and the so-called dissolution apparatus has been automated to various extents [21].

Due to the design of running six or seven experiments at the same time, dissolution testing lends itself readily to parallel automation. An illustration of one parallel automation dissolution experiment is shown in Fig. 8 [22]. This system employs a central Zymark XP robot coupled to a diode-array spectrophotometer, with fiber-optic probes of various path lengths. Although several ancillary tasks associated with the dissolution experiment (cleaning and loading the vessels with fresh dissolution media or adding samples to the vessels) are handled by the robotic arm, one of the most important operations is the movement and positioning of the fiber-optic probe into each of the vessels. Through the use of the appropriate fiber-optic probe, either absorbance or fluorescence measurements

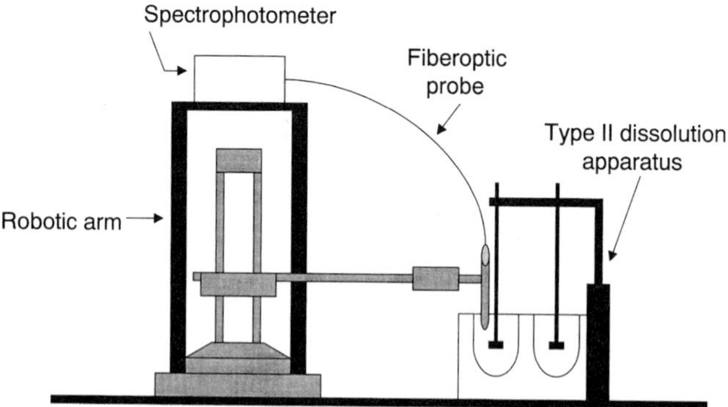

Figure 8 Schematic view of robotic system designed to perform automated dissolution with fiber-optic probe.

can be taken. As only one probe is used for six vessels and the probe is moved to the various vessels, discrete time points are taken instead of continuous measurements.

Other fiber-optic systems using an individual probe for each vessel in the dissolution apparatus have been described, and these provide a true parallel sampling experiment as well as the opportunity to obtain continuous drug release-versus-time profiles for each of the tablets [23]. An advantage of optical measurement is that it eliminates the need for liquid handling and sample collection. But, since fiber-optic probes cannot handle all possible detection needs, the use of liquid sampling is still commonplace. This is usually handled in a parallel automation way by the use of a six-channel parallel ''sipper'' [24] or liquid sampler. Samples obtained from the sipper can be routed to a spectrophotometer with a flow cell or placed into sample collection vessels, such as a 96-well plate, for later quantification.

7 POSTSYNTHESIS IN VIVO DRUG TESTING

After the synthesis and scale up of one or more pharmaceutical compounds, in vivo testing is often conducted to determine their LADMER properties and relative potential as drugs. These LADMER, or liberation, absorption, distribution, metabolism, excretion, and response properties will determine relative and absolute exposure to the drug and clearance rates/pathways, and will greatly influence the overall effectiveness of a drug candidate. For example, if a drug candidate cannot be liberated from the dosage form, absorbed through the route of administration, and distributed so that it reaches the site of action, it is unlikely to be effective.

Many LADMER evaluations can be modeled in vitro. As described above, metabolism can be ascertained through the study of human liver microsomes or hepatocytes, and liberation can be studied through the dissolution-testing model. In vitro modeling is still not sufficiently predictive in enough cases, however, to be able to ascertain with high certainty whether a drug candidate will be successful or not. To gain a closer prediction of this it is still necessary to develop and use animal models, and this requires quantitative determinations of drug substances in biological fluids such as blood plasma, serum, and urine [25].

To quantify drug candidates effectively in biological samples such as plasma, it is necessary to disrupt any drug–protein binding and isolate the drug(s) from as many matrix components as possible. This is done by precipitating the proteins, by solid-phase or liquid–liquid extraction, or through any of several other isolation techniques [11]. Since the advent of LC/MS/MS as a dominant bioanalytical technique [26], biological sample preparation has become one of the rate-limiting steps of postsynthesis in vivo testing. To remove this bottleneck, a number of automation approaches have been invented to improve the throughput and save analyst time. Initially many of these approaches were serial in nature, and a number of serial approaches survive today. It is becoming increasingly apparent, however, that parallel automation processes are more efficient and have the potential to achieve much higher throughput without adversely affecting the quality of the results [27].

One effective approach for parallel automation involves a holistic approach to sample collection, processing, and analysis [28,29]. This process, a representation of which is shown in Fig. 9, requires that samples be collected and placed in a parallel automation accessible format such as a 96-well tube plate [28]. Traditionally in drug LADMER studies, biological fluids are collected in individual capped tubes or bottles. Each tube is labeled individually with the collection time and animal number prior to dosing. The volumes of samples often vary between the individual tubes. As for the generation of samples in the 96-well plate format, which allows them to be manipulated by a parallel workstation, the individual samples can be (1) transferred manually, (2) transferred by a liquid handling workstation, or (3) initially generated and delivered directly to the 96-well format. The last of these approaches seems to have the least amount of transfer overhead and is beginning to be adopted in the pharmaceutical industry. Plastic tubes can be used to contain individual samples within the 96-well format. From this common format, parallel sample processing workstations, such as the Tomtec Quadra 96, can be used to transfer samples and liquids to and from the deep-well tubes. Samples can be moved from old to new tubes as necessary, but once placed into the 96-well format, they never leave it.

An important feature of this integrated sample handling process is that it

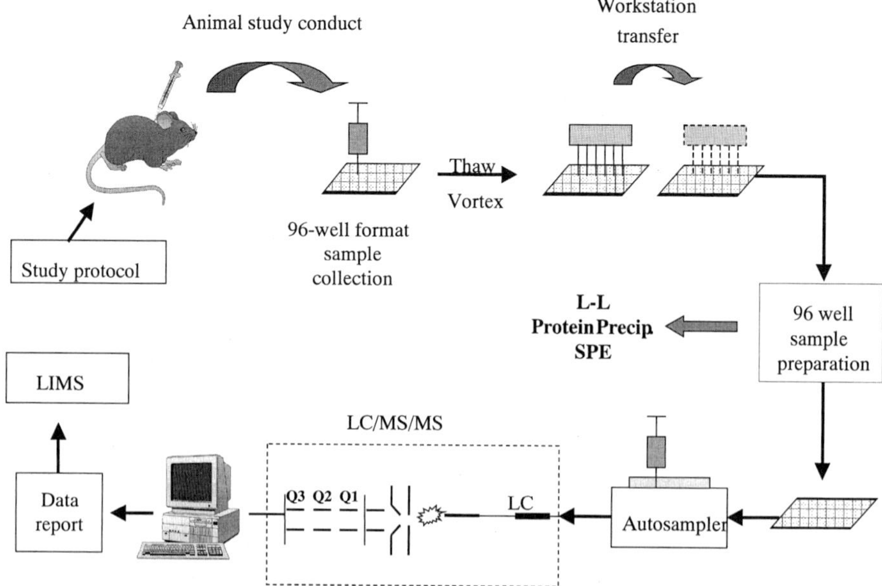

Figure 9 Conceptualized diagram of integrated sample handling process for discovery bioanalysis.

ties directly into the information management process. A study protocol and a sample collection paradigm are created within the framework of an associated laboratory information management system (LIMS). This LIMS allows information created by study participants from different groups to be shared electronically, thereby eliminating the need for redundancy in data entry for all studies. Entries are made only once, and individual sample labeling is replaced by bar coding of an entire 96-well rack of samples. An important component of the LIMS functionality is the generation and universal availability of a plate map that identifies the contents of each tube in the deep-well plate; sample identification is thereby done electronically. Drug concentration data is collected by a data acquisition system and electronically conveyed to the LIMS. Once in the LIMS, the data can be reprocessed, graphically displayed, and reported in any of several different ways, such as a spreadsheet table, a graph, or a hard copy report.

From a laboratory automation standpoint, this holistic view of study conduct allows a parallel sample-preparation approach to be selected and readily inserted into the study flow. An example of this is the use of 96-well liquid–liquid extraction [29].

Liquid extraction has not been easy to automate because it was difficult to conceive and implement the liquid-phase separation. Several unlikely approaches have been proposed over the past decade, including passing the biphasic liquid through a porous membrane of silicon-oil impregnated-cellulose or absorbing the water layer with a packed column of diatomaceous earth. More recently, it has been shown that, by using a 96-parallel channel liquid handling workstation and positioning the channel tip near to the phase meniscus (Fig. 10), the phase separation can be achieved with minimal loss of the organic (top) layer. Recoveries of 45 to 60% can be routinely obtained for several drug classes with a solvent such as ethyl acetate. Other solvents such as methyl tertbutyl ether, chloroform, and methyl tertbutyl ether $+5\%$ ethanol are also practical and can help with extractions for drugs of differing polarities.

The overall throughput for this parallel automation approach allows 96 samples to be extracted and prepared for LC/MS/MS in less than 90 minutes. This is a large throughput advantage over either manual (360 minutes) or serial sample preparation (400 to 500 minutes) for a similar size sample set. Two other disadvantages of the on-line serial process are that an expensive LC/MS/MS instrument is tied up for a longer time while sample preparation is being conducted, and that sample degradation can occur in samples at the end of the queue unless steps are taken to stabilize them. Although, with certain on-line serial processing systems (such as the Prospekt solid-phase extraction system), LC/MS/MS can be running on sample A while sample B is being processed, the overall per-sample-time is about twice as long with this system than it would be for off-line parallel processing. Method development time is proportional as well. Other parallel processing systems for automated solid-phase extraction [30,31]

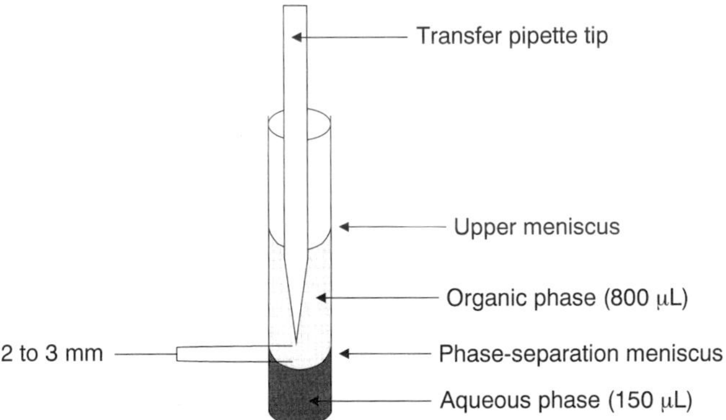

Figure 10 The positioning of a liquid handling workstation tip into a tube containing a biphasic liquid. This approach can be used for rapid phase separation in liquid–liquid extraction experiments, using a 96-well tube format.

and automated solid-phase extraction method development [32] have also been described. A dedicated workstation to do parallel automated solid-phase extraction (Cyclone) is available from Zymark (Hopkington, MA, USA).

8   AUTOMATION AND PARALLEL SEPARATIONS

Over the past decade, many advances in analytical separations have taken place. Although most of these advances have been driven by the need for separating and characterizing mixtures of biomolecules such as proteins and oligonucleotides, the approaches taken have started to spill over into separations for nonbiological mixtures as well. One such area has been the development of instrumentation for multicapillary electrophoresis (MCE) [33]. This approach uses capillary electrophoretic separations in a novel, parallel way to increase rapidly the sample throughput relative to serial systems. The approach, while requiring fairly complex instrumentation, has proven especially useful when relatively high-resolution separations are needed and sufficient throughput cannot be gained by simply shortening the run time.

A schematic representation for such a system is shown in Fig. 11. In this instrument, a stacker (or sample queue) containing plates of samples is located

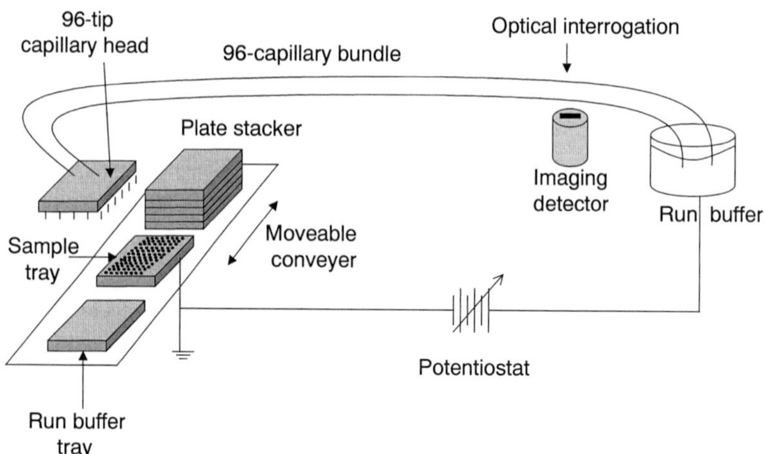

Figure 11   Schematic diagram of an automated multicapillary electrophoresis (MCE) instrument, showing a 96-capillary bundle attached to a 96-capillary sampling head. A movable conveyer allows the capillary head to move between the sampling position and the run (buffer) position. Fresh sample plates can be moved from the plate stacker to the sampling position as needed.

on a movable conveyer (position A). One of these plates (the first in the queue) has been moved to the in-process position (position B) where it is sampled by an array of 96 capillaries, the tips of which are held in position in the standard 8 by 12 format. After a brief (2 to 30 s) electrokinetic injection (by applying a potential difference of up to 400 V/cm of capillary length) of the samples from the sample tray, the capillary tips are moved to a position in a container of separation buffer, at a third position on the conveyer (position C). This separation buffer provides support electrolyte to each capillary before and after sampling to allow completion of the separations.

The samples are held at ground electrical potential during the sampling process, and the sampling head, containing the 96 capillary tips, is positioned so that only one capillary will sample from each of the 96 wells. As is typical of all capillary electrophoretic separations, sample components enter the separation capillaries and are separated based on their relative electrophoretic mobility toward an electrode of opposite polarity. The polarity of the opposite electrode may be switched before the separation to positive or negative, depending on need. Neutral compounds and components that are the same polarity as the counterelectrode may be swept into the capillary and separated because of the electroosmotic flow of ions [34].

A complex and technically interesting aspect of this separation approach is the detection strategy. Although the details for the various approaches used here are beyond the scope of this chapter on parallel automation, two predominant techniques have arisen. The capillary array consists of 96 capillaries of approximately 60 to 100 cm in length, with detection windows set approximately 2/3 that distance from the injection end. Optical windows are made at the detection site by removal of the outer capillary protective coating. This coating is usually polyimide and can be removed by application of a flame, followed by methanol rinsing, to create a window in each capillary, where some means of optical detection can be applied.

All detection schemes reported for MCE so far have been optically based and therefore consist of an illumination source, a light collection system, a spectral discrimination device, and a detector. Both fluorescence and absorbance detection approaches have been used with some success, and each has advantages. The fluorescence system is more sensitive but requires precolumn derivatization of analytes to ensure that a fluorophore is present in the molecule. Such treatment adds an extra step and decreases selectivity, especially for separations involving small molecules, by making all analytes more similar in structure. Absorption detection is simpler to implement but provides less sensitivity and is limited in this regard. Although yet to be demonstrated, a third approach to detection in the MCE experiment is the possibility of multiplexing the output of the capillaries to one or more mass spectrometers. Time-of-flight mass spectrometry would be an effective system because of its high duty cycle and capabilities for rapid spectral scanning. As with the other detection approaches, the complexity of such a MCE/MS system would be high.

The MCE system, while providing two orders of magnitude higher throughput, is proportionately more complex, especially at the detection end. The potential for between-channel interferences, whether arising from chemical crosstalk between capillaries or from optical crosstalk generated by stray light, is a problem that limits the overall performance. The automation requirements for such a system are, however, relatively simple and effective. The approach illustrates how the use of specific and dedicated devices, such as a 96-capillary head and capillary bundles, allows the implementation of parallel automation into a 96- or 384-well format and greatly increases throughput of an analytical instrument system.

9 FUTURE PROSPECTS IN PARALLEL AUTOMATION

There are a number of opportunities for development of parallel automation processes in chemical analysis. As progress is made, it is apparent that parallel automation processes will become more common, especially in those areas that require high or ultrahigh throughput. A few suggested areas are described below.

Conversion of serial to parallel automation. As suggested earlier in this chapter, serial automation processes are useful where sample load is moderate to light (such as in-process testing for large-scale processes), where a few unique experiments are being conducted or where expensive or hard-to-obtain pieces of instrumentation would constrain the deployment of parallel operation. Examples of this later area could be NMR spectroscopy or one-of-a-kind immunoaffinity chromatography systems.

Situations where parallel automation is called for include those with a high sample load and a redundancy of repetitive operations. At the current stage of automation development, microscale (micrograms or microliters) operations would appear to be better candidates than larger scale operations. In situations that call for parallel automation, such as biological sample preparation, it is likely that most serial sample preparation, serial experimentation combined with chemical analysis or serial sampling from multiple studies, will be supplanted by parallel processing within a few years. The cost of not doing so is too great for industrial companies.

In-process control of parallel synthesis. For real-time in-process control of a single industrial batch process, such as the content of a single component in a large batch process, on-line serial sampling seems logical. If near real-time testing of many small batches (parallel combinatorial synthesis) is ever to become practical, however, it is likely that parallel sampling and chemical analysis will be conducted. One simple way to look for the formation of products in a parallel combinatorial synthesis would be to obtain a small sample in parallel from a number of combinatorial wells. These samples could be pooled and injected into an LC/MS/MS or CE/MS/MS system, using the mass spectrometer to deconvolute the results by differing mass to charge ratios. In this way, relative responses of the various products could be monitored, thereby defining the time courses of many reactions in near-real time. Alternatively, after samples are harvested from many wells, they could be injected into an instrument sequentially. The former strategy seems more efficient, assuming that the products are of sufficiently different masses to allow deconvolution on a mass spectrometer.

Economy of scale—more channels in parallel. A trend that will become apparent over time will be to have a greater degree of parallelism in automation. Already this trend is becoming apparent as systems go from 4 to 8 to 12 to 96 to 384 channels in parallel for many automation applications. This economy of scale will allow more samples to be processed or more experiments to be run per unit of time or unit cost. With the increase in samples there will be greater challenges in system complexity and greater need for specialized apparatus, such as 96- or 384-channel heads for liquid transfer. The ultimate limit for the number of channels to be run in parallel will be the sample demands of the experiment. For example, a large amount of LADMER information such as absolute bioavailability can be obtained with a relatively small sample number, so that 50 or 60 parallel channels would be adequate for many investigations, and 96 channel processing still has a built in comfort margin.

Miniaturized parallel experiments. The great frontier for parallel automation, and many aspects of analytical chemistry, is the challenge of miniaturization, or the so-called "lab on a chip" concept [35]. Using this concept, many visionary analytical scientists have asked how could it be possible to run 50, 100, or 250 thousand experiments in parallel. The answer could lie in microfabricated devices that would be analogous to a laboratory beaker in the same way that a very-large-scale integrated (VLSI) circuit is analogous to a single resistor or transistor. Figure 12 illustrates this concept for a capillary electrophoresis separation system. In this figure, all the components required to perform a complete capillary separation, including sample, run buffer, waste vesicles and separation capillary, are diminished in size and etched onto a silica surface through standard photolithography techniques. The silica plate is covered with a polymeric material, and the contacts necessary to perform electrically activated switching are integrated into the design as well. Due to the miniature size, a number of discrete capillary electrophoresis experiments could be placed onto a single one-inch chip.

Much of the technology needed to perform this type of microfluidic manipulation is commercially available [36] at a density of nearly 1000 valves per square inch. If four valves are required to make an operational capillary electropherograph, this density could allow for nearly 250 separate electrophoretic separa-

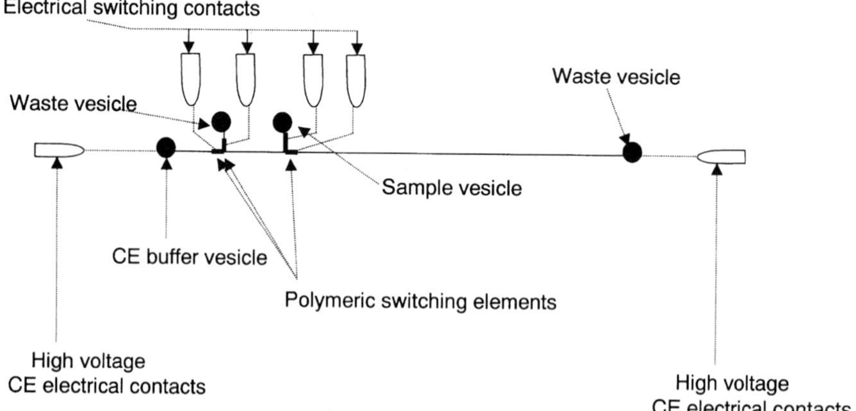

Figure 12 Pictorial diagram of a miniaturized capillary electrophoresis unit, as etched on a silica surface. The actual size of this unit would be approximately 1 inch in length. Detection for this separation, an ongoing challenge, would be conducted by MALDI or by optical means.

tion experiments per square inch. A major challenge associated with this technology continues to be designing detection systems sensitive enough so that adequate detection limits can be achieved. Once this type of technology becomes widely used for chemical analysis, it can readily be combined with automated processes for sample introduction, preparation, separation, and data reporting. A major challenge at this juncture will become the management and reduction of the associated data and resulting information.

Applications to other areas in the chemical industries. Parallel automation applications in pharmaceutical chemistry have led the way in recent years, possibly due to the high level of competition and the high stakes in this arena. Also, combinatorial synthesis, an invention of the pharmaceutical industry, has required that support functions such as analytical chemistry keep pace. As many of the recent automation applications in combinatorial synthesis spill over into other chemical industries, parallel automation in chemical analysis will follow.

10 CONCLUSION

Due to the ongoing demand for speed in industrial chemical analysis, parallel automation processes are beginning to replace serial automation processes, especially in those areas where high numbers of assays or samples are expected. For in situ or on-line situations where sparse strategic sampling is the norm, serial automation will likely play a continued role. Parallel automation for chemical analysis has recently been implemented on 96-well parallel-processing liquid handling workstations. Expensive robotic arms, while powerful and flexible, do not directly lend themselves to parallel automation systems unless the effort is made to combine them with specialized multichannel hardware. Such hybrid systems are powerful and useful but are extremely complicated to implement and maintain. In the future, it seems likely that automated systems will be miniaturized and that this miniaturized format will lend itself readily to increased use of automation.

REFERENCES

1. RE Dole. Comprehensive survey of chemical libraries yielding enzyme inhibitors, receptor agonists and antagonists, and other biologically active agents: 1992 through 1997. Mol. Divers. 1998; 3(4):199–233.
2. WL Fitch. Analytical methods for quality control of combinatorial libraries. Mol. Divers. 1998–1999; 4(1):39–45.
3. DT Rossi, MW Sinz. Mass Spectrometry in Drug Discovery. Marcel Dekker, New York, 2001, Chap. 7.
4. CM Sun. Recent advances in liquid-phase combinatorial chemistry. Comb. Chem. High Throughput Screen. 1999; 2(6):299–318.
5. HP Nestler, R Liu. Combinatorial libraries: studies in molecular recognition. Comb. Chem. High-Throughput Screen. 1998; 1(3):113–126.
6. RA Felder, JC Boyd, K Margrey, W Holman, J Savory. Robotics in the medical laboratory. Clin-Chem. 1990; 36(9):1534–1543.
7. JC Berridge. Advances in automation of pharmaceutical analysis. J-Pharm-Biomed-Anal. 1989; 7(12):1313–1321.
8. GA Smith, TL Lloyd. Automated Solid-Phase Extraction and Sample Preparation-Finding the Right Solution for your Laboratory. LC-GC, Current Trends and Developments in Sample Preparation, May 1998, S22–31.
9. H Fouda, RP Schneider. Robotics for the bioanalytical laboratory: a flexible system for the analysis of drugs in biological fluids. T.R.A.C. 1987; 6:139–147.
10. DT Rossi, MW Sinz. Mass Spectrometry in Drug Discovery. Marcel Dekker, New York, 2001, Chap. 12.
11. DT Rossi, MW Sinz. Mass Spectrometry in Drug Discovery. Marcel Dekker, New York, 2001, Chap. 6.
12. TD Parker, DS Wright, DT Rossi. Design and evaluation of an automated solid-phase extraction method development system for use with biological fluids. Anal. Chem. 1996; 68:2437–2441.
13. K Zheng, X Liang, DT Rossi, GD Nordblom, CM Barksdale, D Lubman. On-line analysis of affinity bound analytes by capillary LC-MS. Rapid Comm. Mass Spectrom. 2000; 14:261–269.

14. OH Chan, DT Rossi, KL Hoffman, LA Stilgenbauer, BH Stewart. High-throughput screening method for determining membrane partitioning potential using immobilized artificial membrane chromatography. American Association of Pharmaceutical Scientists National Meeting, San Francisco, October 1998.

15. T Wang, J Cohen, DB Kassel, L Zeng. A multiple electrospray interface for parallel mass spectrometric analysis of compound libraries. Comb. Chem. High Throughput Screen. 1999; 2(6):327–374.

16. L Zeng, L Burton, K Yung, B Shushan, DB Kassel. Automated analytical/preparative HPLC-MS system for the rapid characterization and purification of compound libraries. J. Chromatogr. 1998; 794:3–13.

17. L Zeng, DB Kassel. Development of a fully automated parallel HPLC/MS system for the analytical characterization and preparation of combinatorial libraries. Anal. Chem. 1998; 70:4380–4388.

18. DA Lake, MV Johnson, CN McEwen, BS Larson. Sample preparation for high throughput accurate mass analysis by matrix-assisted laser desorption/ionization time-of-flight mass spectrometry. Rapid Comm. Mass Spectrom. 2000; 14:1008–1013.

19. GJ Dear, RS Plumb, BC Sweatman, IM Ismail, J Ayerton. Tandem mass spectrometry linked fractionation collection from the isolation of drug metabolites from biological matrices. Rapid Comm. Mass Spectrom. 1999; 13:886–894.

20. PG Welling. Pharmacokinetics: Processes and Mathematics. American Chemical Society, Washington, DC, 1986.

21. E Lamparter, CH Lunkenheimer. The automation of dissolution testing of solid oral dosage forms. J. Pharm. Biomed. Anal. 1992; 10(10–12):727–733.

22. P Rogers, PA Hailey, GA Johnson, VA Dight, C Read, A Shingler, P Savage, T Roche, J Mondry. A comprehensive and flexible approach to the automated-dissolution testing of pharmaceutical drug products incorporating direct UV-Vis fiber-optic analysis, on-line fluorescence analysis and off-line storage options. Lab. Auto. Robotics 2001; 12:12–22.

23. I Nir, BD Johnson, J Johansson, C Schatz. Application of fiber-optic dissolution testing for actual products. Pharm. Tech. 2001; 2–6.

24. A Papas, MY Alpert, SM Marchese, JW Fitzgerald, MF Delany. Anal. Chem. 1985; 1408–1411.

25. M Jemal. High-throughput quantitative bioanalysis by LC/MS/MS. Biomed. Chromatogr. 2000; 14:422–429.

26. DT Rossi, MW Sinz. Mass Spectrometry in Drug Discovery. Marcel Dekker, New York, 2001, Chap. 2.

27. DT Rossi, N Zhang. Automated solid-phase extraction: current aspects and future prospects. J. Chromatogr. 2000; 885:97–113.

28. N Zhang, K Rogers, K Gajda, JR Kagel, DT Rossi. Integrating sample collection and handling for drug discovery bioanalysis. J. Pharm. Biomed. Analysis 2000; 23:551–560.

29. N Zhang, KL Hoffman, W Li, DT Rossi. Semi-automated 96 well liquid-liquid extraction for quantitation of drugs in biological fluids. J. Pharm. Biomed. Analysis 2000; 22:131–138.

30. KL Hoffman, LD Andress, TD Parker, RJ Guttendorf, DT Rossi. Automated deter-

mination of a novel anti-inflammatory drug in plasma using batch robotic sample preparation and HPLC. Lab. Robotics Auto. 1996; 8:237–242.

31. DT Rossi, KL Hoffman, NJ Dolphin, H Bockbrader and TD Parker. Tandem-in-time mass spectrometry as a quantitative bioanalytical tool. Anal. Chem. 1997; 69: 4519–4523.

32. DT Rossi. Automating solid-phase extraction method development for biological fluids. LC-GC, 17, 4s (1999) s4–s8.

33. S Behr, M Matzig, A Levin, H Eickhoff, C Heller. A fully automated multicapillary electrophoresis device for DNA analysis. Electrophoresis 1999; 20:1492–1507.

34. P Camilleri. In Capillary Electrophoresis: Theory and Practice (Camilleri, P., ed.) CRC Press: Boca Raton, FL, 1993.

35. CL Colyer, T Tang, N Chiem, DJ Harrison. Clinical potential of microchip capillary electrophoresis systems. Electrophoresis 1997; 18(10):1733–41.

36. G Worthington. Active micro-fluidic devices for integration of biological assays. 12th International Symposium on Pharmaceutical and Biomedical Analysis, Monterey, CA, May 2001, abstract L13.

Index